Lecture Notes in Bioinformatics 11490

Subseries of Lecture Notes in Computer Science

More information about this series at http://www.springer.com/series/5381

Zhipeng Cai · Pavel Skums ·
Min Li (Eds.)

Bioinformatics Research and Applications

15th International Symposium, ISBRA 2019
Barcelona, Spain, June 3–6, 2019
Proceedings

Editors
Zhipeng Cai
Georgia State University
Atlanta, GA, USA

Pavel Skums
Georgia State University
Atlanta, GA, USA

Min Li ⓘ
Central South University
Changsha, China

ISSN 0302-9743 ISSN 1611-3349 (electronic)
Lecture Notes in Bioinformatics
ISBN 978-3-030-20241-5 ISBN 978-3-030-20242-2 (eBook)
https://doi.org/10.1007/978-3-030-20242-2

LNCS Sublibrary: SL8 – Bioinformatics

This Springer imprint is published by the registered company Springer Nature Switzerland AG
The registered company address is: Gewerbestrasse 11, 6330 Cham, Switzerland

Preface

On behalf of the Program Committee, we would like to welcome you to the proceedings of the 15th edition of the International Symposium on Bioinformatics Research and Applications (ISBRA 2019), held in Barcelona, Spain, June 3–6 2019. The symposium provides a forum for the exchange of ideas and results among researchers, developers, and practitioners working on all aspects of bioinformatics and computational biology and their applications. This year we received 95 submissions in response to the call for extended abstracts. The Program Committee decided to accept 22 of them for full publication in the proceedings and oral presentation at the symposium. We also accepted 23 for oral presentation. Furthermore, we received 20 submissions in response to the call for short abstracts.

The technical program invited keynote talks were given by Prof. Niko Beerenwinkel from ETH Zurich, Prof. Martin Vingron from Max Planck Institute for Molecular Genetics, and Prof. Roderic Guigó from Pompeu Fabra University. We would like to thank the Program Committee members and the additional reviewers for volunteering their time to review and discuss symposium papers. We would like to extend special thanks to the steering and general chairs of the symposium for their leadership, and to the finance, publicity, and local organization chairs for their hard work in making ISBRA 2019 a successful event. Last but not least, we would like to thank all authors for presenting their work at the symposium.

April 2019

Zhipeng Cai
Pavel Skums
Min Li

Organization

Steering Committee

Yi Pan (Chair)	Georgia State University, USA
Dan Gusfield	University of California Davis, USA
Ion Mandoiu	University of Connecticut, USA
Marie-France Sagot	Inria, France
Ying Xu	University of Georgia, USA
Aidong Zhang	University of Virginia, USA

General Chairs

Gabriel Valiente	Technical University of Catalonia, Spain
Alexander Zelikovsky	Georgia State University, USA

Local Chair

Gabriel Valiente	Technical University of Catalonia, Spain

Program Chairs

Zhipeng Cai	Georgia State University, USA
Pavel Skums	Georgia State University, USA
Min Li	Central South University, China

Publicity Chairs

Olga Glebova	Georgia State University, USA
Mingon Kang	Kennesaw State University, USA
Fa Zhang	Chinese Academy of Science, China

Program Committee

Kamal Al Nasr	Tennessee State University, USA
Max Alekseyev	George Washington University, USA
Mukul S. Bansal	University of Connecticut, USA
Paola Bonizzoni	Universita di Milano-Bicocca, Italy
Zhipeng Cai	Georgia State University, USA
Hongmin Cai	South China University of Technology, China
Xing Chen	China University of Mining and Technology, China
Xuefeng Cui	Tsinghua University, China
Ovidiu Daescu	University of Texas at Dallas, USA

Lei Deng	Central South University, China
Pufeng Du	Tianjin University, China
Oliver Eulenstein	Iowa State University, USA
Xin Gao	King Abdullah University of Science and Technology, Saudi Arabia
Xuan Guo	University of North Texas, USA
Olga Glebova	Georgia State University, USA
Zengyou He	Dalian University of Technology, China
Steffen Heber	North Carolina State University, USA
Jinling Huang	East Carolina University, USA
Mingon Kang	Kennesaw State University, USA
Wooyoung Kim	University of Washington, USA
Danny Krizanc	Wesleyan University, USA
Xiujuan Lei	Shaanxi Normal University, China
Shuai-Cheng Li	City University of Hong Kong, SAR China
Yaohang Li	Old Dominion University, USA
Jing Li	Case Western Reserve University, USA
Min Li	Central South University, China
Xiaowen Liu	Indiana University-Purdue University Indianapolis, USA
Bingqiang Liu	Shandong University, China
Ion Mandoiu	University of Connecticut, USA
Igor Mandric	University of California, USA
Serghei Mangul	University of California, USA
Fenglou Mao	National Institute of Health, USA
Yuri Orlovich	Belarusian State University, Belarus
Andrei Paun	University of Bucharest, Romania
Nadia Pisanti	Università di Pisa, Italy and Erable Team, Italy; Inria, France
Yuri Porozov	ITMO University, Russia
Russell Schwartz	Carnegie Mellon University, USA
Joao Setubal	University of Sao Paulo, Brazil
Yi Shi	Shanghai Jiaotong University, China
Xinghua Shi	University of North Carolina at Charlotte, USA
Dong Si	University of Washington, USA
Pavel Skums	Georgia State University, USA
Ileana Streinu	University of Massachusetts Amherst, USA
Emily Su	Taipei Medical University, China
Shiwei Sun	Institute of Computing Technology, Chinese Academy of Sciences, China
Sing-Hoi Sze	Texas A&M University, USA
Weitian Tong	Georgia Southern University, USA
Gabriel Valiente	Technical University of Catalonia, Spain
Jianxin Wang	Central South University, China
Li-San Wang	University of Pennsylvania, USA
Guohua Wang	Harbin Institute of Technology, China

Seth Weinberg	Virginia Commonwealth University, USA
Yubao Wu	Georgia State University, USA
Fangxiang Wu	University of Saskatchewan, Canada
Yufeng Wu	University of Connecticut, USA
Zeng Xiangxiang	Xiamen University, China
Guoxian Yu	Southwest University, China
Ning Yu	Georgia State University, USA
Alex Zelikovsky	Georgia State University, USA
Le Zhang	Sichuan University, China
Xuegong Zhang	Tsinghua University, China
Xing-Ming Zhao	Fudan University, China
Leming Zhou	University of Pittsburgh, USA
Shuigeng Zhou	Fudan University, China
Quan Zou	Tianjin University, China

Webmasters

Filipp Rondel	Georgia State University, USA
Sergey Knyazev	Georgia State University, USA

Additional Reviewers

Adrian Caciula	Maria Atamanova
Alexander Artyomenko	Marmar Moussa
Alexey Markin	Mercè Llabrés
Andrii Melnyk	Mohammed Alser
Anton Nekhai	Mohammad Masum
Chong Chu	Murray Patterson
David Koslicki	Nagakishore Jammula
Deyvid Amgarten	Natalia Khanzhina
Dmitry Antipov	Nathan LaPierre
Ekaterina Gerasimov	Pavel Avdeyev
Fahad Alqahtani	Pavel Skums
Fatemeh Mohebbi	Pelin Burcak Icer
Feng Shi	Raffaella Rizzi
Francesco Masulli	Sanjiv Dinakar
German Demidov	Sergey Aganezov
Huidong Chen	Sergey Malov
Igor Mandric	Seyed Hosseini
Ivan Smetannikov	Tom Hartmann
James Lara	Tejaswini Mallavarapu
Jie Hao	Vadim Nazarov
Kiril Kuzmin	Viachaslau Tsyvina
Lin Yang Yang	Wei Li
Marco Previtali	Yan Wang
Marek Cygan	Zhongpai Gao

Contents

Computational Proteomics

Machine and Deep Learning

Data Analysis and Methodology

Genome Analysis

Computing a Consensus Phylogeny
via Leaf Removal

Zhi-Zhong Chen[1(✉)], Shohei Ueta[1], Jingyu Li[2], and Lusheng Wang[2]

[1] Division of Information System Design, Tokyo Denki University,
Hatoyama, Saitama 350-0394, Japan
zzchen@mail.dendai.ac.jp
[2] Department of Computer Science, City University of Hong Kong,
Tat Chee Avenue, Kowloon, Hong Kong SAR
cswangl@cityu.edu.hk

Abstract. Given a set $\mathcal{T} = \{T_1, T_2, \ldots, T_m\}$ of phylogenetic trees with the same leaf-label set X, we wish to remove some leaves from the trees so that there is a tree T with leaf-label set X displaying all the resulting trees. One objective is to minimize the total number of leaves removed from the trees, while the other is to minimize the maximum number of leaves removed from an input tree. Chauve *et al.* [6] refer to the problem with the first (respectively, second) objective as *AST-LR* (respectively, *AST-LR-d*), and show that both problems are NP-hard. They further present algorithms for the parameterized versions of both problems, but it seems that their algorithm for the parameterized version of AST-LR is flawed [7]. In this paper, we present a new algorithm for the parameterized version of AST-LR and also show that Chauve *et al.*'s algorithm for the parameterized version of AST-LR-*d* can be sped up by an exponential factor. We further design heuristic integer-linear programming (ILP for short) models for AST-LR and AST-LR-*d*. Our experimental results show that the heuristic models can be used to significantly speed up solving the exact models proposed in [7].

1   Introduction

When studying the evolutionary history of a set X of existing species, one can obtain a phylogenetic tree with leaf set X with high confidence by looking at a segment of sequences or a set of genes [11]. When looking at different segments of sequences, different phylogenetic trees with leaf set X can be obtained with high confidence, too. In order to facilitate the comparison of the resulting trees, a number of distance metrics have been proposed in the literature [4,6,10,14,15]. Among the metrics, the rSPR-distance defined in [5,15] is an important metric that often helps us discover reticulation events. In particular, it provides a lower bound on the number of reticulation events [2,3], and has been regularly used to model reticulate evolution [12,13].

Recently, Chauve *et al.* [6] define two new metrics highly related to the rSPR distance. Basically, to compute the rSPR distance of two phylogenetic trees,

© Springer Nature Switzerland AG 2019
Z. Cai et al. (Eds.): ISBRA 2019, LNBI 11490, pp. 3–15, 2019.
https://doi.org/10.1007/978-3-030-20242-2_1

we are allowed to delete *any* edges from the input trees so that the resulting forests become topologically identical, as long as the number of deleted edges is minimized. In contrast, the two new metrics defined in [6] require that any edge removed from the input trees must be *incident to a leaf*. Moreover, the new metrics are defined for any number of phylogenetic trees and also allow different leaves to be removed from different trees. More specifically, given a set of phylogenetic trees, the two new metrics ask us to remove leaves from the input trees so that there is a single tree *displaying* all the resulting trees, where a tree T *displays* another tree T' if and only if T' is topologically identical to a subtree of T. One of the metrics requires that the total number of leaves removed from the input trees is minimized, while the other requires that the maximum number of leaves removed from an input tree is minimized. The problem of computing the former metric is denoted by AST-LR, while the latter is denoted by AST-LR-d.

As easily observed in [6], the special cases of AST-LR and AST-LR-d where there are only two input trees are basically the maximum agreement subtree problem and hence can be solved in $O(n \log n)$ time [8]. However, Chauve *et al.* [6] show that both AST-LR and AST-LR-d are NP-hard in general. They then present an algorithm for the parameterized version of each of the problem. Their algorithm for the parameterized version of AST-LR (respectively, AST-LR-d) runs in $O\left(12^q mn^3\right)$ (respectively, $O\left(1152^d d^{3d}(n^2 + mn \log n)\right)$) time, where m is the number of input trees, n is the number of leaves in each input tree, and q (respectively, d) is the parameter.

Chauve *et al.*'s algorithm [6] for the parameterized version of AST-LR consists of two stages, where the second stage is much more time-consuming than the first. To prove that even the first stage is impractical, Chen *et al.* [7] implemented the first stage and the experimental results in [7] show that it is indeed very slow. Chauve *et al.*'s algorithm [6] for the parameterized version of AST-LR-d is much more complicated and has a much higher time-complexity than the parameterized version of AST-LR. Since both of Chauve *et al.*'s parameterized algorithms look impractical, integer-linear programming (ILP for short) models for AST-LR and AST-LR-d have been proposed in [7]. Experimental results in [7] show that GUROBI (a popular ILP solver) can solve the ILP model for AST-LR within much shorter time than the first stage of Chauve *et al.*'s algorithm for the parameterized version of AST-LR. Another merit of the ILP models is that they can be used to verify if a complicated fixed-parameter algorithm has been correctly implemented.

Unfortunately, it seems that Chauve *et al.*'s algorithm [6] for the parameterized version of AST-LR is flawed [7] roughly for the following reason. Recall that their algorithm consists of two stages. The second stage is based on Theorem 6 in [6] and hence requires that for every set t of three leaf labels, the subtree T_t induced by t in a targeted solution has been fixed in the first stage. However, a counterexample in [7] shows that the requirement is not necessarily always satisfied. More specifically, it is explained in [7] that if the first three trees in Fig. 1 together with the parameter 4 are given as input, then after the first stage, it is possible that only the leaves incident to the four broken edges have been removed and only the triplets in $\{ad|e, ad|j, cd|f\}$ have been explicitly fixed for a targeted solution, but the subtree induced by $\{a, c, d\}$ in the targeted solution has not been fixed and none of the modified input trees contains all the leaves in $\{a, c, d\}$.

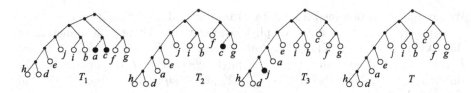

Fig. 1. Three trees and a center tree for them obtained by removing the black leaves.

In this paper, we present a new algorithm for the parameterized version of AST-LR. The algorithm runs in $O\left((4q-2)^q m^2 n^2\right)$ time. We also improve the time complexity of Chauve *et al.*'s algorithm [6] for the parameterized version of AST-LR-d by an exponential factor (namely, roughly 2^d). In more details, our algorithm for the parameterized version of AST-LR-d runs in time $O\left(36^d(4d^3 + 5d^2 + d)^d((n + m)n \log n)\right)$. We further design heuristic ILP models AST-LR and AST-LR-d and discuss how to use them to speed up solving the exact ILP models proposed in [7]. Our experimental results show that the heuristic ILP models lead to significant speedup of the exact ILP models.

Our program is available at http://rnc.r.dendai.ac.jp/consensusTree.html.

2 Preliminaries

Throughout this paper, a phylogenetic tree always means a rooted tree whose leaves are distinctively labeled. Unless stated otherwise, a phylogenetic tree is always binary, i.e., each non-leaf vertex has exactly two children in the tree.

Let T be a phylogenetic tree. For each vertex v of T, the subtree rooted at v is called a *pendant subtree* of T. We use $X(T)$ to denote the leaf-label set of T. For each $x \in X(T)$, we use x^T to denote the leaf of T labeled x. Moreover, for a subset Y of $X(T)$, we use Y^T to denote $\{x^T \mid x \in Y\}$. If T is clear from the context, we simply write x and Y instead of x^T and Y^T, respectively. Moreover, for a subset Y of $X(T)$, we use $T - Y$ to denote the phylogenetic tree obtained from T by first removing the leaves in Y^T and further repeatedly removing an unlabeled leaf or contracting an edge leaving a *unifurcate* vertex (i.e., vertex with only one child) until no such leaf or edge exists. We use $T|_Y$ to denote $T - (X(T) \setminus Y)$. T *displays* another phylogenetic tree T' if $X(T') \subseteq X(T)$ and $T' = T|_{X(T')}$.

A *leaf-prune-and-regraft (LPR) operation* on T is the operation of replacing T by another phylogenetic tree T' with $X(T) = X(T')$ such that T and T' are different but $T - \{x\}$ and $T' - \{x\}$ are identical, In other words, T' is obtained from T as follows.

1. Choose a leaf x and an edge $e = (u, v)$ of T such that v is neither the sibling nor the parent of x in T.
2. Remove the edge entering x and contract the edge leaving p, where p was the parent of x before the removal.
3. Replace e by three edges (u, w), (w, v), and (w, x), where w is a new vertex.

We say that T' is obtained from T by *pruning* x and *regrafting* it on e.

Let $(T_1, \ldots, T_m)$ be a list of phylogenetic trees, where it is unnecessary that $X(T_1) = \cdots = X(T_m)$. A *leaf-disagreement* for $(T_1, \ldots, T_m)$ is a list $L = (Y_1, \ldots, Y_m)$ such that there is a phylogenetic tree T displaying $T_1 - Y_1$ through $T_m - Y_m$. The *size* of L is $\sum_{i=1}^m |Y_i|$, the *radius* of L is $\max_{i=1}^m |Y_i|$, and T is called a *center tree* witnessing L. For example, for the first three trees T_1, T_2, and T_3 in Fig. 1, $L = (\{a, c\}, \{c\}, \{j\})$ is a leaf-disagreement of size 4 and radius 2 and the last tree T in the figure witnesses L. For an integer k, a *size-k* (respectively, *radius-k*) center tree for $\{T_1, \ldots, T_m\}$ is a center tree witnessing a leaf-disagreement of size (respectively, radius) at most k for $(T_1, \ldots, T_m)$. T_1, $\ldots$, T_m are *compatible* if $(\emptyset, \ldots, \emptyset)$ is a leaf-disagreement for $(T_1, \ldots, T_m)$.

Given a set $\{T_1, \ldots, T_m\}$ of phylogenetic trees with $X(T_1) = \cdots = X(T_m)$, *AST-LR* (respectively, *AST-LR-d*) is the problem of computing a leaf-disagreement for $(T_1, \ldots, T_m)$ whose size (respectively, radius) is minimized over all leaf-disagreements for $(T_1, \ldots, T_m)$.

Let T_1 and T_2 be two phylogenetic trees with $X(T_1) = X(T_2)$. A leaf-disagreement (Y_1, Y_2) for (T_1, T_2) is *1-sided* if $Y_2 = \emptyset$. As easily observed in [6], the following hold:

– For every leaf-disagreement (Y_1, Y_2) for (T_1, T_2), $(Y_1 \cup Y_2, \emptyset)$ is a 1-sided leaf-disagreement for (T_1, T_2).
– For every 1-sided leaf-disagreement $(Y, \emptyset)$ for (T_1, T_2) and for every $Y_1 \subseteq Y$, $(Y_1, Y \setminus Y_1)$ is a leaf-disagreement for (T_1, T_2).

For simplicity, we use Y to denote a 1-sided leaf-disagreement $(Y, \emptyset)$ for (T_1, T_2). Y is *minimal* if for every $y \in Y$, $Y \setminus \{y\}$ is not a 1-sided leaf-disagreement for (T_1, T_2). We use $d_{LR}(T_1, T_2)$ to denote the minimum size of a 1-sided leaf-disagreement Y for (T_1, T_2). Given T_1 and T_2, $d_{LR}(T_1, T_2)$ can be computed in $O(n \log n)$ time, where $n = |X(T_1)|$ [8]. Moreover, by the above observation, $d_{LR}(T_1, T_2)$ (respectively, $\lceil \frac{1}{2} d_{LR}(T_1, T_2) \rceil$) is the minimum size (respectively, radius) of a leaf-disagreement for (T_1, T_2). In other words, if we require that there are only two trees in the input, then AST-LR and AST-LR-d become basically the same problem and can be solved in $O(n \log n)$ time.

3   Parameterized Algorithm for AST-LR

A *triplet* is a phylogenetic tree with exactly 3 leaves. We use $xy|z$ to denote the triplet t such that $X(t) = \{x, y, z\}$ and x and y are siblings in t. Two triplets t_1 and t_2 *conflict* if $t_1 = xy|z$ and $t_2 \in \{xz|y, yz|x\}$. We say that $xy|z$ is a *triplet of* a phylogenetic tree T if $\{x, y, z\} \subseteq X(T)$ and $T|_{\{x,y,z\}} = xy|z$. For a phylogenetic tree T, we use $tr(T)$ to denote the set of triplets in T. Moreover, for a set $\mathcal{T}$ of phylogenetic trees, we use $tr(\mathcal{T})$ to denote $\bigcup_{T \in \mathcal{T}} tr(T)$, and use $X(\mathcal{T})$ to denote $\bigcup_{T \in \mathcal{T}} X(T)$. A *full set of triplets* on a set $\mathcal{X}$ of labels is a set $\mathcal{S}$ of triplets such that $\mathcal{X} = X(\mathcal{S})$ and $\mathcal{S}$ contains exactly one of $xy|z$, $xz|y$, and $yz|x$ for each triple $\{x, y, z\} \subseteq \mathcal{X}$. A triple $\{x, y, z\}$ is *conflicting* in $\mathcal{T}$ if $tr(\mathcal{T})$ has two conflicting triplets with label set $\{x, y, z\}$.

3.1   Deciding the Existence of a Center Tree

In this subsection, we sketch how to apply Aho *et al.*'s polynomial-time algorithm [1] to deciding whether there is a phylogenetic tree displaying a given set $\tilde{\mathcal{T}} = \{\tilde{T}_1, \ldots, \tilde{T}_\ell\}$ of phylogenetic trees. If $|X(\tilde{T}_i)| \leq 2$ for some $\tilde{T}_i \in \tilde{\mathcal{T}}$, then it is easy to see that there is a phylogenetic tree displaying the trees in $\tilde{\mathcal{T}}$ if and only if there is a phylogenetic tree displaying the trees in $\tilde{\mathcal{T}} \setminus \{\tilde{T}_i\}$. So, we may assume that $|X(\tilde{T}_i)| \geq 3$ for every $\tilde{T}_i \in \tilde{\mathcal{T}}$. For convenience, we use $\tilde{T}_{i,l}$ (respectively, $\tilde{T}_{i,r}$) to denote the pendant subtree of $\tilde{T}_i$ rooted at the left (respectively, right) child of the root of $\tilde{T}_i$.

Recall the well-known fact that each phylogenetic tree T is the unique phylogenetic tree displaying the triplets in $tr(T)$. So, to decide whether there is a phylogenetic tree displaying the trees in $\tilde{\mathcal{T}}$, it suffices to decide whether there is a phylogenetic tree displaying the triplets in $tr(\tilde{\mathcal{T}})$. Note that a triplet $xy|z$ here has the same meaning as the constraint $(x, y) < (x, z)$ in [1]. So, for our purpose, we can call Aho *et al.*'s algorithm on input $(X(\tilde{\mathcal{T}}), tr(\tilde{\mathcal{T}}))$.

Consider the call of Aho *et al.*'s algorithm on input $(X(\tilde{\mathcal{T}}), tr(\tilde{\mathcal{T}}))$. The algorithm is recursive and actually returns a not-necessarily-binary phylogenetic tree displaying the triplets in $tr(\tilde{\mathcal{T}})$ if one exists. We claim that there is a not-necessarily-binary phylogenetic tree displaying the triplets in $tr(\tilde{\mathcal{T}})$ if and only if there is a phylogenetic tree displaying the triplets in $tr(\tilde{\mathcal{T}})$. The "if" part is clear. To see the "only-if" part, assume that T is a non-binary phylogenetic tree displaying the triplets in $tr(\tilde{\mathcal{T}})$. Let v be a vertex with more than two children in T. We modify T by choosing two arbitrary children v_1 and v_2 of v in T, adding a new vertex u, and replacing the edges (v, v_1) and (v, v_2) with the edges (v, u), (u, v_1), (u, v_2). The modification decreases the degree of v by 1 and one can easily see that the modified T still displays the triplets in $tr(\tilde{\mathcal{T}})$. So, we can repeat the modification until T has no vertex with more than two children, while always ensuring that T displays the triplets in $tr(\tilde{\mathcal{T}})$.

We next detail Aho *et al.*'s recursive algorithm. In the base case where $tr(\tilde{\mathcal{T}}) = \emptyset$, the algorithm returns "yes" together with an arbitrary phylogenetic tree whose leaf-label set is $X(\tilde{\mathcal{T}})$. On the other hand, in case $tr(\tilde{\mathcal{T}}) \neq \emptyset$, it obtains a partition P of $X(\tilde{\mathcal{T}})$ by first initializing $P = \{\{x\} \mid x \in X(\tilde{\mathcal{T}})\}$ and then repeatedly replacing two subsets $S_1 \in P$ and $S_2 \in P$ with $S_1 \cup S_2$ if there is a $xy|z \in tr(\tilde{\mathcal{T}})$ with $x \in S_1$ and $y \in S_2$. We claim that for every $\tilde{T}_i \in \tilde{\mathcal{T}}$, $X(\tilde{T}_{i,l})$ is a subset of some set in P and so is $X(\tilde{T}_{i,r})$. This is true because if x and y are two leaves in $\tilde{T}_{i,l}$ (respectively, $\tilde{T}_{i,r}$), then for each leaf z in $\tilde{T}_{i,r}$ (respectively, $\tilde{T}_{i,l}$), $xy|z$ is a triplet in $tr(\tilde{\mathcal{T}})$. Now, if $|P| = 1$, the algorithm returns "no". Otherwise, for each set $S \in P$, it computes $\tilde{\mathcal{T}}_S = \{\tilde{T}_i \in \tilde{\mathcal{T}} \mid X(\tilde{T}_i) \subseteq S\} \cup \{\tilde{T}_{i,l} \mid \tilde{T}_i \in \tilde{\mathcal{T}},$ $X(\tilde{T}_{i,l}) \subseteq S$, and $X(\tilde{T}_{i,r}) \nsubseteq S\} \cup \{\tilde{T}_{i,r} \mid \tilde{T}_i \in \tilde{\mathcal{T}}, X(\tilde{T}_{i,r}) \subseteq S$, and $X(\tilde{T}_{i,l}) \nsubseteq S\}$, and makes a recursive call on input $(X(\tilde{\mathcal{T}}_S), tr(\tilde{\mathcal{T}}_S))$. If at least one of the $|P|$ recursive calls returns "no", the algorithm returns "no"; otherwise, the algorithm receives a not-necessarily-binary phylogenetic tree T_S from the recursive call on input $(X(\tilde{\mathcal{T}}_S), tr(\tilde{\mathcal{T}}_S))$ for each $S \in P$, combine the $|P|$ trees T_S into a single not-necessarily-binary phylogenetic tree T by adding a new root and connecting it to the root of T_S for each $S \in P$, and further returns "yes" together with T.

For convenience, we refer to P as the *label-partition* for $\tilde{\mathcal{T}}$. Alternatively, we can obtain P as follows. First, we construct an auxiliary graph $H = (V_1 \cup V_2, E_1 \cup E_2 \cup E_3)$ from $\tilde{\mathcal{T}}$, where

- V_1 consists of the leaves (together with their labels) in the trees in $\tilde{\mathcal{T}}$;
- for each $\tilde{T}_i \in \tilde{\mathcal{T}}$, V_2 contains two vertices $v_{i,1}$ and $v_{i,2}$;
- for each $\tilde{T}_i \in \tilde{\mathcal{T}}$ and for each leaf x of $\tilde{T}_i$, if x is a descendant of the left (respectively, right) child of the root in $\tilde{T}_i$, then E_1 (respectively, E_2) contains the edge $\{x, v_{i,1}\}$ (respectively, $\{x, v_{i,2}\}$);
- for every two vertices x and y in V_1, if x and y have the same label, then E_3 contains the edge $\{x, y\}$.

Let $K_1, \ldots, K_h$ be the connected components of H. For each $i \in \{1, \ldots, h\}$, let X_i be the set of all $x \in X(\tilde{\mathcal{T}})$ such that x is the label of some vertex in $V(K_i) \cap V_1$. Then, one can easily see that $P = \{X_1, \ldots, X_h\}$. This new computation of P is more efficient because it uses the trees in $\tilde{\mathcal{T}}$ directly rather than using the triplets in $tr(\tilde{\mathcal{T}})$. Indeed, there is an even more efficient way of computing P. To see this, first note that for each $x \in X(\tilde{\mathcal{T}})$, the vertices of H with label x form a clique C_x. Suppose that we modify H by contracting C_x to a single vertex (still with label x) for each $x \in X(\tilde{\mathcal{T}})$. The modified H has the same number of connected components as before. Moreover, if we compute P from the connected components of the modified H as before, then P should be the same as before. Furthermore, instead of constructing H and then modifying it, we can construct the modified H from the trees in $\tilde{\mathcal{T}}$ directly in $O(\ell|X(\tilde{\mathcal{T}})|)$ time. In this way, P can be computed in $O(\ell|X(\tilde{\mathcal{T}})|)$ time because the modified graph H has $O(\ell + |X(\tilde{\mathcal{T}})|)$ vertices and $O(\ell|X(\tilde{\mathcal{T}})|)$ edges. The reason why we prefer H to the modified H is that H makes our analysis in Sect. 3.2 easier.

3.2   A New Algorithm

In this subsection, we present a new algorithm for the parameterized version of AST-LR. So, consider an instance $(q, \{T_1, \ldots, T_m\})$ of the problem. Note that $X(T_1) = \cdots = X(T_m)$. For each $i \in \{1, \ldots, m\}$, we call Cole *et al.*'s algorithm [8] to compute $d_{LR}(T_i, T_j)$ for each $j \in \{1, \ldots, m\}$, and then check if $\sum_{j=1}^{m} d_{LR}(T_i, T_j) \leq q$. If $\sum_{j=1}^{m} d_{LR}(T_i, T_j) \leq q$ for at least one $i \in \{1, \ldots, m\}$, then we are done by returning "yes". The total time taken by the calls of Cole *et al.*'s algorithm is $O(m^2 n \log n)$, where n is the number of leaves in each of $T_1, \ldots, T_m$. So, we may assume that there is no

$i \in \{1, \ldots, m\}$ with $\sum_{j=1}^{m} d_{LR}(T_i, T_j) \leq q$. Then, for each phylogenetic tree T with $\sum_{j=1}^{m} d_{LR}(T, T_j) \leq q$, we have $d_{LR}(T, T_i) \geq 1$ for every $i \in \{1, \ldots, m\}$. Consequently, $m \leq q$.

Since our algorithm will be recursive, we need to consider a call originated from the *root call* (i.e., the call on input $(q, \{T_1, \ldots, T_m\})$ after zero or more subsequent calls. Let $(\hat{q}, \{\hat{T}_1, \ldots, \hat{T}_k\})$ be the input to the call. We will maintain the invariant that for each $i \in \{1, \ldots, k\}$, there is a $j \in \{1, \ldots, m\}$ with $\hat{T}_i = T_j|_{X(\hat{T}_i)}$. So, $k \leq m$. However, it is not necessarily true that $X(\hat{T}_1) = \cdots = X(\hat{T}_k)$. If there is an $i \in \{1, \ldots, k\}$ with $|X(\hat{T}_i)| \leq 2$, then we can remove $\hat{T}_i$ from the input because there is a leaf-disagreement of size at most $\hat{q}$ for $(\hat{T}_1, \ldots, \hat{T}_k)$ if and only if there is a leaf-disagreement of size at most $\hat{q}$ for $(\hat{T}_1, \ldots, \hat{T}_{i-1}, \hat{T}_{i+1}, \ldots, \hat{T}_k)$. So, we may assume that $|X(\hat{T}_i)| \geq 3$ for every $i \in \{1, \ldots, k\}$.

Let $\hat{\mathcal{T}} = \{\hat{T}_1, \ldots, \hat{T}_k\}$. We next detail our algorithm on input $(\hat{q}, \hat{\mathcal{T}})$. In the base case where $\hat{q} < 0$, the algorithm returns "no". So, assume that $\hat{q} \geq 0$. Our algorithm first calls Aho *et al.*'s algorithm to decide if there is a phylogenetic tree displaying the trees in $\hat{\mathcal{T}}$. If the call returns "yes", then our algorithm returns "yes". Otherwise, as sketched in the last subsection, the call returns "no" because it has found a set $\tilde{\mathcal{T}} = \{\tilde{T}_1, \ldots, \tilde{T}_\ell\}$ of two or more phylogenetic trees satisfying the following conditions:

C1. Each $\tilde{T}_i \in \tilde{\mathcal{T}}$ is a pendant subtree of some $\hat{T}_{j_i} \in \hat{\mathcal{T}}$ with $|X(\tilde{T}_i)| \geq 3$.
C2. If i and i' are different integers in $\{1, \ldots, \ell\}$, then $j_i \neq j_{i'}$.
C3. The partition P of $X(\tilde{\mathcal{T}})$ constructed from $tr(\tilde{\mathcal{T}})$ is $X(\tilde{\mathcal{T}})$.

By Conditions C1 and C2, $\ell \leq k \leq m \leq q$.

Consider the auxiliary bipartite graph $H = (V_1 \cup V_2, E_1 \cup E_2 \cup E_3)$ constructed from $\tilde{\mathcal{T}}$ as in Sect. 3.1. Since $P = X(\tilde{\mathcal{T}})$, H is connected. Our algorithm constructs a vertex-induced subgraph H' of H as follows. Initially, H' is a copy of H. Then, as long as H' has a vertex $x \in V_1$ such that removing x from H' does not disconnect H', we keep modifying H' by removing x. Suppose that we have finished modifying H' in this way. Let $V_1' = \{x \in V_1 \mid x \text{ still remains in } H'\}$.

Lemma 1. $|V_1'| \leq 2|V_2| - 2 = 4\ell - 2$.

For each $i \in \{1, \ldots, \ell\}$, let $V_{1,i}' = \{x \in V_1' \mid x \text{ is a leaf of } \tilde{T}_i\}$ and $\tilde{T}_i' = \tilde{T}_i|_{V_{1,i}'}$.

Lemma 2. *There is no phylogenetic tree displaying* $\tilde{\mathcal{T}}' = \{\tilde{T}_i' \mid V_{1,i}' \neq \emptyset\}$.

By Conditions C1 and C2, $\tilde{T}_i' = \hat{T}_{j_i}|_{V_{1,i}'}$ as well for each $\tilde{T}_i' \in \tilde{\mathcal{T}}'$. Thus, by Lemma 2, we need to delete at least one leaf of V_1' from some $\hat{T}_{j_i}$ with $\tilde{T}_i' \in \tilde{\mathcal{T}}'$, in order to make the trees in $\{\hat{T}_{j_i} \mid \tilde{T}_i' \in \tilde{\mathcal{T}}'\}$ compatible. For convenience, we refer to V_1' as a *small witness* for the incompatibility of the trees in $\hat{\mathcal{T}}$. Now, for each $i \in \{1, \ldots, \ell\}$ and for each leaf x in $\tilde{T}_i$, our algorithm makes a recursive call on input $(\hat{q} - 1, \{\hat{T}_1, \ldots, \hat{T}_{j_i-1}, \hat{T}_{j_i} - \{x\}, \hat{T}_{j_i+1}, \ldots, \hat{T}_k\})$. If at least one of the calls returns "yes", then the algorithm returns "yes"; otherwise, it returns "no".

To compute V_1' from $\tilde{\mathcal{T}}$ efficiently, we can proceed as follows.

1. Try to find an x such that x is a proper leaf descendant of a child of the root of some $\tilde{T}_i$ with $i \in \{1, \ldots, \ell\}$ and the label-partition for $\{\tilde{T}_1, \ldots, \tilde{T}_{i-1}, \tilde{T}_i - \{x\}, \tilde{T}_{i+1}, \ldots, \tilde{T}_\ell\}$ is the same as that for $\{\tilde{T}_1, \ldots, \tilde{T}_\ell\}$. (*Comment:* As noted in Sect. 3.1, the label-partition can be computed in $O(\ell |X(\tilde{T})|)$ time.)
2. If x is found in Step 1, then remove it from $\tilde{T}_i$ and go to Step 1. Otherwise, set V_1' to be the set of leaves in $\tilde{T}_1, \ldots, \tilde{T}_\ell$.

So, V_1' can be computed in $O(\ell^2 |X(\tilde{T})|^2)$ time.

Theorem 1. *The parameterized AST-LR for input $(q, \{T_1, \ldots, T_m\})$ can be solved in $O\left((4q-2)^q m^2 n^2\right)$ time, where n is the number of leaves in each input tree.*

4   Parameterized Algorithm for AST-LR-d

We first review Chauve *et al.*'s algorithm [6] for the parameterized version of AST-LR-d. An instance of the parameterized version consists of an integer d and a set $\{T_1, \ldots, T_m\}$ of phylogenetic trees with $X(T_1) = \cdots = X(T_m)$, and the objective is to decide whether there is a leaf-disagreement of radius at most d for $(T_1, \ldots, T_m)$. Chauve *et al.*'s algorithm for the problem actually solves a more general problem. More specifically, other than d and $\{T_1, \ldots, T_m\}$, the input to their algorithm also includes an integer d' and the algorithm is supposed to return "yes" if and only if T_1 can be transformed into a radius-d center tree for $\{T_2, \ldots, T_m\}$ by performing at most d' LPR operations. Obviously, to solve the parameterized version for $(d, \{T_1, \ldots, T_m\})$, it suffices to solve the generalized problem for $(d, (T_1, \ldots, T_m), d')$ with $d' = d$.

So, consider the call of Chauve *et al.*'s algorithm on input $(d, \{T_1, \ldots, T_m\}, d')$. Since the algorithm is recursive, we need to consider a call originated from the *root call* (i.e., the call on input $(d, \{T_1, \ldots, T_m\}, d)$) after zero or more subsequent calls. Let $(d, \{\tilde{T}_1, T_2, \ldots, T_m\}, d')$ be the input to the call. In one base case where $d' < 0$, the algorithm returns "no". In another base case where $d' \geq 0$ and $\tilde{T}_1$ is a radius-d center tree for $\{T_2, \ldots, T_m\}$, it returns "yes". So, assume that neither of the base cases occurs. Then, there must exist an $i \in \{2, \ldots, m\}$ such that $d_{LR}(\tilde{T}_1, T_i) > d$. Without loss of generality, we may assume $i = 2$.

Basically, their algorithm tries to transform $\tilde{T}_1$ into a radius-d center tree for $\{T_2, \ldots, T_m\}$ by performing at most d' LPR operations. There are two main ideas behind their algorithm. One is to compute a set $S \subseteq X(\tilde{T}_1)$ in $O(n^2)$ time such that $|S| \leq 32 \cdot d^2$ and every minimal 1-sided leaf-disagreement Y for $(\tilde{T}_1, T_2)$ with $|Y| \leq d$ is a subset of S. So, in order to transform $\tilde{T}_1$ into a radius-d center tree for $\{T_2, \ldots, T_m\}$, we have to prune at least one $y \in Y$ from $\tilde{T}_1$ and further regraft it on an edge of $\tilde{T}_1$. Their other idea is to find, for each $y \in Y$, a set P_y of edges in $\tilde{T}_1$ with $|P_y| \leq 18(d+d') + 8$ in $O(n \log n + nd)$ time such that y should be regrafted on one of the edges in P_y in order to transform $\tilde{T}_1$ to a radius-d center tree for $\{T_2, \ldots, T_m\}$. Based on the two ideas, their algorithm then makes a recursive call on input $(d, \{\tilde{T}_1', T_2, \ldots, T_m\}, d' - 1)$ for every $y \in Y$ and every edge $e_y \in P_y$, where $\tilde{T}_1'$ is obtained from $\tilde{T}_1$ by pruning y and regrafting it on e_y. It is easy to see that their algorithm takes $O\left(1152^d d^{3d}(n^2 + mn \log n)\right)$ time.

We next describe how to speed up Chauve $et\ al.$'s algorithm. The idea is to avoid computing S. More specifically, instead of S, we obtain a subtree $\hat{T}_1$ of $\tilde{T}_1$ and a subtree $\hat{T}_2$ of $\tilde{T}_2$ as follows.

1. Initially, $\hat{T}_1$ (respectively, $\hat{T}_2$) is a copy of $\tilde{T}_1$ (respectively, T_2).
2. As long as there is an $x \in X(\hat{T}_1)$ with $d_{LR}(\hat{T}_1 - \{x\}, \hat{T}_2 - \{x\}) \geq d + 1$, keep removing x from both $\hat{T}_1$ and $\hat{T}_2$. ($Comment$: $d_{LR}(\hat{T}_1 - \{x\}, \hat{T}_2 - \{x\}) \geq d_{LR}(\hat{T}_1, \hat{T}_2) - 1$.)

Obviously, $d_{LR}(\hat{T}_1, \hat{T}_2) = d + 1$ because $d_{LR}(\tilde{T}_1, T_2) \geq d + 1$. Since $\hat{T}_1$ is a subtree of $\tilde{T}_1$, we must prune at least one leaf y of $\hat{T}_1$ from $\tilde{T}_1$ and regraft it on an edge e_y of $\tilde{T}_1$ in order to satisfy $d_{LR}(\tilde{T}_1, T_2) \leq d$. From Lemma 10 in [6], it is clear that e_y should be an edge in the set P_y. So, we compute P_y as in Chauve $et\ al.$'s algorithm. Now, for each leaf y of $\hat{T}_1$ and for each $e_y \in P_y$, our algorithm makes a recursive call on input $(d, \{\tilde{T}_1', T_2, \ldots, T_m\}, d' - 1)$, where $\tilde{T}_1'$ is obtained from $\tilde{T}_1$ by pruning y and regrafting it on e_y.

To analyze the time complexity of our algorithm, we need to bound $|X(\hat{T}_1)|$.

Lemma 3. $|X(\hat{T}_1)| \leq (d+1)(4d+1)$.

Theorem 2. *The parameterized version of AST-LR-d for input $(d, \{T_1, \ldots, T_m\})$ can be solved in $O\left(36^d(4d^3 + 5d^2 + d)^d((n+m)n \log n)\right)$, where n is the number of leaves in each input tree.*

5   ILP Approach to the Problems

We use the notation $(j_1, \ldots, j_k) \subseteq S$ to denote an ordered subset $(j_1, \ldots, j_k)$ of a set S. Let $\mathcal{T} = \{T_1, \ldots, T_m\}$ be a set of phylogenetic trees not necessarily with $X(T_1) = \cdots = X(T_m)$. A quadruple $(x_j, x_k, x_l, x_h) \subseteq X(\mathcal{T})$ is *conflicting* in $\mathcal{T}$ if $\{x_j x_k | x_l, x_l x_h | x_k, x_k x_h | x_j\} \subseteq tr(\mathcal{T})$ or $\{x_j x_k | x_l, x_l x_h | x_k, x_j x_h | x_k\} \subseteq tr(\mathcal{T})$.

Lemma 4. [9] *A full set $\mathcal{T}$ of triplets is compatible if and only if no quadruple $(x_j, x_k, x_l, x_h) \subseteq X(\mathcal{T})$ is conflicting in $\mathcal{T}$.*

5.1   Exact ILP Models

For each $(j, k, l) \subseteq \{1, \ldots, n\}$, we define $I_{j,k,l,0} = \{i \in \{1, \ldots, m\} \mid T_i|_{\{x_j, x_k, x_l\}} = x_j x_k | x_l\}$, $I_{j,k,l,1} = \{i \in \{1, \ldots, m\} \mid T_i|_{\{x_j, x_k, x_l\}} = x_j x_l | x_k\}$, and $I_{j,k,l,2} = \{i \in \{1, \ldots, m\} \mid T_i|_{\{x_j, x_k, x_l\}} = x_k x_l | x_j\}$. The ILP model for AST-LR-d proposed in [7] is based on Lemma 4 and is as follows.

$$\begin{aligned}
\min \quad & d \\
\text{s.t.} \quad & \forall_{1 \leq i \leq m} \ \textstyle\sum_{j=1}^{n} y_{i,j} \leq d \\
& \forall_{(j,k,l) \subseteq \{1,\ldots,n\}} \ a_{j,k,l} + b_{j,k,l} \leq 1 \\
& \forall_{(j,k,l) \subseteq \{1,\ldots,n\}} \forall_{i \in I_{j,k,l,0}} \ y_{i,j} + y_{i,k} + y_{i,l} \geq a_{j,k,l} + b_{j,k,l} \\
& \forall_{(j,k,l) \subseteq \{1,\ldots,n\}} \forall_{i \in I_{j,k,l,1}} \ y_{i,j} + y_{i,k} + y_{i,l} \geq 1 - b_{j,k,l} \\
& \forall_{(j,k,l) \subseteq \{1,\ldots,n\}} \forall_{i \in I_{j,k,l,2}} \ y_{i,j} + y_{i,k} + y_{i,l} \geq 1 - a_{j,k,l} \\
& \forall_{(j,k,l,h) \subseteq \{1,\ldots,n\}} \ a_{j,k,l} + b_{j,k,l} - a_{k,l,h} - a_{j,k,h} \geq -1 \\
& \forall_{(j,k,l,h) \subseteq \{1,\ldots,n\}} \ a_{j,k,l} + b_{j,k,l} - a_{k,l,h} - b_{j,k,h} \geq -1 \\
& \text{all variables } y_{i,j},\ a_{j,k,l},\ b_{j,k,l} \ : \ \text{binary}
\end{aligned}$$

To obtain an ILP model for AST-LR, we just simply modify the above ILP model for AST-LR-d by deleting the first set of constraints and replacing the objective function d with $\sum_{i=1}^{m} \sum_{j=1}^{n} y_{i,j}$.

5.2 Heuristic ILP Models

The exact ILP models may take long time to be solved by an ILP solver (such as Gurobi and Cplex). So, we here propose heuristic ILP models instead. As before, we first present a heuristic ILP model for AST-LR-d and then modify it into a heuristic ILP model for AST-LR.

The idea is to avoid explicitly computing the triplets $T|_{\{x_j,x_k,x_l\}}$ in the output center tree T. Intuitively speaking, the idea is to remove the variables $a_{j,k,l}$ and $b_{j,k,l}$ from the exact models. Without explicitly computing the triplet $T|_{\{x_j,x_k,x_l\}}$, we have to resort to removing direct conflicts between the input trees. In other words, for every $(j,k,l) \subseteq \{1,\ldots,n\}$ and for every $\{i_1,i_2\} \subseteq \{1,\ldots,m\}$, if $T_{i_1}|_{\{x_j,x_k,x_l\}}$ and $T_{i_2}|_{\{x_j,x_k,x_l\}}$ are different, we add the constraint $y_{i_1,j} + y_{i_1,k} + y_{i_1,l} + y_{i_2,j} + y_{i_2,k} + y_{i_2,l} \geq 1$ to the model. Similarly, for every $(j,k,l,h) \subseteq \{1,\ldots,n\}$ and for every $\{i_1,i_2,i_3\} \subseteq \{1,\ldots,m\}$, if one of the following holds, then we add the constraint $y_{i_1,j} + y_{i_1,k} + y_{i_1,l} + y_{i_2,j} + y_{i_2,k} + y_{i_2,l} + y_{i_3,j} + y_{i_3,k} + y_{i_3,l} \geq 1$ to the model:

- $T_{i_1}|_{\{x_j,x_k,x_l\}} = x_j x_k | x_l$, $T_{i_2}|_{\{x_j,x_k,x_l\}} = x_l x_h | x_k$, and $T_{i_3}|_{\{x_j,x_k,x_l\}} = x_k x_h | x_j$.
- $T_{i_1}|_{\{x_j,x_k,x_l\}} = x_j x_k | x_l$, $T_{i_2}|_{\{x_j,x_k,x_l\}} = x_l x_h | x_k$, and $T_{i_3}|_{\{x_j,x_k,x_l\}} = x_j x_h | x_k$.

Now, we have obtained the following heuristic ILP model for AST-LR-d:

$$\begin{aligned}
\min \quad & d \\
\text{s.t.} \quad & \forall_{1 \leq i \leq m} \ \textstyle\sum_{j=1}^{n} y_{i,j} \leq d \\
& \forall_{(j,k,l) \subseteq \{1,\ldots,n\}} \forall_{\{t_1,t_2\} \subseteq \{0,1,2\}} \forall_{i_1 \in I_{j,k,l,t_1}} \forall_{i_2 \in I_{j,k,l,t_2}} \\
& \quad y_{i_1,j} + y_{i_1,k} + y_{i_1,l} + y_{i_2,j} + y_{i_2,k} + y_{i_2,l} \geq 1 \\
& \forall_{(j,k,l,h) \subseteq \{1,\ldots,n\}} \forall_{i_1 \in I_{j,k,l,0}} \forall_{i_2 \in I_{k,l,h,2}} \forall_{i_3 \in I_{j,k,h,1} \cup I_{j,k,h,2}} \\
& \quad y_{i_1,j} + y_{i_1,k} + y_{i_1,l} + y_{i_2,j} + y_{i_2,k} + y_{i_2,l} + y_{i_3,j} + y_{i_3,k} + y_{i_3,l} \geq 1 \\
& \text{all variables } y_{i,j} \ : \ \text{binary}
\end{aligned}$$

By experiments, we have found that the heuristic model can be solved much faster than its exact counterpart by Gurobi or Cplex. Of course, an optimal solution s^* of the heuristic model may not lead to a correct leaf-disagreement. In more details, even though s^* tells us which leaves should be removed from each T_i, there may not exist a center tree displaying all the resulting trees. Nevertheless, instead of solving the exact model directly, we can first solve the heuristic model to obtain s^* and then proceed to solving the exact model with the help of s^*. The crucial points are the following:

- The value of d in s^* is a lower bound on the optimal objective value of the exact ILP model. We can incorporate this bound into the exact model when solving it with an ILP solver. The bound can help the solver prune a lot of unnecessary branches of the search tree.
- The values of $y_{i,j}$'s in s^* can be used as a starting partial-solution when solving the exact model by an ILP solver. By experiments, we have found that the partial start can often be extended to a feasible (and hence optimal) solution of the corresponding exact model by the solver. As the result, the solver can often solve the exact model within almost the same time as the heuristic model.

As will be seen in Sect. 5.3, using the heuristic model as in the above leads to significant speedup of solving the exact model.

To obtain a heuristic ILP model for AST-LR, we modify the above heuristic ILP model for AST-LR-d as we did in the exact case. In Sect. 5.3, we will see that the heuristic model can be used to speed up solving the exact model.

5.3   Experimental Results

To evaluate our ILP models empirically, we run our program on a Ubuntu (x64) desktop PC with i7-4790K CPU and 31.4 GiB RAM and another CentOS (x64) desktop PC with E5-2687W(v4) CPU and 252.2 GiB RAM. In order to solve our ILP models, we use Gurobi as the solver.

Our models have many constraints and hence it often takes time for Gurobi to get started. Note that most of the constraints are for excluding conflicting quadruples. In our experiments, we use these constraints as *lazy* constraints when using Gurobi to solve the models. In more details, Gurobi will remove these constraints at the beginning, but will add an initially-removed constraint back later if the constraint is found to be violated by the incumbent integral solution. In this way, Gurobi can start up much faster. Moreover, by experiments, we have found that a set of trees without conflicting triples often have no conflicting quadruples. Thus, we can often expect that very few initially-removed constraints are added back and in turn Gurobi can finish within much shorter time.

To test the performance of our models, we generate simulated datasets as follows. First, for each $n \in \{15, 25\}$, we generate a set $\mathcal{S}_n$ of 100 random phylogenetic trees with n leaves using the program of [3]. Then, for each $n \in \{15, 25\}$, each $T \in \mathcal{S}_n$, and each $k \in \{1, 3, 5, 7\}$, we generate 5 trees $T_{n,k,1}, \ldots, T_{n,k,5}$

from T by performing k random Nearest-Neighbor Interchange (NNI) moves on T; the trees $T_{n,k,1}, \ldots, T_{n,k,5}$ together form an instance of AST-LR and AST-LR-d. So, in total, there are 800 instances in our experiment for each of AST-LR and AST-LR-d. Since there are many instances and some of them may take long time to solve, we use the Ubuntu (respectively, CentOS) machine for those instances with $n = 15$ (respectively, $n = 25$).

Our experimental results for AST-LR-d and AST-LR are omitted here for lack of space but are summarized in two tables which are available at

http://rnc.r.dendai.ac.jp/~chen/centerTree_tables.pdf

From the tables, one can see that compared with solving the exact model directly, it is much faster to first solve the heuristic model and then use its output to solve the exact model. Moreover, in case both of the methods fail to find optimal solutions within the time limit, the latter method finds better heuristic solutions.

References

1. Aho, A.V., Sagiv, Y., Szymanski, T.G., Ullman, J.D.: Inferring a tree from lowest common ancestors with application to the optimization of relational expressions. SIAM J. Comput. **10**, 405–421 (1981)
2. Baroni, M., Grunewald, S., Moulton, V., Semple, C.: Bounding the number of hybridisation events for a consistent evolutionary history. J. Math. Biol. **51**, 171–182 (2015)
3. Beiko, R.G., Hamilton, N.: Phylogenetic identification of lateral genetic transfer events. BMC Evol. Biol. **6**, 15 (2006)
4. Bordewich, M., Semple, C.: On the computational complexity of the rooted subtree prune and regraft distance. Ann. Comb. **8**, 409–423 (2005)
5. Buneman, P.: The recovery of trees from measures of dissimilarity. In: Kendall, D., Tauta, P. (eds.) Mathematics in the Archaeological and Historical Sciences, pp. 387–395. Edinburgh University Press, Edinburgh (1971)
6. Chauve, C., Jones, M., Lafond, M., Scornavacca, C., Weller, M.: Constructing a consensus phylogeny from a leaf-removal distance (extended abstract). In: Fici, G., Sciortino, M., Venturini, R. (eds.) SPIRE 2017. LNCS, vol. 10508, pp. 129–143. Springer, Cham (2017). https://doi.org/10.1007/978-3-319-67428-5_12
7. Chen, Z.-Z., Ueta, S., Li, J., Wang, L.: Finding a center tree of phylogenetic trees via leaf removal. In: Proceedings of the 2018 IEEE International Conference on Bioinformatics and Biomedicine (to appear)
8. Cole, R., Farach-Colton, M., Hariharan, R., Przytycka, T.M., Thorup, M.: An $O(n \log n)$ algorithm for the maximum agreement subtree problem for binary trees. SIAM J. Comput. **30**, 1385–1404 (2000)
9. Guillemot, S., Mnich, M.: Kernel and fast algorithm for dense triplet inconsistency. Theor. Comput. Sci. **494**, 134–143 (2013)
10. Li, M., Tromp, J., Zhang, L.: On the nearest neighbour interchange distance between evolutionary trees. J. Theor. Biol. **182**, 463–467 (1996)
11. Ma, B., Wang, L., Zhang, L.: Fitting distances by tree metrics with increment error. J. Comb. Optim. **3**, 213–225 (1999)
12. Maddison, W.P.: Gene trees in species trees. Syst. Biol. **46**, 523–536 (1997)
13. Nakhleh, L., Warnow, T., Lindner, C.R., John, K.S.: Reconstructing reticulate evolution in species - theory and practice. J. Comput. Biol. **12**, 796–811 (2005)

14. Robinson, D., Foulds, L.: Comparison of phylogenetic trees. Math. Biosci. **53**, 131–147 (1981)
15. Swofford, D., Olsen, G., Waddell, P., Hillis, D.: Phylogenetic inference. In: Hillis, D., Moritz, D., Mable, B. (eds.) Molecular Systematics, 2nd edn, pp. 407–514. Sinauer Associates, Sunderiand (1996)

The Review of Bioinformatics Tool for 3D Plant Genomics Research

Xiangyu Yang[1], Zhenghao Li[2], Jingtian Zhao[1], Tao Ma[2],
Pengchao Li[3], and Le Zhang[1(✉)]

[1] College of Computer Science, Sichuan University, Chengdu 610065, China
zhangle06@scu.edu.cn
[2] College of Life Sciences, Sichuan University, Chengdu 610065, China
[3] School of Computer and Information Science, Southwest University,
Beibei District, Chongqing 400715, China

Abstract. Since the genome of the nucleus is a complicated three-dimensional spatial structure but not a single linear structure, biologists consider that 3D structure of plant chromatin is highly correlated with the function of the genome, which can be used to study the regulation mechanisms of genes and their evolutionary process. Because plants are more prone to chromosome structural variation and the 3D structure of plant chromatin are highly correlated with the function of the genome, it is important to investigate the impact of chromosome structural variation on gene expression by analyzing 3D structure. Here, we will briefly review the current bioinformatics tools for 3D plant genome study, which covers Hi-C data processing tools, then are the tools for A and B compartments identification, topologically associated domains (TAD) identification, identification of significant interactions and visualization. And then, we could provide the useful information for the related 3D plant genomics research scientists to select the appropriate tools according to their study. Finally, we discuss how to develop the future 3D genomic plant bioinformatics tools to keep up with the pace of scientific research development.

Keywords: Plant three-dimensional genomes · Hi-C · Bioinformatics tools · Chromatin spatial structure

1 Introduction

Previous Studies indicate that the genome in the nucleus is not a single linear structure, but a complex three-dimensional (3D) architecture that displays a hierarchical pattern [1–6] (see Fig. 1). Since plants usually are polyploid and have a more complex genome structure rather than animals [7], the 3D structure of the plant genome is more complicated than that of animals. Moreover, because plants are more prone to chromosome structural variation [8] and the 3D structure of plant chromatin are highly correlated with the function of the genome [9, 10], it is important to investigate the impact of chromosome structural variation on gene expression by analyzing 3D structure. Now, 3D structure of plant chromatin is mainly used to study the regulation mechanisms of genes and their evolutionary process [11, 12]. High-throughput chromosome conformation

© Springer Nature Switzerland AG 2019
Z. Cai et al. (Eds.): ISBRA 2019, LNBI 11490, pp. 16–27, 2019.
https://doi.org/10.1007/978-3-030-20242-2_2

capture (Hi-C) technique is a main method to probe the three-dimensional architecture of whole genomes [12–15].

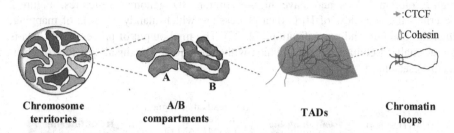

Chromosome territories   A/B compartments   TADs   Chromatin loops

➤:CTCF

◖:Cohesin

Fig. 1. Schematic representation of chromatin hierarchical architecture

The process of Hi-C experiment is crosslinking with formaldehyde, digesting DNA with a restriction enzyme, marking with biotin, ligation, de-crosslinking, shearing the DNA, capturing the biotin-containing fragments with streptavidin beads, purification, and using second-generation high-throughput sequencing for captured fragments [1, 3, 16]. However, the Hi-C data is not only big, the size of which could be dozens of GBs [1, 5], but also is comprised such invalid data that is inevitably generated during the experiment [16, 17]. Therefore, it is necessary to process and analyze the complicated Hi-C data by using bioinformatics tools.

Currently, the 3D plant genome research becomes a hot study area [12, 18–20], which focuses on constructing 3D genome map, A and B compartments identification, topologically associated domains (TADs) identification and identification of significant interactions. For example, Mascher et al. [21] locates the spatial organization of chromatin in the nucleus by using 3D genome map. Also, Wang et al. [12] use 3D maps to identify A/B compartments and significant interactions for cotton. In addition, Liu et al. [15] indicate that thousands of distinct TADs cover about a quarter of the rice genome.

However, since only a few studies comprehensively review the import bioinformatics tools for 3D architecture research of plant genomes, this review firstly introduce the Hi-C data processing tools, then are the tools for A and B compartments identification, topologically associated domains (TAD) identification, identification of significant interactions and visualization. Finally, we discuss the future research direction for 3D genomic plant bioinformatics tools.

2   3D Plant Genomic Bioinformatics Tools

2.1   Hi-C Data Processing Tool

After Hi-C experiment and high-throughput sequencing analysis [1], we obtain raw Hi-C data which are comprised of pair-end reads in FASTQ file format [1, 22]. At present, Hi-C requires high sequencing depth to have high resolution. However, as the sequencing depth increases, more low quality sequences are generated. Therefore, we have to preprocess the raw Hi-C data to receive the clean Hi-C data as following steps.

Firstly, we use FastQC [23] to control the quality of the reads, and then use the filtering tools such as Trimmomatic [24] to filter the invalid data. Moreover, several artificial intelligent algorithms, such as deDoc [25, 26], can use low-resolution Hi-C to predict high-resolution TADs and have high-resolution 3D genomic structures. Figure 2 describes the workflow of Hi-C data processing, which mainly consists of mapping, filtering, binning, and normalization [12, 27]. The final output of process Hi-C data is the normalized contact matrices, which is the basis for hierarchical structure of chromatin research.

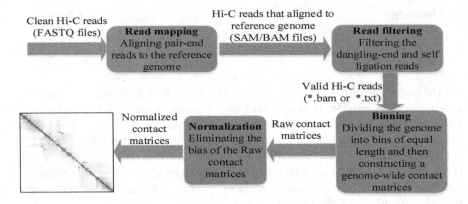

Fig. 2. The workflow for Hi-C data processing

There are many bioinformatics tools developed for Hi-C data preprocessing, which are listed by Table 1. Here, we introduce commonly used HiC-Pro [28], HOMER [29] and Juicer [30] in detail.

Table 1. Integration software for Hi-C data processing

Tools	Mapping, Filtering, Binning, Normalization	URL	Features
HOMER [29], developed by Java, R and Perl in 2010	×, √ √, √, √	http://homer.ucsd.edu/homer/interactions	Providing identification of significant interactions
Hiclib [17], developed by Python in 2012	√, √, √, √	https://bitbucket.org/mirnylab/hiclib	Iterative mapping accounts for the modification of fragments' sequences
HiCUP [31], developed by R and Perl in 2015	√, √, ×, ×	http://www.bioinformatics.babraham.ac.uk/projects/hicup	Mapping data to a specified reference genome and removing artefacts; Providing an easy-to-interpret quality control report
HiCdat [32], developed by R and C++ in 2015	√, √, √, √	https://github.com/MWSchmid/HiCdat	Easy-to-use; Focusing on the analysis of larger chromosomes structural features

(continued)

Table 1. *(continued)*

Tools	Mapping, Filtering, Binning, Normalization	URL	Features
HiC-Pro [28], developed by Python in 2015	√, √, √, √	https://github.com/ nservant/HiC-Pro	Fast iterative correction method; Memory-efficient data format; Distinguishing allele-specific interactions
Juicer [30], developed by Java in 2016	√, √, √, √	https://github.com/ theaidenlab/juicer	Processing terabase scale Hi-C datasets with a single click; Annotating TADs and loops automatically; Being compatible with multiple cluster operating systems
TADbit [33], developed by Python in 2017	√, √, √, √	https://github.com/ 3DGenomes/tadbit	Providing TAD identification and 3D model construction

HiC-Pro is an optimized and flexible pipeline to process Hi-C data from raw reads to normalized contact matrices [28]. It provides a fast implementation of the iterative correction method and a memory-efficient data format, which allows us to analyze allele-specific interaction effect. Though we can use HiC-Pro to do all the steps of Hi-C data processing by one instruction, it is not only hard to install and configure, but also needs a number of preparations before running.

The installation and configuration of HOMER [29] is much simpler than HIC-pro. HOMER will analyze data with overlapping windows when generating Hi-C contact matrices. It not only can identify significant intra-chromosomal interactions and inter-chromosomal interactions, but also does not penalize features that span boundaries [29]. However, when processing a very big Hi-C data, it costs a lot of memory for parallel computing. Moreover, the format of contact matrices file occupies a large storage space.

Juicer is not only an easy-to-use and fully-automated pipeline for the Hi-C data processing, but also can annotate structural features automatically [30]. Moreover, Juicer allows to process datasets at the tera base scale. However, the queues of its running alignments are so complicated that we have to usually modify the script.

2.2 A and B Compartments Identification

Plant chromatin regions can be divided into A and B compartments [12, 34]. The A compartment is usually associated with higher gene density, more active epigenetic marks, and higher transcription activity than B compartment. By analyzing A/B compartments, we can better understand the distribution pattern of euchromatin and heterochromatin in the plant chromosomes [34].

We usually convert the normalized contact matrices into Pearson correlation matrices. Then employ the first principal component (PC1) [35, 36] on Pearson correlation matrices [29, 37] to identify A and B compartments. The area where the value of PC1 is positive represents the A compartments and the negative represents B compartments (see Fig. 3). HOMER [29], HiTC [37] and Juicer [30] provide the runHiCpca.pl, pca.hic and Eigenvector function to identify A/B compartment, respectively.

For example, when using HOMER [29] to identify A/B compartment, the input is a Paired-End Tag Directory and the output is a txt and a bedGraph file of the PC1. The input of HiTC [37] is a normalized contact matrix and the output is a R program object which saves the data of PC1. As for Juicer [30], the input file is a.hic file that stores contact matrices from multiple resolutions and the output is a txt file containing the PC1 data. In summary, Homer's operation and output are better than another two tools.

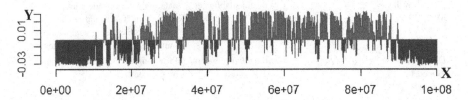

Fig. 3. A/B compartments analysis. The x-axis is the base coordinate in the species reference genome, and the y-axis is the first principal component (PC1). The red histogram represents the A compartment and the blue histogram is the B compartment. (Color figure online)

2.3 TADs Identification

As 3D genome of various plants have been studied, it has been found that TADs are also ubiquitous in plant chromatin [11, 12, 27]. The locations of housekeeping genes [38] are closely associated with cross-tissue conserved TADs. Therefore, the TADs identification helps us understand the spatial relationship of local chromatin regions and provides a potential relationship between regulatory elements and genes [39]. As an example, Fig. 4 describes the TAD schematic diagram by black square.

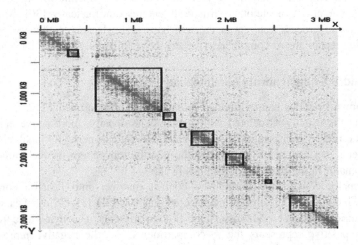

Fig. 4. TAD schematic diagram. Both the x-axis and the y-axis represent coordinates of the reference genome for Populus bolleana Lauche, and the TADs are marked by the black square.

The identification methods for TADs consists two categories [40]. One defines a one-dimensional test statistic from the contact matrices to detect TADs, assuming that TADs are non-overlapping. The other explores the two-dimensional structure of the contact matrices and identifies the TADs hierarchy by optimization or clustering algorithm [41–43]. Table 2 list the commonly used methods. Here, we will introduce TopDom [44], DI-HMM [2] and HiCseg [45] in detail.

TopDom is an efficient and deterministic method [44]. It is fast due to the linear time complexity. Moreover, it can identify the high accurate TADs by depending on the single and intuitive parameter. The input file is the normalized contact matrices, and the output is three files that provide start and end coordinates of the TAD domain on the chromosome.

DI-HMM [2] is to identify TADs based on one-dimensional signal. It devises directionality index (DI) to quantify the degree of upstream or downstream interaction bias for a genomic region, then uses a Hidden Markov model (HMM) based on the DI to identify biased "states" and therefore infers the locations of topological domains in the genome. However, HiCseg [45] identifies TADs based on image segmentation theory and maximum likelihood approach. Compared with DI-HMM, HiCseg and TopDom, TopDom not only runs much faster than DI-HMM and HiCseg, but also can identify more TADs in the same computational environment [44].

Table 2. Methods for identifying TADs

Method	Hierarchical structure	Principle	URL
DI-HMM [2] developed in 2012	NO	Devising directionality index (DI) to quantify the degree of upstream or downstream interaction bias for a genomic region, then use a Hidden Markov model (HMM) to identify biased "states"	https://media.nature.com/original/ nature-asets/nature/journal/v485/ n7398/extref/nature11082-s1.pdf
HiCseg [45] developed in 2014	NO	Performing image segmentation based on maximum likelihood approach	https://cran.r-project.org/web/ packages/HiCseg/index.html
Insulation Score [46] developed in 2015	NO	Devising insulation score to reflect the aggregate of interactions in the interval	https://github.com/dekkerlab/ crane-nature-2015
TopDom [44] developed in 2016	NO	The average contact frequency between regions within a TAD is much higher than between inside and outside regions	http://zhoulab.usc.edu/TopDom/
TADtree [40] developed in 2016	YES	Devising empirical distributions of contact frequencies within TADs, where positions that are far apart have a greater contacts enrichment than close positions	http://compbio.cs.brown.edu/ projects/tadtree/
HiTAD [47] developed in 2017	YES	Optimizing spatial TADs combinations based on the result of improved DI-HMM in space	https://pypi.python.org/pypi/ TADLib
TADbit [33] developed in 2017	NO	Using BIC-penalized likelihood	https://github.com/3DGenomes/ tadbit

2.4   Identification of Significant Interactions

The significant interaction [48, 49] indicates that the contact frequency between two regions of chromatins is significantly greater than in their neighborhood [5, 48, 50]. Significant interactions always are used to investigate the regulation and expression of plant genes [51, 52]. Table 3 lists commonly used methods for significant interactions. Here, we will introduce HOMER [29], Fit-Hi-C [53] and HiCCUPS [5] in detail.

Table 3. The methods for identification of significant interaction

Tools	Model	Features	URL
HOMER [29] developed in 2010	Binomial distribution	Multi-functional integrated software	http://homer.ucsd.edu/homer/interactions
HiCCUPS [5] developed in 2014	Poisson process	Eliminating the effects of TAD structure	https://github.com/aidenlab/juicer/wiki/Download
Fit-Hi-C [53] developed in 2014	Binomial distribution	Fitting contact frequency with the specified genomic distance twice	https://noble.gs.washington.edu/proj/fit-hi-c/
diffHic [54] developed in 2015	Negative binomial distribution	Using edgeR and biological duplication	https://bioconductor.org/packages/release/bioc/html/diffHic.html
HIPPIE [55] developed in 2015	Negative binomial distribution	Resolution based on the length of the restriction enzyme fragments; Providing error checking	http://wanglab.pcbi.upenn.edu/hippie
GOTHiC [48] developed in 2017	Binomial distribution	Calculating expected contact matrices; Provides a statistical framework for further data analysis	http://bioconductor.org/packages/release/bioc/html/GOTHiC.html
HiC-DC [50] developed in 2017	Zero-truncated negative binomial regression distribution	Considering the characteristics of the zero-inflation and over-dispersion of counts in the contact matrices	https://bitbucket.org/leslielab/hic.dc

HOMER [29] can obtain the results of significant interaction with only single instruction. The input file is a Paired-End Tag Directory, and the output is a text file that contains information about significant interactions. Since HOMER has a number of data analysis functions, it is convenient for us to carry out 3D genomes analysis.

Fit-Hi-C [53] uses the spline model to fit such contact frequency and distance for each pairs of loci twice that can identify a large number of significant interactions of chromatins. The analysis results are highly sensitive, low accurate and have high false positives.

HiCCUPS [5] is part of the Juicer [30] software suite, which can identify the significant interactions from the normalized Hi-C contact matrices. Although the number of significant interactions recognized by HiCCUPS is relatively small, the accuracy is higher than both Homer and Fit-Hi-C.

2.5   Visualization

For each aforementioned step, we will have the corresponding output data. However, these corresponding output data are too complicated to be intuitively understand. For example, the.hic file generated by the Juicer to represent the contact matrices infor-mation is in such a compressed binary file format that needs the help of the visual-ization tool to visualize the spatial organization and structure of chromatins. Here, we list commonly used visualization tools for plant Hi-C data by Table 4.

Table 4. The visualization applications

Tools	Year	GUI	URL
HiTC [37]	2012	NO	http://www.bioconductor.org/packages/release/bioc/html/HiTC.html
Juicebox [56]	2016	YES	https://github.com/aidenlab/Juicebox
HiCPlotter [57]	2015	NO	https://github.com/kcakdemir/HiCPlotter

HiTC [37], an R package, is compatible with HiC-Pro, which can visualize A/B compartments and the Hi-C contact matrices with annotation files like TADs. However, it cannot interact and has fewer options for operation. Moreover, since using HiTC requires professional programming skill, the tool is not easy to use.

Juicebox [56] is a visualization tool for the output of Juicer, which can use Hi-C contact matrices to make heat maps with tracks to visualize TADs and loops. In addition, Juicebox offers an interactive graphical user interface (GUI) and online ser-vice rather than HiTC. However, Juicebox can only visualize the low resolution heat map, but not support the high resolution chromatin heat map [30, 56].

HiCPlotter [57] is a Python data visualization tool for Hi-C contact matrices with different data types. It can add multiple types of tracks to the heat map of contact matrices. Compared with HiTC and Juicebox, an important advantage for HiCPlotter is to offer comparison function to compare two different contact matrices. However, the compatible capacity of HiCPlotter still needs to be improved that it cannot accept the contact matrices file produced by Juicer.

3   Summary

Due to the rapid development of high-throughput sequencing technology and Hi-C technology, there has been a surge in three-dimensional plant genome research recently.

Therefore, this study presents commonly used bioinformatics tools for plant 3D genome research from the Hi-C data processing, A/B compartments identification, TADs identification, identification of significant interactions and visualization. This study illustrates the advantages, shortcomings and application scope for the related bioinformatics tools, and tries to provide the useful information for the related 3D plant genomics research scientists to select the appropriate tools according to their study.

Though there are already a number of bioinformatics tools developed for 3D plant genomics study, the tool still should be continually optimized and innovated to keep up with the pace of scientific research development. Thus, we consider the following future direction of the bioinformatics tools developed for 3D plant genomics study. Firstly, we should increase the computational efficiency to support quickly accumulated 3D genomic data. Secondly, we should develop a unified data format and integrate the commonly used functions together to make a friend analytic tool for 3D plant study. Finally, we should improve such the existing tools by adding the comparative analysis function for the data sets from multiple experiments, that can help the researchers investigate the spatial structure similarities and differences in the same kind cells in various life cycles or homologous species.

Acknowledgements. This work was supported by the National Natural Science Foundation of China [61372138], the National Science and Technology Major Project [2018ZX10201002] and the Chinese Chongqing Distinguish Youth Funding [cstc2014jcyjjq40003].

References

1. Erez, L.A., Berkum, N.L., Van, L.W., Maxim, I., Tobias, R., Agnes, T.: Comprehensive mapping of long-range interactions reveals folding principles of the human genome. Science **326**, 289–293 (2009)
2. Dixon, J.R., Siddarth, S., Feng, Y., Audrey, K., Yan, L., Yin, S.: Topological domains in mammalian genomes identified by analysis of chromatin interactions. Nature **485**(7398), 376 (2012)
3. Job, D., Marti-Renom, M.A., Mirny, L.A.: Exploring the three-dimensional organization of genomes: interpreting chromatin interaction data. Nat. Rev. Genet. **14**(6), 390–403 (2013)
4. Doğan, E.S., Chang, L.: Three-dimensional chromatin packing and positioning of plant genomes. Nature Plants (2018)
5. Rao, S.S.P., Huntley, M.H., Durand, N.C., Stamenova, E.K., Bochkov, I.D., Robinson, J.T.: A 3D map of the human genome at kilobase resolution reveals principles of chromatin looping. Cell **159**(7), 1665–1680 (2014)
6. Xianglin, Z., Huan, F., Xiaowo, W.: The advancement of analysis methods of chromosome conformation capture data. Prog. Biochem. Biophys. **45**, 1093–1105 (2018)
7. Liu, B., Wendel, J.F.: Epigenetic phenomena and the evolution of plant allopolyploids. Mol. Phylogenetics Evol. **29**(3), 365–379 (2003)
8. Kellogg, E.A., Bennetzen, J.L.: The evolution of nuclear genome structure in seed plants. Am. J. Bot. **91**(10), 1709–1725 (2004)
9. Spielmann, M., Lupiáñez, D. G., Mundlos, S.: Structural variation in the 3D genome. Nat. Rev. Genet. **19**(7), 453–467 (2018)
10. Mishra, A., Hawkins, R.D.: Three-dimensional genome architecture and emerging technologies: looping in disease. Genome Med. **9**(1), 87 (2017)

11. Li, X., Wu, L., Wang, J., Sun, J., Xia, X., Geng, X.: Genome sequencing of rice subspecies and genetic analysis of recombinant lines reveals regional yield- and quality-associated loci. BMC Biol. **16**(1), 102 (2018)

12. Wang, M., Wang, P., Lin, M., Ye, Z., Li, G., Tu, L.: Evolutionary dynamics of 3D genome architecture following polyploidization in cotton. Nat. Plants **4**(2), 90 (2018)

13. Dudchenko, O., Batra, S.S., Omer, A.D., Nyquist, S.K., Hoeger, M., Durand, N.C.: De novo assembly of the Aedes aegypti genome using Hi-C yields chromosome-length scaffolds. Science **356**(6333), 92 (2017)

14. Wu, P., Li, T., Li, R., Jia, L., Zhu, P., Liu, Y.: 3D genome of multiple myeloma reveals spatial genome disorganization associated with copy number variations. Nat. Commun. **8**(1), 1937 (2017)

15. Liu, C., Cheng, Y.-J., Wang, J.-W., Weigel, D.: Prominent topologically associated domains differentiate global chromatin packing in rice from Arabidopsis. Nat. Plants **3**(9), 742 (2017)

16. Belton, J.M., Mccord, R.P., Gibcus, J.H., Naumova, N., Zhan, Y., Dekker, J.: Hi–C: a comprehensive technique to capture the conformation of genomes. Methods **58**(3), 268–276 (2012)

17. Imakaev, M., Fudenberg, G., Mccord, R.P., Naumova, N., Goloborodko, A., Lajoie, B.R.: Iterative correction of Hi-C data reveals hallmarks of chromosome organization. Nat. Methods **9**(10), 999 (2012)

18. Grob, S., Grossniklaus, U.: Chromosome conformation capture-based studies reveal novel features of plant nuclear architecture. Curr. Opin. Plant Biol. **36**, 149–157 (2017)

19. Rodriguezgranados, N.Y., Ramirezprado, J.S., Veluchamy, A., Latrasse, D., Raynaud, C., Crespi, M.: Put your 3D glasses on: plant chromatin is on show. J. Exp. Bot. **67**(11), 89 (2016)

20. Wang, C., Liu, C., Roqueiro, D., Grimm, D., Schwab, R., Becker, C.: Genome-wide analysis of local chromatin packing in Arabidopsis thaliana. Genome Res. **25**(2), 246–256 (2015)

21. Mascher, M., Gundlach, H., Himmelbach, A., Beier, S., Twardziok, S.O., Wicker, T.: A chromosome conformation capture ordered sequence of the barley genome. Nature **544** (7651), 427 (2017)

22. van Berkum, N.L., Lieberman-Aiden, E., Williams, L., Imakaev, M., Gnirke, A., Mirny, L. A.: Hi-C: a method to study the three-dimensional architecture of genomes. J. Vis. Exp. Jove **39**(39), 292–296 (2010)

23. FastQC. http://www.bioinformatics.babraham.ac.uk/projects/fastqc/. Accessed 21 Mar 2019

24. Bolger, A.M., Marc, L., Bjoern, U.: Trimmomatic: a flexible trimmer for Illumina sequence data. Bioinformatics **30**(15), 2114–2120 (2014)

25. Zhang, Y., An, L., Xu, J., Zhang, B., Zheng, W.J., Hu, M.: Enhancing Hi-C data resolution with deep convolutional neural network HiCPlus. Nat. Commun. **9**(1), 750 (2018)

26. Li, A., Yin, X., Xu, B., Wang, D., Han, J., Yi, W.: Decoding topologically associating domains with ultra-low resolution Hi-C data by graph structural entropy. Nat. Commun. **9** (1), 3265 (2018)

27. Wang, M., Tu, L., Yuan, D., Zhu, D., Shen, C., Li, J.: Reference genome sequences of two cultivated allotetraploid cottons, Gossypium hirsutum and Gossypium barbadense. Nat. Genet. **51**, 224 (2018)

28. Nicolas, S., Nelle, V., Lajoie, B.R., Eric, V., Chen, C.J., Jean-Philippe, V.: HiC-Pro: an optimized and flexible pipeline for Hi-C data processing. Genome Biol. **16**(1), 259 (2015)

29. Heinz, S., Benner, C., Spann, N., Bertolino, E., Lin, Y.C., Laslo, P.: Simple combinations of lineage-determining transcription factors prime -regulatory elements required for macrophage and B cell identities. Mol. Cell **38**(4), 576–589 (2010)

30. Durand, N., Shamim, M., Machol, I., Rao, S.P., Huntley, M., Lander, E.: Juicer provides a one-click system for analyzing loop-resolution Hi-C experiments. Cell Syst. **3**(1), 95–98 (2016)
31. Wingett, S., Ewels, P., Furlan-Magaril, M., Nagano, T., Schoenfelder, S., Fraser, P.: HiCUP: pipeline for mapping and processing Hi-C data. F1000res **4**, 1310 (2015)
32. Schmid, M.W., Grob, S., Grossniklaus, U.: HiCdat: a fast and easy-to-use Hi-C data analysis tool. BMC Bioinform. **16**(1), 1–6 (2015)
33. Serra, F., Baù, D., Goodstadt, M., Castillo, D., Filion, G., Marti-Renom, M.A.: Automatic analysis and 3D-modelling of Hi-C data using TADbit reveals structural features of the fly chromatin colors. PLoS Comput. Biol. **13**(7), e1005665 (2017)
34. Dong, P., Tu, X., Chu, P.-Y., Lü, P., Zhu, N., Grierson, D.: 3D chromatin architecture of large plant genomes determined by local A/B compartments. Mol. Plant **10**(12), 1497–1509 (2017)
35. Zhang, L., Zheng, C.Q., Li, T., Xing, L., Zeng, H., Li, T.T.: Building up a robust risk mathematical platform to predict colorectal cancer. Complexity, 14 (2017)
36. Yu, M., Ren, B.: The three-dimensional organization of mammalian genomes. Annu. Rev. Cell Dev. Biol. **33**(1), 265–289 (2017)
37. Servant, N., Lajoie, B.R., Nora, E.P., Giorgetti, L., Chen, C.J., Heard, E.: HiTC: exploration of high-throughput 'C' experiments. Bioinformatics **28**(21), 2843–2844 (2012)
38. Zhang, L., Xiao, M., Zhou, J., Yu, J.: Lineage-associated underrepresented permutations (LAUPs) of mammalian genomic sequences based on a Jellyfish-based LAUPs analysis application (JBLA). Bioinformatics **34**(21), 3624–3630 (2018)
39. Franke, M., Ibrahim, D.M., Andrey, G., Schwarzer, W., Heinrich, V., Schöpflin, R.: Formation of new chromatin domains determines pathogenicity of genomic duplications. Nature **538**(7624), 265–269 (2016)
40. Weinreb, C., Raphael, B.J.: Identification of hierarchical chromatin domains. Bioinformatics **32**(11), 1601 (2015)
41. Zhang, L., Liu, Y., Wang, M., Wu, Z., Li, N., Zhang, J.: EZH2-, CHD4-, and IDH-linked epigenetic perturbation and its association with survival in glioma patients. J. Mol. Cell Biol. **9**(6), 477–488 (2017)
42. Zhang, L., Qiao, M., Gao, H., Hu, B., Tan, H., Zhou, X.: Investigation of mechanism of bone regeneration in a porous biodegradable calcium phosphate (CaP) scaffold by a combination of a multi-scale agent-based model and experimental optimization/validation. Nanoscale **8**(31), 14877–14887 (2016)
43. Zhang, L., Zhang, S.: Using game theory to investigate the epigenetic control mechanisms of embryo development: Comment on: "Epigenetic game theory: How to compute the epigenetic control of maternal-to-zygotic transition" by Qian Wang et al. Phys. Life Rev. **20**, 140–142 (2017)
44. Shin, H., Shi, Y., Dai, C., Tjong, H., Gong, K., Alber, F.: TopDom: an efficient and deterministic method for identifying topological domains in genomes. Nucleic Acids Res. **44**(7), e70–e70 (2016)
45. Celine, L.L., Delattre, M., Mary-Huard, T., Robin, S.: Two-dimensional segmentation for analyzing Hi-C data. Bioinformatics **30**(17), 386–392 (2014)
46. Emily, C., Qian, B., Rachel, P.M., Lajoie, B.R., Wheeler, B.S., Ralston, E.J.: Condensin-driven remodelling of X chromosome topology during dosage compensation. Nature **523**(7559), 240 (2015)
47. Wang, X.T., Cui, W., Peng, C.: HiTAD: detecting the structural and functional hierarchies of topologically associating domains from chromatin interactions. Nucleic Acids Res. **45**(19), e163 (2017)

48. Mifsud, B., Martincorena, I., Darbo, E., Sugar, R., Schoenfelder, S., Fraser, P.: GOTHiC, a probabilistic model to resolve complex biases and to identify real interactions in Hi-C data. PLoS ONE **12**(4), e0174744 (2017)
49. Forcato, M., Nicoletti, C., Pal, K., Livi, C.M., Ferrari, F., Bicciato, S.: Comparison of computational methods for Hi-C data analysis. Nat. Methods **14**(7), 679 (2017)
50. Carty, M., Zamparo, L., Sahin, M., González, A., Pelossof, R., Elemento, O.: An integrated model for detecting significant chromatin interactions from high-resolution Hi-C data. Nat. Commun. **8**, 15454 (2017)
51. Liu, C., Weigel, D.: Chromatin in 3D: progress and prospects for plants. Genome Biol. **16**(1), 170 (2015)
52. Liu, C., Wang, C., Wang, G., Becker, C., Zaidem, M., Weigel, D.: Genome-wide analysis of chromatin packing in Arabidopsis thaliana at single-gene resolution. Genome Res. **26**(8), 1057 (2016)
53. Ay, F., Bailey, T.L., Noble, W.S.: Statistical confidence estimation for Hi-C data reveals regulatory chromatin contacts. Genome Res. **24**(6), 999 (2014)
54. Lun, A.T.L., Smyth, G.K.: diffHic: a Bioconductor package to detect differential genomic interactions in Hi-C data. BMC Bioinform. **16**, 1258 (2015)
55. Hwang, Y.C., Lin, C.F., Valladares, O., Malamon, J., Kuksa, P.P., Zheng, Q.: HIPPIE: a high-throughput identification pipeline for promoter interacting enhancer elements. Bioinformatics **31**(8), 1290–1292 (2015)
56. Durand, N., Robinson, J., Shamim, M., Machol, I., Mesirov, J., Lander, E.: Juicebox provides a visualization system for Hi-C contact maps with unlimited zoom. Cell Syst. **3**(1), 99–101 (2016)
57. Akdemir, K.C., Chin, L.: HiCPlotter integrates genomic data with interaction matrices. Genome Biol. **16**(1), 198 (2015)

Sorting by Reversals, Transpositions, and Indels on Both Gene Order and Intergenic Sizes

Klairton Lima Brito[1]([✉])(iD), Géraldine Jean[2](iD), Guillaume Fertin[2](iD),
Andre Rodrigues Oliveira[1](iD), Ulisses Dias[3](iD), and Zanoni Dias[1](iD)

[1] Institute of Computing, University of Campinas, Campinas, Brazil
{klairton,andrero,zanoni}@ic.unicamp.br
[2] LS2N, UMR CNRS 6004, University of Nantes, Nantes, France
{geraldine.jean,guillaume.fertin}@univ-nantes.fr
[3] School of Technology, University of Campinas, Limeira, Brazil
ulisses@ft.unicamp.br

Abstract. During the evolutionary process, the genome is affected by various genome rearrangements, which are events that modify large stretches of the genetic material. In the literature, several models were designed to estimate the number of events that occurred during the evolution, but these models represent genomes as a sequence of genes, overlooking the genetic material between consecutive genes. However, recent studies show that taking into account the genetic material present between consecutive genes can be more realistic. Reversal and transposition are genome rearrangements widely studied in the literature. A reversal inverts a segment of the genome while a transposition swaps the positions of two consecutive segments. Genomes also undergo non-conservative events (events that alter the amount of genetic material) such as insertion and deletion, which insert and remove genetic material from intergenic regions of the genome, respectively. We study problems considering both gene order and intergenic regions size. We investigate the reversal distance between two genomes in two scenarios: with and without non-conservative events. For both problems, we show that they belong to NP-hard problems class and we present a 4-approximation algorithm. We also study the reversal and transposition distance between two genomes (and the variation with non-conservative events) and we present a 6-approximation algorithm.

Keywords: Genome rearrangements · Intergenic regions · Approximation algorithms

1 Introduction

Genome rearrangements are events that insert, remove or change the position and/or orientation of large stretches of the genetic material. When we compare

Z. Cai et al. (Eds.): ISBRA 2019, LNBI 11490, pp. 28–39, 2019.
https://doi.org/10.1007/978-3-030-20242-2_3

the genomes of two individuals, one of the main goals is to estimate the sequence of rearrangement events that transformed one genome into the other.

A model $\mathcal{M}$ determines the set of rearrangement events allowed to modify the genome. The size of the smallest sequence of rearrangement events in a model $\mathcal{M}$ capable of transforming a genome into another is called *rearrangement distance*. When we assume no duplicated gene, we can map every gene to a unique number to represent genomes as permutations of integer elements. Usually, rearrangement problems are treated as sorting problems and the goal is to turn any genome π into a specific genome $\iota = (1, \ldots, n)$, which is called *identity permutation*. If the orientation of the genes is known, a positive or negative sign is assigned to each element. Otherwise, signs are omitted.

When the gene orientations are unknown, Caprara [6] proved that the Sorting Permutations by Reversals problem belongs to NP-hard problems class. The best algorithm has an approximation factor of 1.375 and was presented by Berman *et al.* [1]. Sorting Permutations by Transpositions problem also belongs to NP-hard problems class [4] and the best algorithm has an approximation factor of 1.375 [7]. When we consider a model allowing reversals and transpositions we have the Sorting Permutations by Reversals and Transpositions problem. The best algorithm for this problem has an approximation factor of $2.8334 + \epsilon$, for any $\epsilon > 0$ [10] and its complexity is unknown.

The representation of a genome only as a gene sequence is a technique very useful, but all the information that is not present on the genes are discarded that implies in loss of information. In particular, the intergenic regions between consecutive genes are not taken into account by these representations. Recently, authors have suggested that incorporating this information may improve the distance estimate from an evolutionary point of view [2,3]. Thus, it is justified to perform investigations considering the order of the genes and the size of the intergenic regions. Works considering gene order and intergenic sizes have already been presented. A model allowing Double-Cut and Join (DCJ) operation was presented together with the NP-hard proof and a 4/3-approximation algorithm [8]. The DCJ [11] is a rearrangement event that cuts the genome into two points and the extremities of the resulting segments are reassembled following specific criteria. A model allowing DCJs, insertions, and deletions also was investigated and an exact polynomial time algorithm was designed [5] when insertions and deletions act only on intergenic regions. Besides, Oliveira *et al.* [9] presented a model that allows the use of only super short reversals (reversals that affect at most two genes) also considering intergenic regions size.

In this paper, we consider that gene orientations are unknown. We investigate four problems to estimate the distance between genomes also taking into account intergenic regions, two of them allowing just conservative events of reversal and transposition: *Sorting by Intergenic Reversals (SbIR)* and *Sorting by Intergenic Reversals and Transpositions (SbIRT)*. We also investigate two other problems that allow non-conservative events of insertion and deletion: *Sorting by Intergenic Reversals, Insertions, and Deletions (SbIRID)* and *Sorting by Intergenic Reversals, Transpositions, Insertions, and Deletions (SbIRTID)*.

This manuscript is organized as follows. Section 2 provides definitions that are used throughout the paper. Section 3 presents the complexity proofs for SbIR and SbIRID problems and an approximation algorithm for each problem addressed in this manuscript. Section 4 concludes the paper.

2 Basic Definitions

Given a genome $\mathcal{G}$ as a sequence of n genes $g_1, \ldots, g_n$ and a sequence of $n + 1$ intergenic regions $r_1, \ldots, r_{n+1}$, each gene is surrounded by two intergenic regions so that: $\mathcal{G} = (r_1, g_1, r_2, g_2, \ldots, r_n, g_n, r_{n+1})$. Our model assumes that (i) genes orientation is unknown, (ii) there are no duplicate genes, and (iii) the considered genomes share the same set of genes. In this way, we can assign each gene a unique value and map the gene sequence as a permutation $\pi = (\pi_1\ \pi_2\ \cdots\ \pi_n)$, $\pi_i \in \mathbb{N}$, $1 \le \pi_i \le n$, and $\pi_i \ne \pi_j$ for all $i \ne j$.

Rearrangement events can split intergenic regions, and each intergenic region has a well-defined amount of nucleotides. Thus, we represent it by the size (amount of nucleotides). The sequence of intergenic regions $\breve{\pi} = (\breve{\pi}_1\ \breve{\pi}_2\ \cdots\ \breve{\pi}_{n+1})$, $\breve{\pi}_i \in \mathbb{N}$, represents the respective intergenic region sizes, such that $\breve{\pi}_i$ is on the left side of π_i, whereas $\breve{\pi}_{i+1}$ is on the right side.

Since we represent genes as a permutation π, we can treat the problem as a sorting problem in which the target permutation is the identity ι. This approach is widely used and it means that if we are able to transform $(\pi, \breve{\pi})$ into $(\iota, \breve{\iota})$ we can also transform $(\alpha, \breve{\alpha})$ into $(\sigma, \breve{\sigma})$. Therefore, an **instance** of our problem is composed by three elements $(\pi, \breve{\pi}, \breve{\iota})$ and the **rearrangement distance** considering a model $\mathcal{M}$ is represented as $d_{\mathcal{M}}(\pi, \breve{\pi}, \breve{\iota})$.

Definition 1. *An intergenic reversal $\rho_{(x,y)}^{(i,j)}$, with $1 \le i \le j \le n$, $0 \le x \le \breve{\pi}_i$, $0 \le y \le \breve{\pi}_{j+1}$, and $\{x, y\} \subset \mathbb{N}$, splits the intergenic regions $\breve{\pi}_i$ and $\breve{\pi}_{j+1}$ into pieces with sizes $(x, x' = \breve{\pi}_i - x)$ and $(y, y' = \breve{\pi}_{j+1} - y)$, respectively. The sequence $(x', \pi_i, \ldots, \pi_j, y)$ is inverted and the pieces (x, y) and (x', y') form new intergenic regions $\breve{\pi}_i$ and $\breve{\pi}_{j+1}$, respectively.*

Definition 2. *An intergenic transposition $\tau_{(x,y,z)}^{(i,j,k)}$, with $1 \le i < j < k \le n + 1$, $0 \le x \le \breve{\pi}_i$, $0 \le y \le \breve{\pi}_j$, $0 \le z \le \breve{\pi}_k$, and $\{x, y, z\} \subset \mathbb{N}$, splits the intergenic regions $\breve{\pi}_i$, $\breve{\pi}_j$, and $\breve{\pi}_k$ into pieces with sizes $(x, x' = \breve{\pi}_i - x)$, $(y, y' = \breve{\pi}_j - y)$, and $(z, z' = \breve{\pi}_k - z)$, respectively. The sequences $(x', \pi_i, \ldots, y)$ and $(y', \pi_j, \ldots, z)$ swap positions without changing orientation, and pieces (x, y'), (z, x'), and (y, z') form new intergenic regions $\breve{\pi}_i$, $\breve{\pi}_{k+i-j}$, and $\breve{\pi}_k$, respectively.*

Figures 1(a) and (b) show a generic example of intergenic reversal and intergenic transposition, respectively.

Definition 3. *An intergenic insertion ϕ_x^i, such that $1 \le i \le (n+1)$, $x > 0$, and $x \in \mathbb{N}$ acts on the intergenic region $\breve{\pi}_i$ inserting an amount x of nucleotides.*

Definition 4. *An intergenic deletion ψ_x^i, such that $1 \le i \le (n+1)$, $0 < x \le \breve{\pi}_i$, and $x \in \mathbb{N}$ acts on the intergenic region $\breve{\pi}_i$ removing an amount x of nucleotides.*

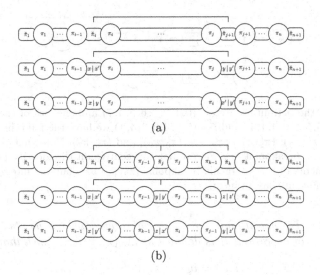

(a)

(b)

Fig. 1. The figure illustrates two intergenic genome rearrangement operations, the reversal (a) and transposition (b).

From now on, we will refer to the operations simply as a reversal, transposition, insertion, and deletion. Since reversal and transposition do not insert or remove genetic material, when dealing solely with conservative events the sum of intergenic regions size in both genomes is the same.

Definition 5. *Given a permutation π with n elements, the extended permutation π' is a permutation with two new elements $\pi'_0 = 0$ and $\pi'_{n+1} = n + 1$.*

From now on, we use the term permutation to denote the extended permutation. We say that an intergenic region is of (in)correct size if it occurs between two elements consecutive both in π' and ι' and whose size is (not) the same in π' and ι'.

Definition 6. *An intergenic breakpoint is a pair of elements π'_i and π'_{i+1} of the permutation π', such that either they are not consecutive in the identity permutation ι', or they are consecutive and the intergenic region between them has an incorrect size.*

An *intergenic breakpoint* represents a region that at some point has to be affected by some operation either to fix the gene order or the size of the intergenic region. Figure 2 shows examples of intergenic breakpoints.

Definition 7. *The number of intergenic breakpoints is denoted as $ib(\pi', \breve{\pi}, \breve{\iota})$.*

Remark 1. The instance $(\iota', \breve{\iota}, \breve{\iota})$ has a property that does not occur in any other instance: $ib(\iota', \breve{\iota}, \breve{\iota}) = 0$.

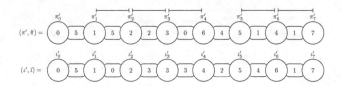

Fig. 2. Given the permutation $\pi' = (0, 1, 2, 3, 6, 5, 4, 7)$ and the size of the intergenic regions $\breve{\pi} = (5, 5, 2, 0, 4, 1, 1)$ and $\breve{\iota} = (5, 0, 3, 3, 2, 4, 1)$ we have the following intergenic breakpoints: (π'_1, π'_2), (π'_2, π'_3), (π'_3, π'_4), (π'_5, π'_6), and (π'_6, π'_7). Note that the elements of the intergenic breakpoint (π'_3, π'_4) are not consecutive on the identity permutation while the elements of the intergenic breakpoint (π'_2, π'_3) are, but the intergenic region between the elements has an incorrect size.

Definition 8. *The variation in the number of intergenic breakpoints after applying a sequence of operations S is denoted as $\Delta_{ib}(\pi', \breve{\pi}, \breve{\iota}, S)$, such that:*

$$\Delta_{ib}(\pi', \breve{\pi}, \breve{\iota}, S) = ib((\pi', \breve{\pi}, \breve{\iota}) \cdot S) - ib(\pi', \breve{\pi}, \breve{\iota}).$$

Definition 9. *A pair (a, b) is a block in a permutation π' if $|a - b| = 1$, the elements a and b are consecutive in the permutation π', and the intergenic region between them has a correct size.*

Definition 10. *Two intergenic breakpoints (π'_i, π'_{i+1}) and (π'_j, π'_{j+1}), such that $i < j$, are connected if the following conditions are fulfilled:*

i. At least one of the pairs (π'_i, π'_{i+1}), (π'_j, π'_{j+1}), (π'_i, π'_j), (π'_i, π'_{j+1}), (π'_{i+1}, π'_j), or (π'_{i+1}, π'_{j+1}) corresponds to two consecutive elements in the identity permutation ι' that do not form a block in the permutation π'.

ii. $\breve{\pi}_{i+1} + \breve{\pi}_{j+1} \geq \breve{\iota}_k$, such that $\breve{\iota}_k$ is the size of the intergenic region between the pair of consecutive elements in the identity permutation ι'.

Connected intergenic breakpoints represent regions that have the potential to remove at least one intergenic breakpoint by placing two consecutive elements and fixing the size of the intergenic region between them. For example, in Fig. 2, the intergenic breakpoints (π'_3, π'_4) and (π'_6, π'_7) are connected while the intergenic breakpoints (π'_2, π'_3) and (π'_3, π'_4) are not.

3   Results

We start this section by showing that problems SbIR and SbIRID belong to the problem class NP-hard. For this, we used a reduction of Sorting by Reversals problem, which does not consider intergenic regions. Then, we present lower bounds (Subsect. 3.1) and approximation algorithms (Subsect. 3.2) for each of the variations of the problems addressed in this work.

The Sorting by Reversals problem (SbR) has already been proven NP-hard [6]. An instance of this problem consists of a permutation δ and a natural number d. The goal is to determine if its possible to transform δ into ι applying at most d reversals.

Lemma 1. *SbIR problem is NP-hard.*

Proof. We can reduce all instances of SbR to instances of SbIR by setting $\pi = \delta$ and $\breve{\pi} = \breve{\iota} = (0\ 0\ ...\ 0)$. Note that it is possible to transform δ into ι applying at most d reversals if and only if $d_{SbIR}(\pi, \breve{\pi}, \breve{\iota}) \leq d$. □

Lemma 2. *SbIRID problem is NP-hard.*

Proof. We can reduce all instances of SbR to instances of SbIRID by setting $\pi = \delta$ and $\breve{\pi} = \breve{\iota} = (0\ 0\ ...\ 0)$. Note that for these instances of the SbIRID problem no insertion and deletion will be applied, otherwise we could get a smaller sequence of reversals just by ignoring the insertions and deletions. That way, it is possible to transform δ into ι applying at most d reversals if and only if $d_{SbIRID}(\pi, \breve{\pi}, \breve{\iota}) \leq d$. □

3.1   Lower Bounds

Following lemmas present lower bounds for each problem we consider.

Lemma 3. $\Delta_{ib}(\pi', \breve{\pi}, \breve{\iota}, \rho) \geq -2$ *for any reversal ρ.*

Proof. Suppose that (π'_{i-1}, π'_i) and (π'_j, π'_{j+1}) are intergenic breakpoints. In this case, the best scenario after applying reversal $\rho^{(i,j)}_{(x,y)}$ removes the intergenic breakpoints (π'_{i-1}, π'_i) and (π'_j, π'_{j+1}), reducing the number of intergenic breakpoints by two. Since any reversal only affects the neighborhood of two pairs of genes and two intergenic regions it is impossible to remove more than two intergenic breakpoints. □

Lemma 4. $\Delta_{ib}(\pi', \breve{\pi}, \breve{\iota}, \tau) \geq -3$ *for any transposition τ.*

Proof. Suppose that (π'_{i-1}, π'_i), (π'_{j-1}, π'_j), and (π'_{k-1}, π'_k) are intergenic breakpoints. In this case, the best scenario is a transposition $\tau^{(i,j,k)}_{(x,y,z)}$ that removes the intergenic breakpoints (π'_{i-1}, π'_i), (π'_{j-1}, π'_j), and (π'_{k-1}, π'_k), reducing the number of intergenic breakpoints by three. Since any transposition only affects the neighborhood of three pairs of genes and three intergenic regions it is impossible to remove more than three intergenic breakpoints. □

Lemma 5. $\Delta_{ib}(\pi', \breve{\pi}, \breve{\iota}, \phi) \geq -1$ *for any insertion ϕ.*

Proof. As an insertion acts in just one intergenic region this means that the best scenario is to remove the intergenic breakpoint (π'_{i-1}, π'_i) after applying an insertion ϕ^i_x, reducing by one the number of intergenic breakpoints. □

Lemma 6. $\Delta_{ib}(\pi', \breve{\pi}, \breve{\iota}, \psi) \geq -1$ *for any deletion ψ.*

Proof. As a deletion acts in just one intergenic region this means that the best scenario is to remove the intergenic breakpoint (π'_{i-1}, π'_i) after applying a deletion ψ^i_x, reducing by one the number of intergenic breakpoints. □

Theorem 1. $d_{SbIR}(\pi, \breve{\pi}, \breve{\iota}) \geq \frac{ib(\pi', \breve{\pi}, \breve{\iota})}{2}$.

Proof. By the Remark 1, we know that $(\iota', \breve{\iota}, \breve{\iota})$ is the only instance with no intergenic breakpoints. To achieve the identity permutation and to fix the intergenic region sizes we need to remove $ib(\pi', \breve{\pi}, \breve{\iota})$ intergenic breakpoints. Also, by Lemma 3, a reversal removes at most two intergenic breakpoints and lemma follows.   □

Theorem 2. $d_{SbIRID}(\pi, \breve{\pi}, \breve{\iota}) \geq \frac{ib(\pi', \breve{\pi}, \breve{\iota})}{2}$.

Proof. Directly by Lemmas 5 and 6, and Theorem 1.   □

Theorem 3. $d_{SbIRT}(\pi, \breve{\pi}, \breve{\iota}) \geq \frac{ib(\pi', \breve{\pi}, \breve{\iota})}{3}$.

Proof. Directly by Remark 1 and Lemmas 3 and 4.   □

Theorem 4. $d_{SbIRTID}(\pi, \breve{\pi}, \breve{\iota}) \geq \frac{ib(\pi', \breve{\pi}, \breve{\iota})}{3}$.

Proof. Directly by Lemmas 5 and 6, and Theorem 3.   □

3.2   Approximation Algorithms

In this subsection, we will present four approximation algorithms. Initially, we will show an algorithm with approximation factor 4 for the SbIR and SbIRID problems. Then, we will present an algorithm with approximation factor 6 for the SbIRT and SbIRTID problems.

Lemma 7. *Let $(\pi, \breve{\pi}, \breve{\iota})$ be an instance such that $\sum_{i=1}^{n+1} \breve{\pi}_i \geq \sum_{i=1}^{n+1} \breve{\iota}_i$ and the number of intergenic breakpoints is greater than one. It is always possible to find at least one pair of intergenic breakpoints that are connected.*

Proof. Since $ib(\pi', \breve{\pi}, \breve{\iota}) > 1$, we can find at least a pair of intergenic breakpoints. We have to show that at least one of those pairs will be connected. Suppose that exists an instance $(\pi, \breve{\pi}, \breve{\iota})$, such that $\sum_{i=1}^{n+1} \breve{\pi}_i \geq \sum_{i=1}^{n+1} \breve{\iota}_i$, $ib(\pi', \breve{\pi}, \breve{\iota}) > 1$, and there is not a pair of intergenic breakpoints that are connected. The possibilities for not finding such a pair of intergenic breakpoints are:

– For all pairs of intergenic breakpoints (π'_i, π'_{i+1}) and (π'_j, π'_{j+1}) the elements (π'_i, π'_{i+1}), (π'_j, π'_{j+1}), (π'_i, π'_j), (π'_i, π'_{j+1}), (π'_{i+1}, π'_j), and (π'_{i+1}, π'_{j+1}) are not consecutive in the identity permutation, but if it is true π cannot be a permutation.
– For all pairs of intergenic breakpoints (π'_i, π'_{i+1}) and (π'_j, π'_{j+1}) we do not have enough intergenic material to remove any intergenic breakpoint $\breve{\pi}_{i+1} + \breve{\pi}_{j+1} < \breve{\iota}_k$, such that $\breve{\iota}_k$ is the size of the intergenic region between the pair of consecutive elements in the identity permutation. If it is true it implies that $\sum_{i=1}^{n+1} \breve{\pi}_i < \sum_{i=1}^{n+1} \breve{\iota}_i$ and that contradicts the initial assumption.   □

Lemma 8. *Let (π'_i, π'_{i+1}) and (π'_j, π'_{j+1}) be intergenic breakpoints that are connected. It is possible to remove at least one intergenic breakpoint after at most two reversals.*

Proof. We analyze the possibilities to remove an intergenic breakpoint based on a pair of connected intergenic breakpoints.

i. (π'_i, π'_j) or (π'_{i+1}, π'_{j+1}) are consecutive in the identity permutation: these cases are symmetric and we need to apply only one reversal $\rho^{(i+1,j)}_{(x,y)}$ to place the element π'_j on the right side of the element π'_i or π'_{i+1} on the left side of π'_{j+1}. As $\breve{\pi}_{i+1} + \breve{\pi}_{j+1} \geq \breve{\iota}_k$, then the x and y parameters can always be chosen properly to fill the intergenic region with the correct size between the consecutive elements generated (Fig. 3(a)).

ii. (π'_i, π'_{j+1}): in this case we apply two consecutive reversals. In this scenario, we need an intergenic breakpoint (π'_k, π'_{k+1}), such that $k < i$ or $k > j$, to apply the sequence of reversals without creating new intergenic breakpoints. We will prove that exists such intergenic breakpoint by contradiction. Suppose that there is no intergenic breakpoint (π'_k, π'_{k+1}) such that $k < i$ or $k > j$. This means that the segments $(\pi'_0, \dots, \pi'_i)$ and $(\pi'_{j+1}, \dots, \pi'_{n+1})$ are composed of consecutive elements with no intergenic breakpoint between them; also we know that π'_i and π'_{j+1} are consecutive elements, but if both statements are true it implies that there are no valid values for the elements π_{i+1} and π_j of the permutation π. If $k < i$ we apply a reversal $\rho^{(k+1,i)}_{(0,\breve{\pi}_{i+1})}$ to obtain the case (i) (Fig. 3(b)). If $k > j$ we apply a reversal $\rho^{(j+1,k)}_{(0,\breve{\pi}_{k+1})}$ to obtain the case (i) (Fig. 3(c)). Note that in both scenarios the intergenic regions sizes remains the same and the case (i) can be applied (Fig. 3(b)).

iii. (π'_{i+1}, π'_j): In this case we apply two consecutive reversals. In this scenario, we need an intergenic breakpoint (π'_k, π'_{k+1}), such that $k > i$ and $k < j$, to apply the sequence of reversals without creating new intergenic breakpoints. We will prove that exists such intergenic breakpoint by contradiction. Suppose that there is no intergenic breakpoint (π'_k, π'_{k+1}) such that $k > i$ and $k < j$. This means that the segment $(\pi'_{i+1}, \dots, \pi'_j)$ is composed of consecutive elements with no intergenic breakpoint between them; also we know that (π'_{i+1}, π'_j) are consecutive elements, but if both statements are true implies that there are no valid values for the elements π_{i+1} and π_j of the permutation π. After identifying the intergenic breakpoint (π'_k, π'_{k+1}) we apply a reversal $\rho^{(i+1,k)}_{(0,\breve{\pi}_{k+1})}$ (Fig. 3(d)) to obtain the case (i).

iv. (π'_i, π'_{i+1}) or (π'_j, π'_{j+1}): these cases are symmetric and we need to apply two consecutive reversals. Initially, we apply a reversal $\rho^{(i+1,j)}_{(0,\breve{\pi}_{j+1})}$ without changing the intergenic regions sizes, as result we obtain the case (i) (Fig. 3(e)).   □

Theorem 5. *SbIR problem is 4-approximable.*

Proof. While the permutation is not sorted and while the permutation has intergenic regions with incorrect size, it is always possible to remove at least one intergenic breakpoint after applying at most two reversals (Lemmas 7 and 8). In the worst case, it gives us a total of $2ib(\pi', \breve{\pi}, \breve{\iota})$ reversals to transform $(\pi, \breve{\pi}, \breve{\iota})$ into $(\iota, \breve{\iota}, \breve{\iota})$. By the Theorem 1, we obtained the lower bound $ib(\pi', \breve{\pi}, \breve{\iota})/2$ and the theorem follows.   □

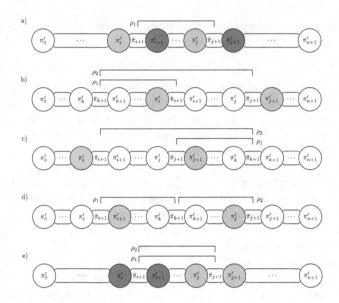

Fig. 3. The possibilities that can be found when we have a pair of connected intergenic breakpoints and the operations of reversal that must be applied to remove at least one intergenic breakpoint. The pair of elements that are consecutive in the identity permutation are represented with a gray scale color.

Lemma 9. *Let* $(\pi, \breve{\pi}, \breve{\iota})$ *be an instance of the Sorting by Intergenic Reversals, Insertions, and Deletions problem, such that* $ib(\pi', \breve{\pi}, \breve{\iota}) > 0$. *It is always possible to find an insertion* ϕ *such that* $\Delta_{ib}(\pi', \breve{\pi}, \breve{\iota}, \phi) \leq 0$.

Proof. Since $ib(\pi', \breve{\pi}, \breve{\iota}) > 0$, then it exists at least one intergenic breakpoint that we can apply an insertion in this region. Therefore, in the worst case the amount of intergenic breakpoints remains the same. □

Lemma 10. *Let* $(\pi', \breve{\pi}, \breve{\iota})$ *be an instance of the Sorting by Intergenic Reversals, Insertions, and Deletions problem, such that* $ib(\pi', \breve{\pi}, \breve{\iota}) = 1$ *and* $\sum_{i=1}^{n+1} \breve{\pi}_i > \sum_{i=1}^{n+1} \breve{\iota}_i$. *It is always possible to find a deletion* ψ *such that* $\Delta_{ib}(\pi', \breve{\pi}, \breve{\iota}, \psi) = -1$.

Proof. Since $ib(\pi', \breve{\pi}, \breve{\iota}) = 1$, then we know that $\pi' = \iota'$, otherwise the number of intergenic breakpoints should be greater than one. Since all the elements of the permutation π' are consecutive and $\sum_{i=1}^{n+1} \breve{\pi}_i > \sum_{i=1}^{n+1} \breve{\iota}_i$, there is an intergenic region $\breve{\pi}_k$, such as $\breve{\pi}_k > \breve{\iota}_k$. Thus the deletion $\psi_{\breve{\iota}_k - \breve{\pi}_k}^k$ removes the intergenic breakpoint (π'_{k-1}, π'_k) and the lemma follows. □

Theorem 6. *SbIRID problem is 4-approximable.*

Proof. We are going to divide the proof into three cases:

  i. $\sum_{i=1}^{n+1} \breve{\pi}_i > \sum_{i=1}^{n+1} \breve{\iota}_i$: Lemmas 7 and 8 remain valid as long as $ib(\pi', \breve{\pi}, \breve{\iota}) > 1$, then we apply only one deletion to remove the last intergenic breakpoint (Lemma 10).

ii. $\sum_{i=1}^{n+1} \breve{\pi}_i = \sum_{i=1}^{n+1} \breve{\iota}_i$: Lemmas 7 and 8 are sufficient to sort the permutation and to fix the sizes of the intergenic regions by applying only reversals.

iii. $\sum_{i=1}^{n+1} \breve{\pi}_i < \sum_{i=1}^{n+1} \breve{\iota}_i$: Initially we apply an insertion to make $\sum_{i=1}^{n+1} \breve{\pi}_i = \sum_{i=1}^{n+1} \breve{\iota}_i$ (Lemma 9). Sequentially, Lemmas 7 and 8 guarantee that the permutation will be sorted and the sizes of the intergenic regions will be fixed applying only reversals. Note that it is not guaranteed that the initial insertion removes any intergenic breakpoint, but then only reversals are applied and the last reversal must remove two intergenic breakpoints. Considering the insertion and the last reversal, on average, we were able to remove one intergenic breakpoint after applying one operation.

Considering the three cases, in the worst scenario, it gives us a total of $2ib(\pi', \breve{\pi}, \breve{\iota})$ operations to transform $(\pi, \breve{\pi}, \breve{\iota})$ into $(\iota, \breve{\iota}, \breve{\iota})$. By Theorem 2, we obtained the lower bound $ib(\pi', \breve{\pi}, \breve{\iota})/2$ and the theorem follows. □

Lemma 11. *Let (π'_i, π'_{i+1}) and (π'_j, π'_{j+1}) be intergenic breakpoints that are connected. It is possible to remove at least one intergenic breakpoint after at most two operations of reversal or transposition.*

Proof. Similar to Lemma 8, we will analyze the possibilities to remove an intergenic breakpoint based on a pair of connected intergenic breakpoints.

i. (π'_i, π'_j) or (π'_{i+1}, π'_{j+1}): We need to apply only one reversal, which is exactly as the procedure shown in case (i) of Lemma 8.

ii. (π'_i, π'_{j+1}): In this case we need to apply one transposition. As shown previously in case (ii) of Lemma 8, we know that it must exist another intergenic breakpoint (π'_k, π'_{k+1}) such that $k < i$ or $k > j$. If $k < i$ we apply a transposition $\tau^{(k+1,i+1,j+1)}_{(x,y,z)}$ to place the element π'_i on the left side of the element π'_{j+1} (Fig. 4(a)). If $k > j$ we apply a transposition $\tau^{(i+1,j+1,k+1)}_{(x,y,z)}$ to place the element π'_{j+1} on the right side of the element π'_i (Fig. 4(b)). In both scenarios, we have that $\breve{\pi}_{i+1} + \breve{\pi}_{j+1} \geq \breve{\iota}_k$, then the x, y, and z parameters always can be chosen properly to fill the intergenic region with the correct size between the consecutive elements generated.

iii. (π'_{i+1}, π'_j): In this case we need to apply one transposition. As shown previously in case (iii) of Lemma 8, we know that it must exist another intergenic breakpoint (π'_k, π'_{k+1}) such that $k > i$ and $k < j$. After identifying the intergenic breakpoint (π'_k, π'_{k+1}) we apply a transposition $\tau^{(i+1,k+1,j+1)}_{(x,y,z)}$ to place the element π'_j on the left side of the element π'_{i+1} (Fig. 4(c)). Since $\breve{\pi}_{i+1} + \breve{\pi}_{j+1} \geq \breve{\iota}_k$, then the x, y and z parameters always can be chosen properly to fill the intergenic region with the correct size between the consecutive elements generated.

iv (π'_i, π'_{i+1}) or (π'_j, π'_{j+1}): We need to apply two reversals exactly as the procedure shown on Lemma 8 case (iv). □

Theorem 7. *SbIRT problem is 6-approximable.*

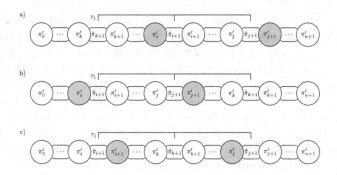

Fig. 4. The possibilities to remove at least one intergenic breakpoint by applying only one transposition. The pair of elements that are consecutive in the identity permutation are represented with the color gray.

Proof. While the permutation is not sorted and with all the intergenic regions with the correct size it is always possible to remove at least one intergenic breakpoint after applying at most two operations (Lemmas 7 and 11). In the worst case, it gives us a total of $2ib(\pi', \breve{\pi}, \breve{\iota})$ operations to transform $(\pi, \breve{\pi}, \breve{\iota})$ into $(\iota, \breve{\iota}, \breve{\iota})$. By Theorem 3, we obtained the lower bound $ib(\pi', \breve{\pi}, \breve{\iota})/3$ and the theorem follows. $\square$

Theorem 8. *SbIRTID problem is 6-approximable.*

Proof. Similar to Theorem 6, we are going to divide the analysis into three cases:

i. $\sum_{i=1}^{n+1} \breve{\pi}_i > \sum_{i=1}^{n+1} \breve{\iota}_i$: Lemmas 7 and 11 remain valid as long as $ib(\pi', \breve{\pi}, \breve{\iota}) > 1$, then we apply only one deletion to remove the last intergenic breakpoint (Lemma 10).

ii. $\sum_{i=1}^{n+1} \breve{\pi}_i = \sum_{i=1}^{n+1} \breve{\iota}_i$: Lemmas 7 and 11 are sufficient to sort the permutation and to fix the sizes of the intergenic regions by applying only reversals and transpositions.

iii. $\sum_{i=1}^{n+1} \breve{\pi}_i < \sum_{i=1}^{n+1} \breve{\iota}_i$: Initially we apply an insertion to make $\sum_{i=1}^{n+1} \breve{\pi}_i = \sum_{i=1}^{n+1} \breve{\iota}_i$ (Lemma 9). Sequentially, Lemmas 7 and 11 guarantee that the permutation will be sorted and the sizes of the intergenic regions will be fixed applying only reversals and transpositions. Note that it is not guaranteed that the initial insertion removes any intergenic breakpoint, but then only reversals and transpositions are applied and the last operation (reversal or transposition) must remove at least two intergenic breakpoints. Considering the insertion and the last operation, on average, we were able to remove one intergenic breakpoint after applying one operation.

Considering the three cases, in the worst scenario, it gives us a total of $2ib(\pi', \breve{\pi}, \breve{\iota})$ operations to transform $(\pi, \breve{\pi}, \breve{\iota})$ into $(\iota, \breve{\iota}, \breve{\iota})$. By Theorem 4, we obtained the lower bound $ib(\pi', \breve{\pi}, \breve{\iota})/3$, and the theorem follows. $\square$

4   Conclusion

We proved that the problems of Sorting by Intergenic Reversals and Sorting by Intergenic Reversals, Insertions, and Deletions belong to NP-hard problems class. Besides, we presented for both problems an algorithms with approximation factor 4. We also investigate the Sorting by Intergenic Reversals and Transpositions problem and the variation with non-conservative events of insertion and deletion. For both problems, we designed approximation algorithms of factor 6.

As future works, we intend to improve the approximation factors of the algorithms and develop cost functions that consider the likelihood of each operation.

Acknowledgments. This work was supported by the National Council for Scientific and Technological Development - CNPq (grants 400487/2016-0, 425340/ 2016-3, and 140466/2018-5), the São Paulo Research Foundation - FAPESP (grants 2013/08293-7, 2015/ 11937-9, 2017/12646-3, and 2017/16246-0), the Brazilian Federal Agency for the Support and Evaluation of Graduate Education - CAPES, and the CAPES/COFECUB program (grant 831/15).

References

1. Berman, P., Hannenhalli, S., Karpinski, M.: 1.375-approximation algorithm for sorting by reversals. In: Möhring, R., Raman, R. (eds.) ESA 2002. LNCS, vol. 2461, pp. 200–210. Springer, Heidelberg (2002). https://doi.org/10.1007/3-540-45749-6_21

2. Biller, P., Guéguen, L., Knibbe, C., Tannier, E.: Breaking good: accounting for fragility of genomic regions in rearrangement distance estimation. Genome Biol. Evol. **8**(5), 1427–1439 (2016)

3. Biller, P., Knibbe, C., Beslon, G., Tannier, E.: Comparative genomics on artificial life. In: Beckmann, A., Bienvenu, L., Jonoska, N. (eds.) CiE 2016. LNCS, vol. 9709, pp. 35–44. Springer, Cham (2016). https://doi.org/10.1007/978-3-319-40189-8_4

4. Bulteau, L., Fertin, G., Rusu, I.: Sorting by transpositions is difficult. SIAM J. Comput. **26**(3), 1148–1180 (2012)

5. Bulteau, L., Fertin, G., Tannier, E.: Genome rearrangements with indels in intergenes restrict the scenario space. BMC Bioinform. **17**(14), 426 (2016)

6. Caprara, A.: Sorting permutations by reversals and eulerian cycle decompositions. SIAM J. Discrete Math. **12**(1), 91–110 (1999)

7. Elias, I., Hartman, T.: A 1.375-approximation algorithm for sorting by transpositions. IEEE/ACM Trans. Comput. Biol. Bioinform. **3**(4), 369–379 (2006)

8. Fertin, G., Jean, G., Tannier, E.: Algorithms for computing the double cut and join distance on both gene order and intergenic sizes. Algorithms Mol. Biol. **12**(1), 16 (2017)

9. Oliveira, A.R., Jean, G., Fertin, G., Dias, U., Dias, Z.: Super short reversals on both gene order and intergenic sizes. In: Alves, R. (ed.) BSB 2018. LNCS, vol. 11228, pp. 14–25. Springer, Cham (2018). https://doi.org/10.1007/978-3-030-01722-4_2

10. Rahman, A., Shatabda, S., Hasan, M.: An approximation algorithm for sorting by reversals and transpositions. J. Discrete Algorithms **6**(3), 449–457 (2008)

11. Yancopoulos, S., Attie, O., Friedberg, R.: Efficient sorting of genomic permutations by translocation, inversion and block interchange. Bioinformatics **21**(16), 3340–3346 (2005)

Unifying Gene Duplication, Loss, and Coalescence on Phylogenetic Networks

Peng Du, Huw A. Ogilvie, and Luay Nakhleh[✉]

Rice University, Houston, TX 77005, USA
{peng.du,huw.a.ogilvie,nakhleh}@rice.edu

Abstract. Statistical methods were recently introduced for inferring phylogenetic networks under the multispecies network coalescent, thus accounting for both reticulation and incomplete lineage sorting. Two evolutionary processes that are ubiquitous across all three domains of life, but are not accounted for by those methods, are gene duplication and loss (GDL).

In this work, we devise a three-piece model—phylogenetic network, locus network, and gene tree—that unifies all the aforementioned processes into a single model of how genes evolve in the presence of ILS, GDL, and introgression within the branches of a phylogenetic network. To illustrate the power of this model, we develop an algorithm for estimating the parameters of a phylogenetic network topology under this unified model.

We demonstrate the application of the model and the accuracy of the algorithm on simulated as well as biological data.

Our work adds to the biologist's toolbox of methods for phylogenomic inference by accounting for more complex evolutionary processes.

Keywords: Phylogenetic network · Coalescence · Introgression · Gene duplication and loss

1 Introduction

Independently evolving lineages of eukaryotic organisms are typically referred to as *species* (they may also be referred to as *populations* depending on the context and operational definition of those terms). Over evolutionary time scales, species lineages bifurcate to form two descendant species from a single ancestral species. This gives rise to a *species tree*, which is a phylogenetic tree describing the evolutionary history of a set of species.

Estimating a species tree is challenging as gene trees are expected to be discordant with the species tree because of several well known processes. The first

This work was supported in part by NSF grants DBI-1355998, CCF-1302179, CCF-1514177, CCF-1800723, and DMS-1547433.

Z. Cai et al. (Eds.): ISBRA 2019, LNBI 11490, pp. 40–51, 2019.
https://doi.org/10.1007/978-3-030-20242-2_4

process leading to discordance is incomplete lineage sorting (ILS), where multiple versions or *alleles* of a gene persist in a species up through to its ancestral species [9]. The second is horizontal gene transfer (HGT) through hybrid speciation [11], introgression [10], and speciation with gene flow [13]. This can lead to gene coalescent times which are younger than the earliest speciation event separating the corresponding species. The third is gene duplication and loss (GDL), where new copies of a gene are created at new loci in the genome, so that the relationship between sequences from different species at different loci (paralogs) reflects the duplication and loss process rather than the speciation process [4].

ILS has been addressed by years of research into the multispecies coalescent (MSC), a mathematical model which describes the evolution of gene trees within a species tree and naturally accommodates ILS [4]. In the MSC, the relationship between sequences from different species at orthologous (as opposed to paralogous) loci is represented by a gene tree, evolving within a species tree, and constrained so that its coalescent times must be older than the corresponding most recent common ancestors (MRCAs).

More recently, HGT has been addressed by generalizing the MSC model to the multispecies network coalescent (MSNC) model, which represents the evolutionary history of species as a phylogenetic network [22]. This flexible model of reticulate evolution can naturally accommodate hybrid speciation [25] and introgression [21]. Implementations of this method include mcmc_seq in PhyloNet [20,23] and SpeciesNetwork in BEAST [25].

GDL has been addressed by the development of models which add a third layer to the MSC between the species tree and the gene trees. This is known as the locus tree, and it contains vertices encoding duplication events, as well as vertices which directly correspond to the speciation vertices of the species tree [16]. The duplicate copy of a gene is assumed to reside in a new unlinked locus, so that there are multiple copies of a gene present in a single genome. The leaves of a single locus tree can therefore represent multiple loci, and the source of data in this model may be more appropriately termed "gene families" (cf. "genes").

DLCoal, the original implementation of the three-layer model [16], is relatively inflexible. It takes as input a gene tree topology, a species tree fixed in topology, branch lengths and effective population sizes, and rates of gene duplication and loss. From such input data it can estimate the locus tree, the mapping of gene tree coalescent vertices to locus tree branches, and the mapping of speciation vertices in the locus tree to the species tree. DLCoal also relies on the accuracy of the supplied gene tree topology, which may contain errors due to the gene tree inference method or insufficient information in the original multiple sequence alignment (MSA). A later method, DLC-Coestimation [24], avoids that potential issue by jointly estimating the gene tree along with the locus tree and reconciliations and mapping directly from a gene family MSA.

The most recent implementation of the three-layer model jointly estimates the species, locus and gene tree topology and times, as well as general parameters including duplication and loss rates from the MSAs of multiple gene families [5]. In a simulation study, this method was able to successfully infer the species tree topology, and outperformed using the MSC model alone (without accounting for GDL) when estimating species divergence times [5].

While the above methods either account for both ILS and HGT, or for both ILS and GDL, no model has been designed or implemented that accounts for all three processes which generate gene tree discordance; ILS, HGT and GDL. Here we present a new model which extends the MSNC to a three-layer model by adding a locus network between the species network and gene trees. This new model accounts for HGT at the species network level, GDL at the locus network level, and ILS at the gene tree level. We have implemented a maximum *a posteriori* (MAP) search for this model which jointly estimates the speciation times, inheritance probabilities and duplication and loss rates. Using simulation experiments, we show that it can accurately infer the aforementioned parameters.

We also used simulated data and an empirical data set of six yeast species to study the difference in accuracy between our new method and an MSNC method which does not account for GDL. Results from those experiments showed that accounting for GDL in addition to ILS and HGT is particularly important when estimating reticulation times.

2   Methods

Similar to the three-layer model of [16], we develop a three-layer model that uses a locus network (different from the locus tree of [16]) as an intermediate layer between the species network and gene tree. This structure allows for unified modeling of coalescence and GDL, where all coalescence events are captured by the relationship between the gene tree and locus network, and all GDL events are captured by the relationship between the locus network and phylogenetic network. The reticulation events (e.g., introgression) are captured by the fact that the species and locus structures are both networks, rather than trees.

2.1   The Three-Layer Model

A species network $\mathbb{S} = (V(S), E(S), \tau^S)$ is a directed acyclic graph depicting the reticulate evolutionary histories of a set of species where $V(S)$ is the set of vertices in the network, $E(S)$ is the set of edges and τ^S contains the set of branch lengths of the edges. We use S to denote $\{V(S), E(S)\}$. Further, $V = r \cup V_L \cup V_T \cup V_N$ where r is the root of the network, V_L is the set of leaf vertices, V_T denotes the set of tree vertices with two children and one parent and V_N represents the set of reticulation vertices with one child and two parents. The set of all internal vertices is $IV(S) = r \cup V_T \cup V_N$. If vertex u has only one parent, we call this parent $pa(u)$. The set of children of u is denoted as $c(u)$. For each reticulation vertex u with two parents v and w, there is an inheritance probability $\gamma \in [0,1]$ such that the probability of locus u inheriting from v is γ and inheriting from w is $1 - \gamma$. Γ is a vector of all inheritance probabilities for all vertices in V_N, $\Gamma((v,u)) = \gamma$ and $\Gamma((w,v)) = 1 - \gamma$. The population sizes are denoted as N^S and the population size on branch $e(u,v)$ is $N^S((u,v))$.

Locus Networks and Locus-Network-to-Species-Network Reconciliation. A locus network $\mathbb{L} = (V(L), E(L), \tau^L)$ is generated by applying duplication and loss events onto the species network with a top-down birth-death process [1,2,18]. Birth events create new loci by duplicating an existing locus, and death (loss) events eliminate loci so that it will have no sampled descendants. Fully describing the result of this process requires a reconciliation R^L from the locus network to the species network, where the vertices on the locus network can be mapped to either the vertices or the branches of the species network. If $u \in V(L)$ is mapped to a species network vertex, then we call it a speciation vertex; the set of speciation vertices is denoted as $V_S(L)$. If it is mapped to a species network branch; we call it a duplication vertex and the set of duplication vertices is denoted as $V_D(L)$. Branches with no existing leaf vertices are pruned out (Fig. 1). For a duplication, a new locus is generated, so a mapping $\delta(u, v) = 1$ or $\delta(u, v) = 0$ is used to indicate whether (u, v) leads to the new (daughter) locus or if (u, v) is the mother branch where u is the duplication vertex. The population size of branch $e = (u, v)$ in the locus network is the population size of the branch $e' = (w, x)$ on the species network where $R^L(u) = (w, x)$ or $R^L(v) = (w, x)$ or $R^L(u) = w, R^L(v) = x$. Similarly, $\Gamma((u, v)) = \Gamma((w, x))$ where $(w, x) \in E(S)$ if $R^L(u) = (w, x)$ or $R^L(v) = (w, x)$ or $R^L(u) = w, R^L(v) = x$. It is important to note that reticulation edges present in the species network may be deleted from the locus network (as in the (F, X) branch leading to the B2 locus in Fig. 1), so the locus network can be a tree or more tree-like with fewer reticulation vertices than the species network.

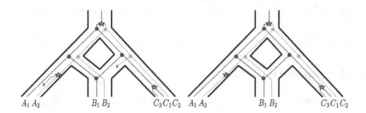

Fig. 1. A gene duplication and loss scenario inside of a species network on three species A, B, and C. (Left) The complete locus network embedded in the species network, produced by a birth-death process, and containing all duplication and loss events. (Right) Lineages in the locus network with no sampled loci due to loss events are pruned from the locus network, resulting in the observed locus network. Extinct lineages are deleted. Duplication, loss, and speciation/hybridization events are represented by $\star$, $\times$, and $\bullet$, respectively. New lineages arising from duplication are colored red and green. (Color figure online)

Gene Trees and Gene-Tree-to-Locus-Network Reconciliation. A gene tree $\mathbb{G} = (V(G), E(G), \tau^G)$ describes the evolution of lineages and the definitions of vertices are similar with those in the species network and locus network. The reconciliation from the gene tree to the locus network is denoted by R^G. The two reconciliations R^L and R^G are collectively denoted by R. For each locus network branch $e = (u, w)$ with $\delta((u, w)) = 1$, the coalescent time of every gene vertex mapped to the leaf vertices under w must be more recent than u. Also, we define M as the mapping from the gene tree leaf vertex-set to the locus network leaf vertex-set. M indicates what gene is from what locus in the locus network. Figure 2 shows the reconciliations from the gene tree to the locus network (R^G) and from the locus network to the species network (R^L).

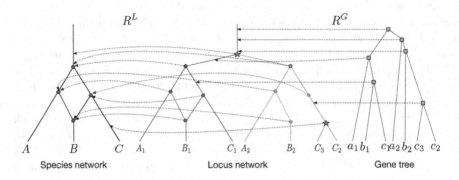

Fig. 2. Gene duplication/loss events are obtained by mapping the nodes of the locus network onto the branches of the species network, via reconciliation R^L (the dotted arrows from the locus network to the species network). Coalescence events are obtained by mapping the gene tree nodes onto the branches of the locus network, via reconciliation R^G (the dotted arrows from the gene tree to the locus network). Duplication, loss, speciation/hybridization, and coalescence events are represented by $\star$, $\times$, $\bullet$, and $\blacksquare$, respectively.

2.2   Model Assumptions

In this model, we need to make some assumptions as made in [5, 16, 24].

1. After the duplication, the daughter locus becomes totally unlinked and any further evolution of the mother and daughter loci, as well as the coalescent histories of the mother and daughter genes, are independent conditional on the species network topology, times and population sizes. Thus we can calculate the coalescent probabilities separately for each locus, and use the product as the gene family coalescent probability.
2. At a locus level, hemiplasy [3] is assumed to be non existent in this model. In other words, for each duplication and loss event, the resulting addition or deletion of locus will be transmitted universally to all descendent species. This allows us to explain all unobserved loci by means of gene loss.
3. In our present implementation, one individual per species is sampled for each locus.

2.3   Probability Distribution

For a species network $\mathbb{S}$ and a set of gene families $\mathbb{GF}$ with each member $\mathbb{GF}_i = (\mathbb{L}_i, \mathbb{G}_i, R_i, M_i, \delta_i^L)$, and parameters θ, the posterior $p(\mathbb{S}, \mathbb{GF}, \theta | D)$ given observed DNA sequences D is

$$p(\mathbb{S}, \mathbb{GF}, \theta | D) \propto \prod_i p(\mathbb{GF}_i | \mathbb{S}, \theta) \times p(D_i | \mathbb{GF}_i) \times p(\theta)$$

where D_i is the DNA sequences for $\mathbb{GF}_i$ and $\theta = \{\mu, \lambda, \Gamma, N^S\}$ which are the duplication rate, loss rate, substitution rate, inheritance probabilities and population size respectively. The term $p(\mathbb{GF}_i | \mathbb{S}, \theta)$ can be decomposed into (we drop the subscript i for readability) the product $p(G, \tau^G, R^G | L, \tau^L, \delta^L, M, \Gamma, N^S) \times p(M | L, \tau^L, R^L, \delta^L) \times p(L, \tau^L, R^L, \delta^L | S, \tau^S, \mu, \lambda)$, and we have $p(D | \mathbb{GF}_i) = p(D | G, \tau^G)$. The term $p(G, \tau^G, R^G | L, \tau^L, \delta^L, M, \Gamma, N^S)$ is the probability of the gene tree coalescing in the locus network under a bounded coalescence model where gene lineages originated from gene duplication events must coalesce earlier than the duplication event. The bounded coalescence model is extended from [16] and gains the capacity to handle hybridization events. The details are in [6].

The term $p(M | L, \tau^L, R^L, \delta^L)$ is the probability of the map of gene tree leaves to locus network leaves. Since we assume no prior knowledge of locus information of each sampled gene copy from a certain species, the mapping has a uniform distribution based on the number of possible permutations:

$$p(M | L, \tau^L, R^L, \delta^L) = \prod_{x \in L(S)} \frac{1}{|u : R^L(u) = x|!}. \tag{1}$$

The number of permutations is constant for a given data set D, so for identification of the MAP topology or algorithms like MCMC which use unnormalized posterior probabilities not scaled by $1/P(D)$, the calculation of this prior is unnecessary. The term $p(L, \tau^L, R^L, \delta^L | S, \tau^S, \mu, \lambda)$ is the probability of the locus network generated inside of the species network with duplication rate μ and loss rate λ and is also derived in [6]. The term $p(\mathbb{S})$ is the prior of the species network which is a compound prior with uniform prior on the topology and exponential prior on divergence times as in [7,19].

2.4   MAP Inference of the Parameters of a Fixed Network Topology

Our goal is to find the maximum *a posteriori* (MAP) estimate of the parameters; that is,

$$(\mathbb{S}^*, \mathbb{GF}^*, \theta^*) = \text{argmax}_{(\mathbb{S}, \mathbb{GF}, \theta)} p(\mathbb{S}, \mathbb{GF}, \theta | D). \tag{2}$$

In this present work, we will focus on inferring the species network parameters—times, population sizes and inheritance probabilities—with the topology being fixed, as well as locus networks, gene trees and reconciliations between them. General parameters such as duplication and loss rates are also inferred. Because of the hierarchical nature of the generative model, changes on higher level components will influence lower level components as well. For example, changing the heights of the species network vertices will also change the heights of corresponding locus network vertices. We developed four groups of operators, each working on different levels of the model, which we describe in detail in [6]. The first group makes changes to the species network and can also alter the locus networks and gene trees. The second group changes the locus network and can also alter the gene trees. The third group makes changes to the gene trees alone, while the fourth applies to the macroevolutionary rates, which in our implementation is limited to the duplication and loss rates.

2.5   Results and Discussion

2.6   Performance on Simulated Data

Simulation Setup. We simulated DNA sequence data for multiple gene families with our gene tree simulator and Seq-Gen [14]. Our gene tree simulator employs the hybridization-duplication-loss-coalescence model and operates in two phases. First, it generates the locus network within a predefined species network by simulating duplications and losses.

Then the gene tree is simulated under a coalescence model along the locus network. If the gene lineages could not coalesce before the duplication event backward in time, it will be rejected and retried until it coalesces after the event, up to 10^8 attempts, beyond which the locus network will be rejected and regenerated. Locus networks with fewer than 3 extant species will be rejected. Once the gene trees were generated, the program Seq-Gen [14] was used to simulate the evolution of DNA sequences down the gene trees under a specified model of evolution. In all simulations reported here, we used the Jukes-Cantor model of evolution [8] to generate 1000 bp long DNA sequences. For Experiments 1 to 3, we used the network of Fig. 3 as

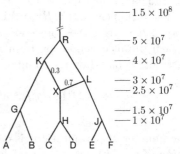

Fig. 3. The model network used to simulate data for Experiments 1 to 3. The values on the right correspond to divergence times of the nodes in number of generations. The inheritance probability values are shown on the reticulation edges.

the model species network. Population sizes are given as the number of diploid individuals, and specified duplication/loss rates and population sizes were set to be the same across all branches of the model networks. A mutation rate of 10^{-9} was used for all simulation experiments.

Experiment 1: Testing the Effects of GDL Rate and Population Size. In this experiment, different settings of duplication/loss rates and population sizes were used to test how these parameters would affect the accuracy of inferences. The duplication and loss rates (both were equal) used were 5×10^{-10}, 10^{-9} and 2.5×10^{-9} and the population sizes were 10^6, 4×10^6, and 8×10^6. For each of the 9 different settings of duplication/loss rates and population size, we generated 10 replica each with 50 gene families and ran 15 million iterations for each data set. First, we calculated the average difference of the estimated divergence times and the true values in population mutation rate units for the 9 settings based on the 10 replica for each setting. Most estimates of the divergence times are accurate across different settings, with the exception of the reticulation time "X" which appears to be less identifiable at smaller population sizes (see [6]). We calculated the difference between the estimated inheritance probabilities and the true value (0.3) on (K, X) (see [6]). No consistent trend was observed in the accuracy of inheritance probability estimates over the ranges of population size and duplication and loss rates studied. To assess the accuracy of locus networks and gene trees, we calculated the topological error in metrics developed by [12] between the estimated and true locus networks and RF distance between true and estimated gene trees [17]. Overall our method shows very good accuracy (indicated by topological distances close to 0). The average distance for the locus networks increases as the duplication and loss rate increases, but it appears invariant to varying population size. This makes sense because the locus networks are determined by the duplication and loss events not the ILS events. If ILS, GDL and HGT are absent all gene tree topologies will identical to the species tree topology, and be perfectly accurate when the species tree topology is fixed at the truth. However in our model gene trees can vary because of all three processes. The prevalence of ILS is partly dependent on population sizes, and therefore it is unsurprising that we show gene tree topological error consistently increasing as population sizes get larger (see [6]). Finally, we assessed the method's performance in terms of estimating the duplication and loss rates. As the results show, the method performs well at estimating both rates under the range of population sizes and duplication and loss rates studied (see [6]).

Experiment 2: Testing the Effect of the Number of Gene Families. In order to determine how our method performs given larger or smaller data sets, we varied the number of gene families (5, 10, 25, and 50) under one setting of duplication/loss rate (2.5×10^{-9}) and population size (4×10^6). 10 replica for each number of gene families were simulated and 15 million iterations were run for each data set. Results (see [6]) show that even for 5 gene families, a relatively small number, the estimated divergence times are generally accurate especially for H and R. The accuracy and precision improve as more gene families are used for example for nodes G, J and L.

We tested the inference of other parameters. Figure 4(a) shows that both the accuracy and precision of the inheritance probability improved for larger numbers of gene families. The accuracy of the duplication rate both appear to improve slightly with more data. The loss rate, while accurately estimated, did not show any consistent trends. As the results show, the accuracy of the locus networks and of the gene trees seems to be stable across different settings in terms of both mean and standard deviation. As gene tree topologies, while independent, are conditioned on the species network topology, when the species network topology is fixed even without any data there will already be a lot of information in the model on the gene tree topologies. Also, the locus networks are independent for each gene tree conditioned on the species network topology. So increasing the number of gene trees will not improve the overall accuracy of gene tree estimates to the same extant as when jointly estimated with the species tree or network topology (Fig. 4(b)).

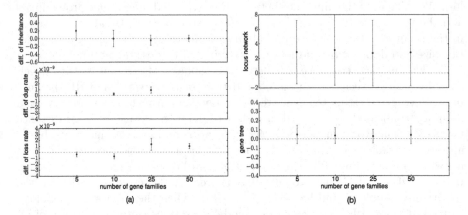

Fig. 4. (a) The difference of estimated parameters from the true values. Top: difference between estimated inheritance probability and the true value (0.3) on (H, X). Middle: difference of estimated duplication rate and true value. Bottom: difference of estimated loss rate and true value. The number of gene families used as input to the inference method is shown on the x-axis. Standard deviations are represented by vertical bar. (b) The average topological distances between the inferred and true networks or trees. Top: Locus network difference. Bottom: Gene tree difference. The number of gene families used as input to the inference method is shown on the x-axis. Standard deviation is represented as vertical bar.

Experiment 3: Comparing Inference With and Without GDL. In this experiment we set out to test how a method that accounts only for incomplete lineage sorting but ignores duplication and loss would perform as compared to our model here. To achieve this, we ran our method and a Bayesian MCMC species network inference method (the mcmc_seq command, with the species network fixed, and using the MAP estimation) in PhyloNet [23] which implements the method of [20]. We simulated 10 replica under duplication and loss rates of 2.5×10^{-9} and population size 10^7 and 50 gene families for each data set. For each gene family we randomly selected one gene copy for each species if there was at least one. As a result, around half of the sequences in the gene families were kept after this pruning of the data sets. We fed the sequences to both methods and ran them both for 15 million iterations. Our results show that our method, which accounts for gene duplication and loss even with a single sampled locus per species, more accurately estimated speciation and reticulation times. This was particularly true of the reticulation vertex, where mcmc_seq dramatically underestimated the reticulation time (see [6]). Also, we have a better estimation of the inheritance probabilities than mcmc_seq. Our estimation is 0.268 and mcmc_seq had estimation of 0.464 where the true value is 0.3.

2.7   Biological Data

We used the yeast genome data set with duplications reported on http://compbio.mit.edu/dlcoal/ and randomly selected two data sets restricted to six genomes. One consists of 100 gene families each with exactly one copy for each species with alignments 1000nt–2000nt in length; the other consists of 100 gene families with possibly multiple or 0 copies for each species with alignments 1000nt–2000nt in length. We used 10^{-10} as duplication and loss rate and 4×10^{-10} as mutation rate and 10^7 as population size which are comparable with the settings used in [15]. Then we fed mcmc_seq with the first data set and ran the command 10 times for 15 million iterations each with the maximum number of reticulation vertex set to be one. The most prevalent topology is shown in Fig. 5(a) and appeared in 7 of the 10 runs. We then fed our method with the second data set and run 7 times each with 15 million iterations. A table of the average estimated divergence times is given in Fig. 5(b). We can see that most of the divergence times are similar and the only significant differences are at the divergence times of vertices J, X and L. Given our method is better at estimating divergence times given results from Experiment 3, it seems that the ones obtained by our method here are probably more accurate estimations.

The inheritance probability on branch (R, X) estimated by mcmc_seq was 0.503 ± 0.147 while the value estimated by our method was 0.461 ± 0.091. The error is the standard deviation among runs, and shows that the estimated inheritance probabilities of the two methods are very close.

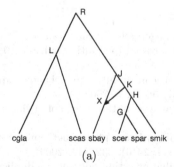

		Our method	mcmc_seq
G		0.04068 ± 0.001	0.0401 ± 0.001
H		0.0708 ± 0.001	0.0677 ± 0.002
X		0.02851 ± 0.015	0.0626 ± 0.0292
K		0.0937 ± 0.003	0.0911 ± 0.004
J		0.1187 ± 0.005	0.1227 ± 0.004
L		0.2321 ± 0.002	0.1877 ± 0.007
R		0.2415 ± 0.002	0.1955 ± 0.008

(a)   (b)

Fig. 5. (a) The yeast species network topology inferred by mcmc_seq on the 100 gene families. (b) Mean and standard deviation of the estimated divergence times of mcmc_seq and our method (std's smaller than 0.001 are rounded to 0.001).

3   Conclusions

In this work, we developed a probabilistic model that simultaneously accounts for hybridization, gene duplication, loss and ILS. We also devised a stochastic search algorithm for parameterizing phylogenetic networks based on this model. This algorithm provides estimates of evolutionary parameters, as well as gene histories and their reconciliations. Results based on simulation studies show good performance of the algorithm as well as insights obtained by employing the new model as compared with existing models that exclude gene duplication and loss.

We identify three natural directions for future research. First, while in this work we assumed a fixed phylogenetic network topology, in most empirical studies such a topology is not given or known. Developing a method that infers the phylogenetic network, along with all the parameters that the current method estimates, is essential for proper application of the model. Second, while this work focused on obtaining point estimates of the phylogenetic network's parameters, developing a method that estimates a posterior distribution on the space of phylogenetic networks and their parameters would provide additional information, including assessment of statistical significance and the uniqueness and distinguishability of optimal solutions. Third, the computational bottleneck in this domain stems from the time it takes to compute the likelihood of a given point in the parameter space as well as from the need to walk an enormous and complex space of such parameters. For example, it took between 15 and 20 h for a single run of 15 million iterations on a data set with four or five species and 50 gene families. Developing algorithmic techniques and potentially alternative likelihood functions to speed up these calculations is imperative for this work to be applicable to data sets of the scale that biologists can now generate using the latest sequencing technologies.

References

1. Åkerborg, Ö., Sennblad, B., Arvestad, L., Lagergren, J.: Simultaneous Bayesian gene tree reconstruction and reconciliation analysis. Proc. Natl. Acad. Sci. **106**(14), 5714–5719 (2009)
2. Arvestad, L., Lagergren, J., Sennblad, B.: The gene evolution model and computing its associated probabilities. J. ACM (JACM) **56**(2), 7 (2009)
3. Avise, J.C., Robinson, T.J.: Hemiplasy: a new term in the lexicon of phylogenetics. Syst. Biol. **57**(3), 503–507 (2008)
4. Degnan, J.H., Rosenberg, N.A.: Gene tree discordance, phylogenetic inference and the multispecies coalescent. Trends Ecol. Evol. **24**(6), 332–340 (2009)
5. Du, P., Nakhleh, L.: Species tree and reconciliation estimation under a duplication-loss-coalescence model. In: Proceedings of the 2018 ACM International Conference on Bioinformatics, Computational Biology, and Health Informatics, pp. 376–385. ACM (2018)
6. Du, P., Ogilvie, H., Nakhleh, L.: Unifying gene duplication, loss, and coalescence on phylogenetic networks. bioRxiv (2019). https://doi.org/10.1101/589655
7. Heled, J., Drummond, A.J.: Bayesian inference of species trees from multilocus data. Mol. Biol. Evol. **27**(3), 570–580 (2009)

8. Jukes, T.H., Cantor, C.R., et al.: Evolution of protein molecules. Mamm. Protein Metab. **3**(21), 132 (1969)
9. Maddison, W.P.: Gene trees in species trees. Syst. Biol. **46**(3), 523–536 (1997)
10. Mallet, J.: Hybridization as an invasion of the genome. Trends Ecol. Evol. **20**(5), 229–237 (2005)
11. Mallet, J.: Hybrid speciation. Nature **446**(7133), 279 (2007)
12. Nakhleh, L.: A metric on the space of reduced phylogenetic networks. IEEE/ACM Trans. Comput. Biol. Bioinform. **7**(2), 218–222 (2010)
13. Nosil, P.: Speciation with gene flow could be common. Mol. Ecol. **17**(9), 2103–2106 (2008). https://doi.org/10.1111/j.1365-294X.2008.03715.x
14. Rambaut, A., Grass, N.C.: Seq-gen: an application for the Monte Carlo simulation of DNA sequence evolution along phylogenetic trees. Bioinformatics **13**(3), 235–238 (1997)
15. Rasmussen, M., Kellis, M.: A Bayesian approach for fast and accurate gene tree reconstruction. Mol. Biol. Evol. **28**(1), 273–290 (2011)
16. Rasmussen, M.D., Kellis, M.: Unified modeling of gene duplication, loss, and coalescence using a locus tree. Genome Res. **22**(4), 755–765 (2012)
17. Robinson, D.F., Foulds, L.R.: Comparison of phylogenetic trees. Math. Biosci. **53**(1–2), 131–147 (1981)
18. Sjöstrand, J., Sennblad, B., Arvestad, L., Lagergren, J.: DLRS: gene tree evolution in light of a species tree. Bioinformatics **28**(22), 2994–2995 (2012)
19. Than, C., Ruths, D., Nakhleh, L.: PhyloNet: a software package for analyzing and reconstructing reticulate evolutionary relationships. BMC Bioinform. **9**(1), 1 (2008)
20. Wen, D., Nakhleh, L.: Co-estimating reticulate phylogenies and gene trees from multi-locus sequence data. Syst. Biol. **67**(3), 439–457 (2018)
21. Wen, D., Yu, Y., Hahn, M.W., Nakhleh, L.: Reticulate evolutionary history and extensive introgression in mosquito species revealed by phylogenetic network analysis. Mol. Ecol. **25**(11), 2361–2372 (2016). https://doi.org/10.1111/mec.13544
22. Wen, D., Yu, Y., Nakhleh, L.: Bayesian inference of reticulate phylogenies under the multispecies network coalescent. PLoS Genet. **12**(5), e1006006 (2016)
23. Wen, D., Yun, Y., Zhu, J., Nakhleh, L.: Inferring phylogenetic networks using PhyloNet. Syst. Biol. **67**(4), 735–740 (2018)
24. Zhang, B., Wu, Y.-C.: Coestimation of gene trees and reconciliations under a duplication-loss-coalescence model. In: Cai, Z., Daescu, O., Li, M. (eds.) ISBRA 2017. LNCS, vol. 10330, pp. 196–210. Springer, Cham (2017). https://doi.org/10.1007/978-3-319-59575-7_18
25. Zhang, C., Ogilvie, H.A., Drummond, A.J., Stadler, T.: Bayesian inference of species networks from multilocus sequence data. Mol. Biol. Evol. **35**(2), 504–517 (2018). https://doi.org/10.1093/molbev/msx307

The Cluster Affinity Distance
for Phylogenies

Jucheol Moon[1] and Oliver Eulenstein[2(✉)]

[1] Department of Computer Engineering and Computer Science,
California State University Long Beach, Long Beach, CA 90840, USA
jucheol.moon@csulb.edu
[2] Department of Computer Science, Iowa State University, Ames, IA 50011, USA
oeulenst@iastate.edu

Abstract. Studying phylogenetic trees is fundamental to biology and benefitting a vast variety of other research areas. Comparing such trees is essential to such studies for which a growing and diverse collection of tree distances are available. In practice, tree distances suffer from problems that can severely limit their applicability. Notably, these distances include the cluster matching distance that is adapted from the Robinson-Foulds distance to overcome many of the drawbacks of this traditional measure. However, at the same time, the cluster matching distance is much more confined in its application than the Robinson-Foulds distance and makes sacrifices for satisfying the properties of a metric. Here, we propose the cluster affinity distance, a new tree distance that is adapted from the cluster matching distance but has not its drawbacks. Nevertheless, as we show, the cluster affinity distance preserves all of the properties that make the matching distance appealing.

Keywords: Phylogenies · Cluster matching distance ·
Cluster affinity distance

1 Introduction

Phylogenetic trees depict the phylogenetic relationships of entities, like molecular sequences, genomes, or species, and can be of enormous size. For researchers, phylogenetic trees are full of complexities and present a primary tool for studying how entities have evolved the way they are today. Potential applications of such studies are widespread and affecting a vast variety of fundamental research areas such as biology, ecology, epidemiology, and conservation biology.

Studying phylogenetic trees entails the comparative evaluation of their differences and similarities [8,20]. To compare phylogenetic trees a large variety of measures has been considered and analyzed (e.g., [11,12]). However, all of these measures are prone to shortcomings or weaknesses that can severely limit their usefulness in practice, ranging from intrinsic exponential time-complexities (e.g., [1,5,6]) to negatively skewed distributions [21], and several measures suffer from

© Springer Nature Switzerland AG 2019
Z. Cai et al. (Eds.): ISBRA 2019, LNBI 11490, pp. 52–64, 2019.
https://doi.org/10.1007/978-3-030-20242-2_5

topology biases [23] or do not satisfy the properties of a metric (e.g., [16]). For example, the problems of computing the distance for a pair of trees under various tree edit operations are NP-hard, such as the traditional tree edit operations nearest neighbor interchange (NNI), subtree pruning and regrafting (SPR), and tree bisection and reconnection (TBR) [1,6]. Gene tree parsimony costs that rely on evolutionary models to compare trees do not satisfy the properties of a metric [16] and suffer from topology biases [23]. Widely-used is the classic Robinson-Foulds (RF) distance [19], perhaps, because it is not confined by an evolutionary model and linear-time computable [7]. In practice, however, phylogenetic trees are prone to small error, and the RF distance is not sufficiently "robust" to compensate for such errors. A tree distance is considered to be *robust* when a *small error* in the compared trees that is modeled by successive tree edit operations will not cause abrupt distance changes (e.g., [15]). Another significant drawback of the RF distance is its negatively skewed distribution [21] where most distances are close to the maximum possible distance, called *RF diameter*.

Matching distances present a significant leap in comparative phylogenetics. These distances can be seen as weighted adaptions of the classic RF distance, and are designed to overcome its shortcomings. The matching distances, similar to the RF distance, are defined for comparing rooted tree-pairs [4] and unrooted tree-pairs [15]. Here, we focus on the rooted version of the matching distance, which is called the cluster matching distance.

The *cluster matching distance* is defined for a pair of rooted and binary input trees. This distance is based on matching the clusters (or clades) between the two trees perfectly, i.e., every cluster in each tree is paired with exactly one cluster in the other tree. For each cluster pair, the *cluster distance* is defined as the count of the symmetric difference of the involved clusters. Now, the cluster matching distance is the score of a minimum-weight perfect matching; that is, a perfect matching with their overall minimum distance of its cluster pairs.

Despite the desirable properties of the cluster matching distance, its applicability is severely limited when compared to the RF distance. Unlike the RF distance that is also defined for non-binary trees, the matching distance requires binary input trees to establish a perfect matching. In practice, however, trees inferred from biological data are often non-binary and have various multifurcated vertices that allow representing the uncertainty of their actual binary resolution (e.g., [8]). While the perfect matching of the cluster matching distance establishes its metric properties, this matching is prone to skew the actual "similarity" of the input trees. For example, minimum weight perfect matchings can enforce that most clusters are not matched with their smallest possible cluster distance, and thus overestimating largely the actual minimum cluster distances between the input trees. Thus, the same cluster matching distance can describe tree pairs that are quite different in terms of their minimum cluster distances.

Here, provided with the great template of the cluster matching distance, we introduce a new adapted distance, referred to as the *cluster affinity distance*, that is generally defined for multi-furcated and binary trees, and where each cluster in a tree is matched with the smallest cluster distance. At the same time, as we prove, the cluster affinity distance is not jeopardizing any of the properties that make the cluster matching distance appealing.

Related Work. The need to compare phylogenetic trees has given rise to the proliferation of various measures for the pairwise comparison of such trees [11]. Here we describe distance-based measures for the pairwise comparison of trees over the same label set that are closely related to the presented work and discuss their advantages and shortcomings.

While all of the presented measures induce a metric on the tree-space, which is not true for measures that rely on biological models (e.g., [16]), they widely differ in their asymptotic computation times and distributions. Also, these measures vary in terms of their diameters, and gradients regarding the classic tree edit operations NNI, SPR, and TBR. The *diameter* of a measure for the tree space of all n taxa trees is the maximum distance between any pair of trees in this space [11]. In practice, such diameters are often used to normalize their corresponding measures to compare them when analyzing tree distances [10]. The *gradient* of a tree edit operation for a given distance metric is the maximum distance between all tree pairs that can be transformed into each other by one edit operation. Errors in trees can be expressed in terms of the tree edit operations [24], and thus the gradient of an edit operation for a measure can be used to describe the robustness regarding the error of this measure [15]. In the following, we overview the measures of interest for this work, which are (i) tree edit measures, (ii) the RF measure, and (iii) the cluster matching distance.

Tree edit measures. Maybe the most natural tree measures are based on the traditional tree edit operations that are informally described for an unrooted and full-binary tree T over n taxa as follows.

Nearest neighbor interchange (NNI): This operation selects an internal edge in T (i.e., an edge that is not incident to a leaf), and exchanges a subtree on one side of the selected edge with a subtree on the other side of the edge.

Subtree prune and regraft (SPR): This operation prunes a subtree from a tree T by cutting an edge and redrafts the subtree to a new vertex obtained by subdividing an edge of the edited tree.

Tree bisection and reconnection (TBR): This operation divides a tree T into two subtrees by removing an edge, and then reconnects these subtrees by creating a new edge between the midpoints of edges in them.

The *NNI, SPR, and TBR measures* are defined to count the minimal number of corresponding edit operations required to change a given pair of trees into each other. The NNI distance has been introduced independently by DasGupta [6] and Li et al. [13], and computing this distance is NP-hard [14]. Later on, the SPR distance and TBR distance were introduced for unrooted and rooted trees, and their NP-hardness was shown eventually [1,5]. All of these measures induce metrics or distances on the space of trees [11]. The diameter of the NNI distance is $\Theta(n \log n)$ [6,13], and the diameters for the SPR and TBR distance are $\Theta(n)$ [1].

The RF distance. The RF distance [19] is a popular and widely used measure [7, 17,19], which can be computed in linear time in the size of the compared tree pair [7]. However, the RF distance has a negatively skewed distribution, and in practice, this distance is mostly useful when the compared trees are "very similar" [21]. Further, the RF distance is not robust towards small changes, such as topological error, since re-attaching a single leaf elsewhere in one of the compared trees can maximize the distance.

The cluster matching distance. The cluster matching distance has been introduced by Bogdanowicz and Giaro [4] for the comparison of a pair of rooted binary trees on the same leaves and is using a minimum-weight perfect matching to compare the clusters of these trees. Let n be the number of leaves of the trees that are compared. The cluster matching distance can be computed in $O(n^{2.5} \log n)$ time [15], and its diameter is bound by $\Theta(n^2)$ [4]. Further, the gradients of the cluster matching distance under the rooted versions of the traditional edit operations NNI, SPR, and TBR are bound by $\Theta(n)$, $\Theta(n^2)$, and $\Theta(n^2)$, respectively [18].

Contribution. We introduce the *cluster affinity distance* for two rooted, and not necessarily binary, input trees. This distance is based on the *directed cluster distance* for an ordered tree pair that pairs each cluster in the first tree with a cluster in the second tree with the minimum cluster distance, and scores the overall cluster distances of these pairs. Now, the cluster affinity distance is the average of the directed cluster distances for each ordered pair of the input trees.

Following the outline of Lin et al. [15], we first show that the cluster affinity distance is a metric. Then, we prove that the diameter of the cluster affinity distance is bound by $\Theta(n^2)$. Moreover, we prove that the gradients of the cluster affinity distance under the rooted versions of the edit operations NNI, SPR, and TBR are bound by $\Theta(n)$, $\Theta(n^2)$, and $\Theta(n^2)$, respectively. By n we refer to the number of leaves of the compared trees.

In an experimental study, we show that the cluster affinity distance, like the cluster matching distance, overcomes major drawbacks of the RF distance. First, we demonstrate the distribution of the cluster affinity distance between randomly generated binary trees using Yule-Harding model and birth-death process model. For both models, the cluster affinity distance is more broadly distributed in the form of a bell-shape and has a broader range than the RF distance. Then, we show how the cluster affinity distance and the RF distance are correlated with the number of classical tree edit operations. When compared to the RF distance, the cluster affinity distance is gradually saturated towards its maximum value.

2 Preliminaries and Basic Definitions

A *(phylogenetic) tree* T is a connected acyclic graph that has exactly one distinguished vertex of degree two, called the *root* of T, and where all of the remaining vertices are either of degree three or one. The vertices of degree larger than one are the *internal* vertices of T, and the remaining vertices are the *leaves* of T. For a tree T, we denote its vertex set, edge set, leaves, internal vertices, and root, by $V(T), E(T), \mathcal{L}(T), V_{int}(T)$, and $r(T)$, respectively.

In the following we introduce needed terminology relating to the semi-order represented by T. We define $\leq_T$ to be the partial order on $V(T)$, where $x \leq_T y$ if y is a vertex on the path between $r(T)$ and x. If $x \leq_T y$, we call x a *descendant* of y, and y an *ancestor* of x. We also define $x <_T y$, if $x \leq_T y$ and $x \neq y$. If $\{x, y\} \in E(T)$ and $x \leq_T y$, then we call y the *parent* of x and x a *child* of y. For a node $x \in V(T)$, the subtree of T rooted at x is denoted by $T(x)$, the *cluster* of x is defined by $\mathcal{C}_T(x) := \mathcal{L}(T(x))$, and the set of all clusters of T is defined by $\mathcal{H}(T) = \bigcup_{x \in V(T)} \mathcal{C}_T(x)$. $X \in \mathcal{H}(T)$ is called a *trivial* cluster if $X = \mathcal{L}(T)$ or $|X| = 1$, it is called *non-trivial* otherwise. The set of non-trivial clusters is defined by $\mathcal{H}'(T) = \bigcup_{x \in V_{int}(T) \setminus r(T)} \mathcal{C}_T(x)$. The symmetric difference of two sets X and Y is defined as $X \ominus Y = (X \setminus Y) \cup (Y \setminus X)$. Let T_1 and T_2 be trees with same leaves, then the *(rooted) Robinson-Foulds (rRF) distance* [19] is defined as $RF(T_1, T_2) := \frac{1}{2}|\mathcal{H}(T_1) \ominus \mathcal{H}(T_2)|$.

Let T be a tree and $\phi(T)$ be the set of trees derived by applying the edit operation ϕ to T, then $\phi(T)$ is called the *(local) neighborhood* of T under ϕ [20]. We provide the definitions for the classic rooted tree edit operations.

rNNI [5]: Let $T_2 \in rNNI(T_1)$. An internal vertex u of a rooted binary tree T_1 has two incident edges that connects its children l and r. A rooted binary tree T_2 is obtained from T_1 by deleting $e = \{u, l\}$ (or $e' = \{u, r\}$), adding the edge between l (or r) and the vertex subdivides the edge that is incident with $Pa_{T_1}(u)$ and u's sibling, and then suppressing any degree-two vertices.

rSPR [5]: Let $T_2 \in rSPR(T_1)$, $e = \{u, v\}$, $u \leq_{T_1} v$. A rooted binary tree T_2 is obtained from T_1 by deleting e, adding the edge between u and the vertex that subdivides the edge of $T_1 \setminus e$, and suppressing degree-two vertices.

rTBR [5]: Let $T_2 \in rTBR(T_1)$. Analogous to rSPR, a rooted binary tree T_2 is obtained from T_1 by deleting e, adding an edge between vertices such that each of the vertices subdivides the edge of one and the other component of $T_1 \setminus e$, and then suppressing any degree-two vertices.

3   A New Metric Space

Definition 1 *(Cluster Affinity Distance).* Let T_1 and T_2 be trees over the same leaves and $f(T_1, T_2) := \Sigma_{X \in \mathcal{H}'(T_1)} (\min_{Y \in \mathcal{H}'(T_2)} |X \ominus Y|)$. The cluster affinity (CA) distance is defined as $CA(T_1, T_2) := \frac{1}{2}(f(T_1, T_2) + f(T_2, T_1))$.

Proposition 1. *The CA distance is $\mathcal{O}(n^2)$ time computable.*

Proof. Let T_1 and T_2 be trees over n taxa. Computing $f(T_1, T_2)$ requires $\mathcal{O}(n^2)$ time by using the n-bit binary vector representation of clusters. For all $X \in \mathcal{H}'(T_1)$, $Y \in \mathcal{H}'(T_1)$, the clusters X and Y are represented by the n-bit binary vectors that map each leaf to 1 if the cluster contains the leaf, and 0 otherwise. Let b_X and b_Y be the vector representations of the clusters X and Y, then

$|X \ominus Y| = |\boldsymbol{b}_X \vee \boldsymbol{b}_Y|^2 - |\boldsymbol{b}_X \wedge \boldsymbol{b}_Y|^2$, where $\vee$ and $\wedge$ denote the bitwise operations OR and AND, respectively. Thus, the CA distance is $\mathcal{O}(n^2)$ time computable.

Proposition 2. The CA distance is a metric.

Proof. Observe that non-negativity and symmetry properties follow directly from definition of the CA distance (i.e., Definition 1).

To show the identity property, suppose that there are trees T_i and T_j where $CA(T_i, T_j) = 0$. Since $f(T_i, T_j) \geq 0$ and $f(T_j, T_i) \geq 0$, $CA(T_i, T_j) = 0$ follows that $f(T_i, T_j) = f(T_j, T_i) = 0$. It follows that $\mathcal{H}'(T_i) \subseteq \mathcal{H}'(T_j)$ and $\mathcal{H}'(T_j) \subseteq \mathcal{H}'(T_i)$. Hence, $\mathcal{H}'(T_i) = \mathcal{H}'(T_j)$ and $T_i = T_j$. The opposite direction is trivial.

To show the triangle inequality property, for a cluster $X_s \in \mathcal{H}'(T_i)$, suppose that $\{Y_s\} = \arg\min_{Y \in \mathcal{H}'(T_j)} |X_s \ominus Y|$ and $\{Z_s\} = \arg\min_{Z \in \mathcal{H}'(T_k)} |Y_s \ominus Z|$. (For simplicity, assume $|\{Y_s\}| = |\{Z_s\}| = 1$). By the property of the symmetric difference operation, $|X_s \ominus Z_s| \leq |X_s \ominus Y_s| + |Y_s \ominus Z_s|$. Because $\min_{Z \in \mathcal{H}'(T_k)} |X_s \ominus Z| \leq |X_s \ominus Z_s|$, $\min_{Y \in \mathcal{H}'(T_j)} |X_s \ominus Y| = |X_s \ominus Y_s|$, and $\min_{Z \in \mathcal{H}'(T_k)} |Y_s \ominus Z| = |Y_s \ominus Z_s|$, it follows that $\min_{Z \in \mathcal{H}'(T_k)} |X_s \ominus Z| \leq |X_s \ominus Y_s| + |Y_s \ominus Z_s|$. Since $f(T_i, T_k) \leq f(T_i, T_j) + f(T_j, T_k)$ and $f(T_k, T_i) \leq f(T_k, T_j) + f(T_j, T_i)$, it follows that $CA(T_i, T_k) \leq CA(T_i, T_j) + CA(T_j, T_k)$.

Definition 2 (Diameter). Let $\mathfrak{T}(n)$ be the space of all trees on the identical n leaves. Then the diameter of $\mathfrak{T}(n)$ with respect to a distance metric D on $\mathfrak{T}(n)$ is defined as $\Delta(D, n) := \max\{D(T_1, T_2) \mid T_1, T_2 \in \mathfrak{T}(n)\}$.

Proposition 3. $\Delta(CA, n) = \Theta(n^2)$.

Proof. Let $T_1 = (\ldots (1, 2), 3), \ldots n-1)$ and $T_2 = (\ldots (n, n-1), n-2), \ldots 1)$ be two caterpillars over the same leaf set.

Let $v_l \in V(T_2)$ such that $\mathcal{C}_{T_2}(v_l) = \{n, n-1, \cdots, \frac{3}{4}n\}$. For all $u \in V_{int}(T_1) \setminus \{r(T_1)\}$, $|\mathcal{C}_{T_1}(u) \ominus \mathcal{C}_{T_2}(v_l)| > \frac{n}{4}$ since $|\{1, 2\} \ominus \{n, n-1, \cdots, \frac{3}{4}n\}| = \frac{n}{4} + 3$, $|\{1, 2, 3\} \ominus \{n, n-1, \cdots, \frac{3}{4}n\}| = \frac{n}{4} + 4, \cdots$, and so on. Similarly, let $v_h \in V(T_2)$ such that $\mathcal{C}_{T_2}(v_h) = \{n, n-1, \ldots, \frac{n}{4}\}$. For all $u \in V_{int}(T_1) \setminus \{r(T_1)\}$, $|\mathcal{C}_{T_1}(u) \ominus \mathcal{C}_{T_2}(v_h)| \geq \frac{n}{4}$ since $|\{1, 2, \cdots, n-1\} \ominus \{n, n-1, \cdots, \frac{n}{4}\}| = \frac{n}{4}$, $|\{1, 2, \cdots, n-2\} \ominus \{n, n-1, \cdots, \frac{n}{4}\}| = \frac{n}{4} + 1$, and so on. For any vertex v such that $v_l \leq_{T_2} v \leq_{T_2} v_h$, $|\mathcal{C}_{T_1}(u) \ominus \mathcal{C}_{T_2}(v)| > \frac{n}{4}$ for all $u \in V_{int}(T_1) \setminus \{r(T_1)\}$. Because there are $\frac{n}{2}$ such vertices in $V(T_2)$, we have $f(T_2, T_1) > \frac{n}{4} \times \frac{n}{2} = \Omega(n^2)$.

Let $u_l \in V(T_1)$ such that $\mathcal{C}_{T_1}(u_l) = \{1, 2, \cdots, \frac{n}{4}\}$. For all $v \in V_{int}(T_2) \setminus \{r(T_2)\}$, $|\mathcal{C}_{T_2}(v) \ominus \mathcal{C}_{T_1}(u_l)| > \frac{n}{4}$ since $|\{n, n-1\} \ominus \{1, 2, \cdots, \frac{n}{4}\}| = \frac{n}{4} + 2$, $|\{n, n-1, n-2\} \ominus \{1, 2, \cdots, \frac{n}{4}\}| = \frac{n}{4} + 3, \cdots$, and so on. Similarly, let $u_h \in V(T_1)$ such that $\mathcal{C}_{T_1}(u_h) = \{1, 2, \ldots, \frac{3}{4}n\}$. For all $v \in V_{int}(T_2) \setminus \{r(T_2)\}$, $|\mathcal{C}_{T_2}(v) \ominus \mathcal{C}_{T_1}(u_h)| > \frac{n}{4}$ since $|\{n, n-1, \cdots, 2\} \ominus \{1, 2, \cdots, \frac{3}{4}n\}| = \frac{n}{4} + 1$, $|\{n, n-1, \cdots, 3\} \ominus \{1, 2, \cdots, \frac{3}{4}n\}| = \frac{n}{4} + 2$, and so on. For any vertex u such that $u_l \leq_{T_1} u \leq_{T_1} u_h$, $|\mathcal{C}_{T_1}(u) \ominus \mathcal{C}_{T_2}(v)| > \frac{n}{4}$ for all $v \in V_{int}(T_2) \setminus \{r(T_2)\}$. Because there are $\frac{n}{2}$ such vertices in $V(T_1)$, $f(T_1, T_2) > \frac{n}{4} \times \frac{n}{2} = \Omega(n^2)$.

For the upper bound, consider two trees T_1 and T_2 on n leaves. For any pair of $u \in V_{int}(T_1) \setminus \{r(T_1)\}$ and $v \in V_{int}(T_2) \setminus \{r(T_2)\}$, $|\mathcal{C}_{T_1}(u) \ominus \mathcal{C}_{T_2}(v)| \leq n-3$. Hence, $f(T_1, T_2) = \mathcal{O}(n^2)$ and $f(T_2, T_1) = \mathcal{O}(n^2)$. Finally, $\Delta(CA, n) = \Theta(n^2)$.

4   Gradients for the Tree Edit Operations

Definition 3 *(Gradient).* *The* gradient *of a tree edit operation ϕ with respect to a distance D on $\mathfrak{T}(n)$ is $\mathcal{G}(T(n), D, \phi) := \max\{D(T_1, T_2) \mid T_1, T_2 \in \mathfrak{T}(n) \wedge T_2 \in \phi(T_1)\}$* [15].

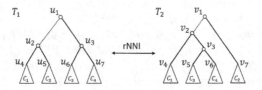

Fig. 1. An rNNI operation: $T_1 \in rNNI(T_2)$, $T_2 \in rNNI(T_1)$, and $CA(T_1, T_2) = \frac{n}{4}$.

Proposition 4. $\mathcal{G}(n, CA, rNNI) = \Theta(n)$.

Proof. Consider the trees T_1 and T_2 with the same n leaves as shown in Fig. 1. Suppose that T_1 and T_2 are well-balanced trees such that $|C_1| = |C_2| = |C_3| = |C_4| = \frac{n}{4}$. Note that $\mathcal{C}_{T_1}(u_1) = C_1 \cup C_2 \cup C_3 \cup C_4$, $\mathcal{C}_{T_1}(u_2) = C_1 \cup C_2$, $\mathcal{C}_{T_1}(u_3) = C_3 \cup C_4$, $\mathcal{C}_{T_2}(v_1) = C_1 \cup C_2 \cup C_3 \cup C_4$, $\mathcal{C}_{T_2}(v_2) = C_1 \cup C_3 \cup C_4$, and $\mathcal{C}_{T_2}(v_3) = C_3 \cup C_4$. It follows that $f(T_1, T_2) = \frac{n}{4}$ since $|\mathcal{C}_{T_1}(u_1) \ominus \mathcal{C}_{T_2}(v_1)| = 0$, $|\mathcal{C}_{T_1}(u_2) \ominus \mathcal{C}_{T_2}(v_4)| = |\mathcal{C}_{T_1}(u_2) \ominus \mathcal{C}_{T_2}(v_7)| = \frac{n}{4}$, $|\mathcal{C}_{T_1}(u_3) \ominus \mathcal{C}_{T_2}(v_3)| = 0$. Similarly, it follows that $f(T_2, T_1) = \frac{n}{4}$ since $|\mathcal{C}_{T_2}(v_2) \ominus \mathcal{C}_{T_1}(u_3)| = |\mathcal{C}_{T_2}(v_2) \ominus \mathcal{C}_{T_1}(u_1)| = \frac{n}{4}$. Therefore, $\mathcal{G}(T(n), CA, rNNI) = \Theta(n)$.

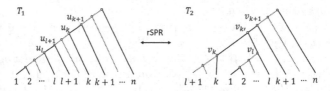

Fig. 2. An rSPR operation: $T_1 \in rSPR(T_2)$, $T_2 \in rSPR(T_1)$, and $CA(T_1, T_2) = \Theta(n^2)$.

Proposition 5. $\mathcal{G}(n, CA, rSPR) = \Theta(n^2)$

Proof. Consider the caterpillar trees T_1 and T_2 with the same n leaves as shown in Fig. 2. By the rSPR operation from T_1 to T_2, the edge $\{u_l, u_{l+1}\}$ is deleted, and the subtree $T_1(u_l)$ is grafted between u_k and u_{k+1} where $1 < l < k < n$. It follows that $|\mathcal{C}_{T_1}(u_l) \ominus \mathcal{C}_{T_2}(v_l)| = 0$, $|\mathcal{C}_{T_1}(u_k) \ominus \mathcal{C}_{T_2}(v_{k'})| = 0$, and $|\mathcal{C}_{T_1}(u_{l+1}) \ominus \mathcal{C}_{T_2}(v_l)| = 1$. Suppose that $\frac{n}{8} < l < \frac{n}{4}$ and $k \geq \frac{3}{4}n$, then $\Sigma_{u_l <_{T_1} u <_{T_1} u_k} |\mathcal{C}_{T_1}(u) \ominus \mathcal{C}_{T_2}(v_l)| = \Sigma_{u_l <_{T_1} u <_{T_1} u_k} |\mathcal{C}_{T_1}(u) \ominus \mathcal{C}_{T_2}(v_{k'})| = 1 + 2 + \cdots + m = \frac{m(m+1)}{2} \geq \frac{3}{8}n^2$ since $m = k - l - 1 \geq \frac{n}{2}$. It follows that $f(T_1, T_2) = \Omega(n^2)$, hence $\mathcal{G}(n, CA, rNNI) = \Omega(n^2)$. The upper bound is trivial by Proposition 3. Therefore, $\mathcal{G}(n, CA, rSPR) = \Theta(n^2)$.

Corollary 1. $\mathcal{G}(n, CA, rTBR) = \Theta(n^2)$.

5   Experiments

We study the characteristics of the distances rRF and CM in comparison with the CA distance using simulated datasets. First, we compare the distances of tree pairs under these measures when randomly sampled under two classic models. Second, we study how the distances rRF, CM, and CA are correlated with the number of consecutive tree edit operations. For the experiments, we define a *profile* to be a tuple of trees over the same leaf set.

Table 1. Descriptive statistics of the distances rRF, CM, CA, between a pair of randomly generated binary trees under the Yule-Harding model and the birth-death process model on 100 and 1000 leaves. (SD: standard deviation, CV: coefficient of variation)

Leaves		100			1000		
Model		rRF	CM	CA	rRF	CM	CA
Yule-Harding	Mean	97.77	891.60	626.31	997.78	17659.27	11090.88
	SD	0.48	38.28	26.58	0.48	423.3	311.87
	CV	0.49%	4.29%	4.24%	0.04%	2.39%	2.81%
	Median	98	889	625	998	17637	11070
	Min	94	760	529	994	16253	10066
	Max	98	1123	808	998	20031	12861
Birth-death	Mean	81.37	891.6	283.2	837.87	17659.27	4764.95
	SD	4.16	38.28	36.89	12.84	423.3	380.38
	CV	5.11%	4.29%	13.02%	1.53%	2.39%	7.98%
	Median	82	889	282	838	17637	4754
	Min	58	760	148	782	16253	3357
	Max	95	1123	458	891	20031	6563

5.1   Distribution of the Tree Distance Metrics

We compared the distance distributions under rRF, CM, and CA for randomly sampled trees under the Yule-Harding model [9] and birth-death process model [2].

Dataset. We generated the profiles $P_k := \{p_1, \ldots, p_l\}$ and $Q_k := \{q_1, \ldots, q_l\}$ of random trees over k leaves for each $k \in \{100, 1000\}$, and $l := 100000$ separately under each of the two models as follows.

Yule-Harding Model. The following procedure [3] is sampling trees for each k.

  i. an initial list of k single-vertex trees is generated.
 ii. two randomly chosen trees are merged into a new tree by making the roots of these trees the children of a new root.
iii. this process is repeated until the list contains only one tree.

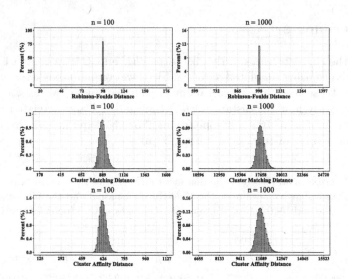

Fig. 3. Distributions of the distances rRF, CM, and CA between a pair of randomly generated binary trees by using the Yule-Harding model on 100 and 1000 leaves. The ranges of the x-axes are $\mu \pm 0.8 \times \mu$ for $n = 100$ and $\mu \pm 0.4 \times \mu$ for $n = 1000$.

Birth-Death Process Model. The software DendroPy version 3.10 [22] was used with the parameters 0.1 and 0 for the birth rate and death rate, respectively.

Experimental Setting. For the profiles P_k and Q_k ($k \in \{100, 1000\}$) generated under each of the two models, we computed the distances rRF and CA for each pair p_i and q_i, where $1 \leq i \leq l$.

Results and Discussion. We discuss the results for each of the two models.

Yule-Harding Model. The distributions of the distances rRF, CM, and CA between the pairs of randomly generated trees are depicted in Fig. 3, and the corresponding descriptive statistics are shown in Table 1. The rRF distances show very narrow distributions for the sampled tree pairs with 100 and 1000 leaves, and thus the coefficient of variation (CV) of these distances are relatively minimal. Also, the minimum value and mean value for both of these distributions are very close to the theoretical maximum values (i.e., diameters). Further, the standard deviation and the range of the rRF distances are similar for tree pairs that have between 100 and 1000 leaves, suggesting that they are not proportional to the number of leaves. In contrast, the distances CM and CA are more broadly and bell-shape like distributed, which also have a much larger CV than the corresponding rRF distributions for 100 and 1000 leaves.

Birth-Death Process Model. Table 1 summarizes the descriptive statistics and Fig. 4 shows the distributions of the distances rRF, CM, and CA between a pair of randomly generated trees. Unlike the Yule-Harding model, the distributions of the distances rRF, CM and CA are all in the form of a bell-shape. However, the CA distance is more widely and evenly dispersed than the CM distance.

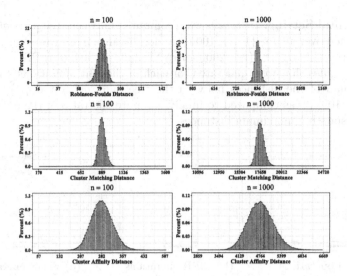

Fig. 4. Distribution of the distances rRF, CM, and CA between a pair of randomly generated binary trees under the birth-death process model on 100 and 1000 leaves. The ranges of the x-axes are $\mu \pm 0.8 \times \mu$ for $n = 100$ and $\mu \pm 0.4 \times \mu$ for $n = 1000$.

Hence, the distribution of the CA distance shows a much larger CV than the distribution of the distances rRF and CM, for both 100 and 1000 leaves.

5.2   Distance Metrics Under the Tree Edit Operations

We show (i) how the distances rRF, CM, and CA correlate with the number of consecutive tree edit operations, and (ii) that the rRF distance is expected to be saturated faster than the distances CM and CA under repeated edit operations.

Dataset. We generated a profile P consisting of 1000 random trees on 500 leaves, where each tree in P was sampled under the Yule-Harding model using the procedure described in Sect. 5.1. For each of the rooted tree edit operations rNNI, rSPR, and rTBR we generated the profiles $Q(i) := \{q(i)_1, \ldots, q(i)_{1000}\}$ for (i) every $i \in \{1, \ldots, 2000\}$ for the rNNI operation, and (ii) every $i \in \{1, \ldots, 500\}$ for the rSPR and rTBR operations. The profiles were generated as follows.

i. Given a tree edit operation, the initial profile $Q(1)$ is set to profile P. If this operation is rNNI, the range r is set to 2000. Otherwise, r is set to 500.
ii. For each $i \in \{2, \ldots, r\}$ the profile $Q(i + 1)$ is generated from profile $Q(i)$. The tree $q(i + 1)_j$ is created by applying the input tree edit operation to tree $q(i + 1)_j$ (for each $j \in \{1, \ldots, 1000\}$) , where the selection of edges in $q(i)_j$ that is needed to specify the operation is chosen randomly. E.g., for the rSPR operation, two edges (possibly including a root edge) of tree $q(i)_j$ are randomly chosen, where the first edge determines the pruning location of the subtree, and the second edge the regrafting location of the subtree.

Experimental Setting. Distances were computed between tree pairs $q(1)_j$ and $q(i)_j$ averaged over all $j \in \{1, \ldots, 1000\}$, under distance measures rRF, CM, and CA for every $i \in \{1, \ldots, r\}$ ($r = 2000$ for rNNI, and $r = 500$ for rSPR and rTBR) using the profiles that were generated for each of the edit operations. Similarly, the maximum of the distances between the tree pairs $q(1)_j$ and $q(i)_j$ over all $j \in \{1, \ldots, 1000\}$ was computed to finally compute the ratio of the averages to their corresponding maximum distances.

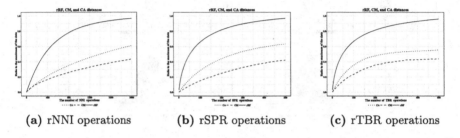

    (a) rNNI operations   **(b)** rSPR operations   **(c)** rTBR operations

Fig. 5. Distances by tree edit operations: average rRF and CA distances of 1000 trees on 500 leaves as a function of the number of consecutive tree edit operations.

5.3 Results and Discussion

Figure 5(a) depicts the average for the distances rRF, CM, and CA between the initial tree and rNNI operation applied trees. The gradient of the rRF distance curve is very steep between $0 \sim 1200$ operations, and the inclination of the curve is gradual after 1600 operations. However, after 1600 operations the distances CM and CA still have increasing trends. Figure 5(b) shows the average of the distances rRF, CM, and CA between the initial tree and rSPR operation applied trees. While the gradient of the rRF distance curve is gradual after 300 operations, the gradients of the distances CM and CA are in increasing trends. Figure 5(c) shows the average for the distances rRF, CM, and CA between the initial tree and rTBR operation applied trees. Unlike the rSPR operation, the gradients of the curves for the distances rRF, CM, and CA are all steep between 0–200 operations, but they are gradual after 300 operations.

6   Conclusion

There may not be an optimal tree comparison measure, and one or more measures may be used by the practitioner depending on the application. However, such choices can be guided by the strengths and weaknesses of such measures. We introduced the CA distance and showed that this distance, like the CM distance, offers a variety of desirable features. In contrast to the CM distance, the CA distance is also applicable to non-binary trees and reflects more precisely the distances between a tree pair in terms of the minimum cluster distances.

Acknowledgements. OE is supported by the National Science Foundation under Grant No. 1617626.

References

1. Allen, B.L., Steel, M.: Subtree transfer operations and their induced metrics on evolutionary trees. Ann. Comb. **5**(1), 1–15 (2001)
2. Arvestad, L., et al.: Gene tree reconstruction and orthology analysis based on an integrated model for duplications and sequence evolution. In: RECOMB, pp. 326–335. ACM (2004)
3. Betkier, A., Szczęsny, P., Górecki, P.: Fast algorithms for inferring gene-species associations. In: Harrison, R., Li, Y., Măndoiu, I. (eds.) ISBRA 2015. LNCS, vol. 9096, pp. 36–47. Springer, Cham (2015). https://doi.org/10.1007/978-3-319-19048-8_4
4. Bogdanowicz, D., Giaro, K.: On a matching distance between rooted phylogenetic trees. Int. J. Appl. Math. Comput. **23**(3), 669–684 (2013)
5. Bordewich, M., Semple, C.: On the computational complexity of the rooted subtree prune and regraft distance. Ann. Comb. **8**(4), 409–423 (2005)
6. DasGupta, B., et al.: On distances between phylogenetic trees. In: SODA 1997, pp. 427–436 (1997)
7. Day, W.H.E.: Optimal algorithms for comparing trees with labeled leaves. J. Classif. **2**(1), 7–28 (1985)
8. Felenstein, J.: Inferring Phylogenies. Sinauer, Sunderland (2003)
9. Harding, E.F.: The probabilities of rooted tree-shapes generated by random bifurcation. Adv. Appl. Probab. **3**(1), 44–77 (1971)
10. Huber, K.T., et al.: Metrics on multilabeled trees: interrelationships and diameter bounds. IEEE/ACM Trans. Comput. Biol. Bioinform. **8**(4), 1029–40 (2011)
11. Katherine, S.J.: Review paper: the shape of phylogenetic treespace. Syst. Biol. **66**(1), e83–e94 (2017)
12. Kuhner, M.K., Yamato, J.: Practical performance of tree comparison metrics. Syst. Biol. **64**(2), 205–14 (2015)
13. Li, M., Tromp, J., Zhang, L.: On the nearest neighbour interchange distance between evolutionary trees. J. Theor. Biol. **182**(4), 463–7 (1996)
14. Li, M., Zhang, L.: Twist-rotation transformations of binary trees and arithmetic expressions. J. Algorithms **32**(2), 155–166 (1999)
15. Lin, Y., Rajan, V., Moret, B.M.E.: A metric for phylogenetic trees based on matching. IEEE/ACM Trans. Comput. Biol. Bioinform. **9**(4), 1014–1022 (2012)
16. Ma, B., Li, M., Zhang, L.: From gene trees to species trees. SIAM J. Comput. **30**(3), 729–752 (2000)
17. Makarenkov, V., Leclerc, B.: Comparison of additive trees using circular orders. J. Comput. Biol. **7**(5), 731–744 (2000)
18. Moon, J., Eulenstein, O.: Cluster matching distance for rooted phylogenetic trees. In: Zhang, F., Cai, Z., Skums, P., Zhang, S. (eds.) ISBRA 2018. LNCS, vol. 10847, pp. 321–332. Springer, Cham (2018). https://doi.org/10.1007/978-3-319-94968-0_31
19. Robinson, D.F., Foulds, L.R.: Comparison of phylogenetic trees. Math. Biosci. **53**(1–2), 131–147 (1981)
20. Semple, C., Steel, M.A.: Phylogenetics. Oxford University Press, Oxford (2003)
21. Steel, M.A., Penny, D.: Distributions of tree comparison metrics. Syst. Biol. **42**(2), 126–141 (1993)

22. Sukumaran, J., Holder, M.T.: Dendropy: a Python library for phylogenetic computing. Bioinformatics **26**(12), 1569–1571 (2010)
23. Wilkinson, M., et al.: The shape of supertrees to come: tree shape related properties of fourteen supertree methods. Syst. Biol. **54**(3), 419–431 (2005)
24. Wu, Y.-C., et al.: TreeFix: statistically informed gene tree error correction using species trees. Syst. Biol. **62**(1), 110–20 (2013)

Modeling SNP-Trait Associations and Realizing Privacy-Utility Tradeoff in Genomic Data Publishing

Zaobo He[1(✉)] and Jianqiang Li[2(✉)]

[1] Department of Computer Science and Software Engineering,
Miami University, Oxford, OH, USA
hez26@miamioh.edu
[2] Faculty of Information Technology,
Beijing University of Technology, Beijing, China
lijianqiang@bjut.edu.cn

Abstract. Genomic data privacy arises as one of the most important concerns facing the wide commoditization of DNA-genotyping. In this paper, we study the problem of privacy preserved kin-genomic data publishing. The major challenge in protecting kin-genomic data privacy is to protect against powerful attackers with abundant background knowledge. We propose a probabilistic model based on factor graph with the knowledge of publicly available GWAS statistics to reveal the dependency relationship between genotypes and phenotypes. Furthermore, a genomic data sanitization method is proposed to protect against optimal inference attacks launched by powerful attackers.

Keywords: Factor graph · SNP/trait associations · Optimal attack ·
Data sanitization

1 Introduction

With the technical advancement of DNA-genotyping, genomic data publishing and analyzing have become an emerging paradigm, consisting of mass genomic data and diversified service requirements. For example, several commercial platforms have established to offer DNA-sequencing services to more than 900,000 users, such as 23andMe [1], OpenSNP [2] or PatientsLikeMe [3]. One typical service is to enable users to learn their predispositions to certain genetic diseases with genotyped DNA. On the other hand, collected DNA information by researchers is beneficial to establish new method of diagnosing diseases, or new medicines. These services and benefit are fundamentally derived from big data analyzing technologies, with the aid of GWAS (Genome-wide association studies) statistics and SNPs (single-nucleotide polymorphisms)-trait associations. GWAS explore genetic variation from case-control studies to figure out whether certain SNPs appear more frequently in people with a certain disease (or genetic trait).

© Springer Nature Switzerland AG 2019
Z. Cai et al. (Eds.): ISBRA 2019, LNBI 11490, pp. 65–72, 2019.
https://doi.org/10.1007/978-3-030-20242-2_6

GWAS show that various SNPs are associated with various complex genetic traits. Most of these information are publicly available. For example, the GWAS catalog [5] publish the SNPs-trait associations and related GWAS statistics.

As a fundamental data analyzing operation, modeling the association of SNPs and traits has been proved to help the understanding the association between genotypes and phenotypes, which is the basis of promising genetic diagnostics and medicine development. Particularly, probabilistic graphical models have been widely applied for modeling probabilistic dependency relationship between various SNPs and traits. For example, a recent work [18] proposed a Bayesian network model for characterizing SNP-trait association. In their model, each SNP or trait is represented as a network node, and a two-layered Bayesian network is constructed. Their model suffers from a serious limitation that lies that in the constructed Bayesian network, the direction of network arcs contradicts to the actual dependency relationship between SNPs and traits. In GWAS, a trait is usually viewed as a dependent variable whereas a SNP is treated as an independent variable, which implies the arcs should always point from SNP to trait. The current modeling based on probabilistic graphical models exists a major limitation which lies that most of the current works do not consider the kin-genomic dependency relationship.

In this paper, we study how to construct an accurate probabilistic graphical models based on GWAS statistics. Although such modeling is attractive, incorporating complex variable dependency and genetic associations among family members bring two major challenges. The first challenge comes from privacy concern. Genomes from two relatives are highly correlated; consequently, a person's genomic information can be easily leaked by her careless relatives, through a simple data releasing on a personal computer, without any consent from her relatives in advance. Although anonymized data is released by careless relatives, once one family member is identified (many de-anonymization algorithms have been proposed [7,15,17]), all family members privacy are a embarrassing situation. For example, Henrietta Lacks [4], whose DNA was sequenced and published online platform SNPedia without the consent of her family. Just after several minutes of uploading, attackers generate a complete report of genetic information about Henrietta Lacks and her family [14]. Thus, genomic data publishing and analyzing should avoid considerable leakage of sensitive information. The second challenge comes from modeling SNP-trait association for a given family. For a given family, modeling the complex genomic dependency is challenging, which involves computing minor allele frequencies (MAFs) of one SNPs, the fundamental rules of Mendelian inheritance [16], etc.

Although numerous efforts have been made for each challenge respectively [6,9–13,19], it remains an unsolved problem to model SNP-trait association for a family and conduct privacy-preserved genomic data publishing. For this challenge, we devote to modeling SNP-trait association for a family efficiently and privacy preserved publish kin-genomic data.

In this paper, we first propose a model to characterize GWAS statistics based on factor graph through incorporating SNPs, traits and related statistics into it. The model presents the conditional dependency between SNPs and traits, and conditional dependency among any two family numbers. Given the proposed model, statistical inference attack can be efficiently launched to predict hidden variables based on revealed variables and auxiliary knowledge.

To derive privacy-preserved genomic data publishing, we assume the attacker is powerful with abundant background knowledge. To meet such a requirement, we formulate an optimization problem that takes optimal attack as input and outputs an optimal data sanitization method, such that the derived sanitized genomic data satisfies the specified approximation degree and privacy guarantee can be maximized.

Our key contributions are summarized as follows.

- We propose a probabilistic model, in which the dependency relationship between genotypes and phenotypes can be revealed.
- The proposed probabilistic model formulate a framework based on which various statistical attacks can be presented.
- An optimization problem is formulated to achieve genomic privacy-utility tradeoff.

The rest of this paper is organized as follows. Section 2 introduces our system and adversary models, together with privacy and utility metrics. Section 3 presents optimization algorithm to solve optimal privacy preserving method. Section 4 concludes this paper.

2   Problem Formulation

In this section, we introduce the system and adversary model, quantification methods for measuring genomic data privacy and utility.

2.1   Adversary Model

The objective of the attacker is to predict the values of hidden traits and SNPs of members of a target family. To launch the attack, the attacker utilizes some auxiliary knowledge, essentially the SNP/trait associations (released by GWAS) and the fundamental rules of Mendelian inheritance. In addition to these publicly available auxiliary knowledge, the attacker collects a subset of traits and SNPs of several members of given family, typically those who have released their genomic information for service or research purpose.

The attack behavior can be formally formulated as exploring marginal probabilities of unknown SNPs and traits from the global probability distribution $\Pr(X_U | X_K, C)$, where X_U represents unknown SNPs and traits; X_K the observed SNPs and traits released by family members, and C the auxiliary knowledge. We assume the attacker has wide range of auxiliary knowledge, so that robust privacy preservation methods must be built. We assume the attacker knows the prior probability of SNPs and traits X of each family member, $\psi(X)$. Clearly, for a family member, $\sum \psi(X) = 1$ is always held for all her possible X. $\psi(X)$ is expressed as a user's profile. Furthermore, we also assume the attacker knows our privacy preservation method. With such open design, this inference attack can be launched efficiently by executing belief propagation on factor graph. (see Sect. 3.2 of [8] for more details about the inference attack).

2.2   Privacy and Utility Metrics

We define F to be the number of members in a given family and S to be the set of SNPs, where $|F| = m$ and $|S| = n$. We also denote the value of an arbitrary SNP $j(j \in S)$ for family member $i(i \in F)$ is s_j^i, where $s_j^i \in \{0, 1, 2\}$, where $s_j^i = 0$ represents $s_j^i = BB$ which implies both alleles of i are major alleles; $s_j^i = 1$ represents $s_j^i = Bb$ which implies alleles of i are a major allele and a minor allele; $s_j^i = 2$ represents $s_j^i = bb$ which implies both alleles are minor alleles. We assume an attacker launch attacks with broad auxiliary knowledge, involving intercepted partial SNPs or traits released by some members of given family.

We express the inference for marginal probabilities as $\Pr(\hat{x}_j^i | X_K, C)$, for all $i \in F$, $j \in S$. We quantify data privacy by measuring the attacker's error in predicting target SNPs and traits. Such incorrectness can be measured by exploring the expected distance between the adversary estimation on target variables $\hat{x}_j^i$, and the true value of the corresponding variables, x_j^i. Since the attacker knows the family member's profile $\psi(X)$ and privacy preserving method $f(X'|X)$, she computes the posterior probability of X, conditional on X' with prior knowledge $\psi(X)$ and $f(X'|X)$:

$$\Pr(X|X') = \frac{\Pr(X, X')}{\Pr(X')} = \frac{f(X'|X)\psi(X)}{\sum_X f(X'|X)\psi(X)}$$

Then, for each variable X with posterior probability distribution $\Pr(X|X')$, the attacker infer the family member's hidden SNPs and traits based on X. We represent the predicted hidden variables from $\Pr(X|X')$ as Z_X. The attackers' objective is then to choose a proper $\hat{Z}$ to minimize the family member's conditional expected genomic privacy, conditional on $\Pr(X|X')$. For an $\hat{Z}$, the family member's conditional expected genomic data privacy is

$$\sum_x \Pr(X|X')d_p(Z_X, \hat{Z})$$

where $d_p(Z_X, \hat{Z})$ is the difference between $Z_X(A)$ and $\hat{Z}$.

For the minimized $\hat{Z}$, it is

$$\min_{\hat{Z}} \sum_X \Pr(X|X')d_p(Z_X, \hat{Z}) \tag{1}$$

For a given X', the conditional genomic data privacy is given by 1. However, the probability distribution of X' after carrying out privacy preservation method is $\Pr(X') = \sum_X f(X'|X)\psi(X)$. Thus, the family member's unconditional expected genomic privacy is

$$\sum_{X'} \psi(X') \min_{\hat{Z}} \sum_X Pr(X|X') d_p(Z_X, \hat{Z})$$
$$= \sum_{X'} \min_{\hat{Z}} \sum_X \psi(X) f(X'|X) d_p(Z_X, \hat{Z})$$

Then the genomic data privacy loss can be formally formulated as follows:

Definition 2.1 Genomic Data Privacy Loss. *Genomic data privacy loss measures how much privacy loses of a family member in the inference attack issued launch by a attacker, which is qualified by the family member's unconditional expected privacy:*

$$\sum_{X'} \min_{\hat{Z}} \sum_X \psi(X) f(X'|X) d_p(Z_X, \hat{Z})$$

We define

$$P_{X'} = \min_{\hat{Z}} \sum_X \psi(X) f(X'|X) d_p(Z_X, \hat{Z}). \tag{2}$$

Substituting $P_{X'}$ into the above expression, it can be rewritten as

$$\sum_{X'} P_{X'}, \tag{3}$$

which is the optimization objective and an optimal privacy preservation method can maximize it, in order to finding the optimal $f(X'|X)$.

Unfortunately, the problem in (2) is nonlinear due to the minimum operator. To reduce computation overhead, the nonlinear problem can be converted into a set of linear constraints:

$$P_{X'} \le \sum_X \psi(X) f(X'|X) d_p(Z_X, \hat{Z}) \quad \forall \hat{Z} \tag{4}$$

Therefore, the problem of maximizing (3) under constraint (2) can be solved given the solution of solving the optimizing (3) under condition (4).

Utility Metrics: we quantify data utility by measuring the average prediction accuracy for arbitrary SNPs and traits. Such accuracy difference $d_u(X, X')$ can also be measured by exploring the expected distance between the prediction value on target variables $\hat{x}_j^i$, and the true value of the corresponding variables, x_j^i. Given $f(X'|X)$, $\psi(X)$, and $d_u(X, X')$, data utility can be measured as the expectation of $du(X, X')$ over all X and X' for a family member: $PUL_i = \sum_{X,X'} \psi(X) f(X'|X) d_u(X, X') \le \delta$.

Attribute set disparity measurer d_u is determined by data semantics. In different applications, du can be defined as Euclidean, Hamming, or Mahalanobis distance, etc.

3 SNP-Trait Association Modeling

Figure 1 shows a factor graph for a family with three members, assuming each member has 2 traits $T = \{t_1, t_2\}$, 3 SNPs $S = \{s_1, s_2, s_3\}$. For two traits t_1 and t_2, the associated SNPs are $\{s_1\}$ and $\{s_1, s_2, s_3\}$, respectively.

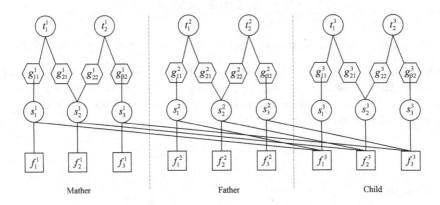

Fig. 1. A factor graph with 2 traits $T = \{t_1, t_2\}$ and 3 SNPs $S = \{s_1, s_2, s_3\}$.

4 Privacy-Utility Tradeoff

The problem of finding optimal privacy preserving genomic data publishing method $f(X'|X)$ can be formulated as follows. A linear program for a family member to find the optimal genomic data sanitization is: choose $f(X'|X)$, $\hat{Z}$, $\forall X, X'$, in order to

Maximize: $\sum_{X'} P_{X'}$

Subject to:

$$P_{X'} \le \sum_{X} \psi(X) f(X'|X) d_p(Z_X(A), \hat{Z}) \quad \forall \hat{Z}$$

$$\sum_{X} \psi(X) \sum_{X'} f(X'|X) d_u(X, X') \le \delta$$

$$f(X'|X) \ge 0 \qquad \forall X, X'$$

$$\sum_{X'} f(X'|X) = 1, \quad \forall X$$

5 Conclusions

We have proposed a probabilistic graphical model for modeling SNPs-trait associations from GWAS statistics, based on factor graph. Furthermore, we have proposed a privacy preserved kin-genomic data publishing method, which is able to protect against inference attacks from powerful attackers.

References

1. https://www.23andme.com/
2. https://opensnp.org/
3. https://www.patientslikeme.com/
4. http://www.nytimes.com/2013/03/24/opinion/sunday/the-immortal-life-of-henrietta-lacks-the-sequel.html?pagewanted=all
5. The NHGRI-EBI catalog of published genome-wide association studies. https://www.ebi.ac.uk/gwas/docs/about
6. Cai, Z., He, Z., Guan, X., Li, Y.: Collective data-sanitization for preventing sensitive information inference attacks in social networks. IEEE Trans. Depend. Secure Comput. **15**(4), 577–590 (2018)
7. Gymrek, M., McGuire, A.L., Golan, D., Halperin, E., Erlich, Y.: Identifying personal genomes by surname inference. Science **339**(6117), 321–324 (2013)
8. He, Z., Yu, J., Li, J., Han, Q., Luo, G., Li, Y.: Inference attacks and controls on genotypes and phenotypes for individual genomic data. IEEE/ACM Trans. Comput. Biol. Bioinform. 1 (2018)
9. He, Z., Cai, Z., Li, Y.: Customized privacy preserving for classification based applications. In: Proceedings of the 1st ACM Workshop on Privacy-Aware Mobile Computing, pp. 37–42. ACM (2016)
10. He, Z., Cai, Z., Yu, J., Wang, X., Sun, Y., Li, Y.: Cost-efficient strategies for restraining rumor spreading in mobile social networks. IEEE Trans. Veh. Technol. **66**(3), 2789–2800 (2017)
11. He, Z., Li, Y., Li, J., Li, K., Cai, Q., Liang, Y.: Achieving differential privacy of genomic data releasing via belief propagation. Tsinghua Sci. Technol. **23**(4), 389–395 (2018)
12. He, Z., Li, Y., Li, J., Yu, J., Gao, H., Wang, J.: Addressing the threats of inference attacks on traits and genotypes from individual genomic data. In: Cai, Z., Daescu, O., Li, M. (eds.) ISBRA 2017. LNCS, vol. 10330, pp. 223–233. Springer, Cham (2017). https://doi.org/10.1007/978-3-319-59575-7_20
13. He, Z., Li, Y., Wang, J.: Differential privacy preserving genomic data releasing via factor graph. In: Cai, Z., Daescu, O., Li, M. (eds.) ISBRA 2017. LNCS, vol. 10330, pp. 350–355. Springer, Cham (2017). https://doi.org/10.1007/978-3-319-59575-7_33
14. Humbert, M., Ayday, E., Hubaux, J.P., Telenti, A.: Addressing the concerns of the lacks family: quantification of kin genomic privacy. In: Proceedings of the 2013 ACM SIGSAC conference on Computer & communications security, pp. 1141–1152. ACM (2013)
15. Humbert, M., Huguenin, K., Hugonot, J., Ayday, E., Hubaux, J.P.: De-anonymizing genomic databases using phenotypic traits. Proc. Priv. Enhancing Technol. **2015**(2), 99–114 (2015)
16. Saghai-Maroof, M.A., Soliman, K.M., Jorgensen, R.A., Allard, R.: Ribosomal dna spacer-length polymorphisms in barley: mendelian inheritance, chromosomal location, and population dynamics. Proc. Natl. Acad. Sci. **81**(24), 8014–8018 (1984)
17. Sweeney, L., Abu, A., Winn, J.: Identifying participants in the personal genome project by name (a re-identification experiment). arXiv preprint arXiv:1304.7605 (2013)

18. Wang, Y., Wu, X., Shi, X.: Using aggregate human genome data for individual identification. In: 2013 IEEE International Conference on Bioinformatics and Biomedicine, pp. 410–415, December 2013
19. Zheng, X., Cai, Z., Li, J., Gao, H.: Location-privacy-aware review publication mechanism for local business service systems. In: IEEE INFOCOM 2017-IEEE Conference on Computer Communications, pp. 1–9. IEEE (2017)

Simultaneous Multi-Domain-Multi-Gene Reconciliation Under the Domain-Gene-Species Reconciliation Model

Lei Li and Mukul S. Bansal[✉]

Department of Computer Science and Engineering, University of Connecticut,
Storrs, CT, USA
{lei.li,mukul.bansal}@uconn.edu

Abstract. The recently developed Domain-Gene-Species (DGS) reconciliation framework, which jointly models the evolution of a domain family inside one or more gene families and the evolution of those gene families inside a species tree, represents one of the most powerful computational techniques for reconstructing detailed histories of domain and gene family evolution in eukaryotic species. However, the DGS reconciliation framework allows for the reconciliation of only a single domain tree (representing a single domain family present in one or more gene families from the species under consideration) at a time, i.e., each domain tree is reconciled separately without consideration of any other domain families that might be present in the gene trees under consideration. However, this can lead to conflicting gene-species reconciliations for gene trees containing multiple domain families.

In this work, we address this problem by extending the DGS reconciliation model to simultaneously reconcile a set of domain trees, a set of gene trees, and a species tree. The new model, which we call the *multi-DGS (mDGS) reconciliation model*, produces a consistent joint reconciliation showing the evolution of each domain tree in its corresponding gene trees and the evolution of each gene tree inside the species tree. We formalize the mDGS reconciliation framework and define the associated computational problem, provide a heuristic algorithm for estimating optimal mDGS reconciliations (both the DGS and mDGS reconciliation problems are NP-hard), and apply our algorithm to a large dataset of over 3800 domain trees and over 7100 gene trees from 12 fly species. Our analysis of this dataset reveals interesting underlying patterns of co-occurrence of domains and genes, demonstrates the importance of mDGS reconciliation, and shows that the proposed heuristic is effective at estimating optimal mDGS reconciliations.

1 Introduction

Most eukaryotic genes are known to contain one or more protein domains [2,4] and it is well understood that the domain content of genes can change over

© Springer Nature Switzerland AG 2019
Z. Cai et al. (Eds.): ISBRA 2019, LNBI 11490, pp. 73–86, 2019.
https://doi.org/10.1007/978-3-030-20242-2_7

time due to evolutionary events such as domain duplications, transfers, or losses [8]. Changes in the domain content of genes have important functional consequences [11,12] and it is therefore important to reconstruct the history of these changes in the evolution of gene families. Several methods have been developed for studying the evolution of domain families (or domain trees), but these methods either do not take gene trees into account [1,13,15] or do not account for the inter-dependence of domain, gene, and species level evolution [10].

The recently developed Domain-Gene-Species (DGS) reconciliation framework [6,7], which jointly models the evolution of a domain family inside one or more gene families and the evolution of those gene families inside a species tree, represents one of the most powerful computational techniques for reconstructing detailed histories of domain and gene family evolution in eukaryotic species. However, the DGS reconciliation framework allows for the reconciliation of only a single domain tree (representing a single domain family present in one or more gene families from the species under consideration) at a time, i.e., each domain tree is reconciled separately without consideration of any other domain families that might be present in the gene trees under consideration. This poses a problem since many gene families (or gene trees) have multiple protein domains; specifically, solving the DGS reconciliation problem on different domain trees that are represented in the same gene tree can yield conflicting reconciliations for that gene tree with the species tree.

Our Contributions. In this work, we address this problem by extending the DGS reconciliation model to simultaneously reconcile a set of domain trees, a set of gene trees, and a species tree. The new model, which we call the *multi-DGS (mDGS) reconciliation model*, produces a consistent joint reconciliation showing the evolution of each domain tree in its corresponding gene trees and the evolution of each gene tree inside the species tree. We formalize the mDGS reconciliation framework and define the associated computational problem, provide a heuristic algorithm for estimating optimal mDGS reconciliations (both the DGS and mDGS reconciliation problems are NP-hard), and apply our algorithm to a large dataset of over 3800 domain trees and over 7100 gene trees from 12 fly species. Our experimental results demonstrate the importance of mDGS reconciliation and show that the proposed heuristic is effective at estimating optimal mDGS reconciliations. We also develop a technique to further improve the accuracy of mDGS reconciliation by using appropriately chosen subsets of the domain and gene trees under consideration and provide a clustering algorithm to find such subsets. An implementation of our heuristic for mDGS reconciliation is available freely from https://compbio.engr.uconn.edu/software/seadog/.

2   Definitions and Preliminaries

We follow the notation and basic definitions from [6,7].

Preliminaries. Throughout this manuscript, the term *tree* refers to rooted binary trees. Given a tree T, we denote its node, edge, and leaf sets by $V(T)$,

$E(T)$, and $Le(T)$ respectively. The root node of T is denoted by $rt(T)$, the parent of a node $v \in V(T)$ by $pa_T(v)$, its set of children by $Ch_T(v)$, and the (maximal) subtree of T rooted at v by $T(v)$. The set of *internal nodes* of T, denoted $I(T)$, is defined to be $V(T) \setminus Le(T)$. We define $\leq_T$ to be the partial order on $V(T)$ where $x \leq_T y$ if y is a node on the path between $rt(T)$ and x. The partial order $\geq_T$ is defined analogously, i.e., $x \geq_T y$ if x is a node on the path between $rt(T)$ and y. We say that y is an *ancestor* of x, or that x is a *descendant* of y, if $x \leq_T y$ (note that, under this definition, every node is a descendant as well as ancestor of itself). We say that x and y are *incomparable* if neither $x \leq_T y$ nor $y \leq_T x$. Given a non-empty subset $L \subseteq Le(T)$, we denote by $lca_T(L)$ the least common ancestor (LCA) of all the leaves in L in tree T; i.e., $lca_T(L)$ is the unique smallest upper bound of L under $\leq_T$.

The input for mDGS reconciliation is a collection of domain trees $\mathcal{D}$, a collection of gene trees $\mathcal{G}$, and a species tree S. The *species tree* is a tree showing the evolutionary history for a chosen set of species. Each *gene tree* is a tree showing the evolutionary history for a set of genes related by common ancestry, called a *gene family*, restricted to the species represented in the species tree. Similarly, a *domain tree* shows the evolutionary history of a set of domains related by common ancestry, called a *domain family*, restricted to the species present in the species tree. For mDGS reconciliation, we require that the collections $\mathcal{D}$ and $\mathcal{G}$ be "complete", in the sense that all gene families represented in any domain tree from $\mathcal{D}$ should be present as a gene tree in $\mathcal{G}$ and all domain families represented in any gene tree of $\mathcal{G}$ should be present as a domain tree in $\mathcal{D}$. We refer to any such "complete" pair of collections $\mathcal{D}$ and $\mathcal{G}$ as a *DG-group*. Essentially, a DG-group can be viewed as a connected component in a bipartite graph where the node set corresponds to all domain families and all gene families present in the species under consideration and an edge connects a domain family node and a gene family node if a domain from that domain family exists in a gene from that gene family.

As in DGS reconciliation [6,7], each leaf in a gene tree is labeled by the species from which that leaf (gene) was sampled. Similarly, each leaf in a domain tree is labeled with the gene from which that leaf (domain) was taken. This defines a leaf-to-leaf mapping from the domain trees to the gene trees, and from the gene trees to the species tree. Since a gene may have multiple domains, there may be multiple domains (possibly from different domain trees) mapping to the same gene. Similarly, since domains from the same domain family may be present in multiple gene families, different leaves of a single domain tree may map to genes from different gene families.

For convenience, we extend the notions of the leaf set, vertex set, and edge set of a tree as follows: $Le(\mathcal{G}) = \cup_{G \in \mathcal{G}} Le(G)$, $V(\mathcal{G}) = \cup_{G \in \mathcal{G}} V(G)$, and $E(\mathcal{G}) = \cup_{G \in \mathcal{G}} E(G)$. And $Le(\mathcal{D}) = \cup_{D \in \mathcal{D}} Le(D)$, $V(\mathcal{D}) = \cup_{D \in \mathcal{D}} V(D)$, and $E(\mathcal{D}) = \cup_{D \in \mathcal{D}} E(D)$.

mDGS Reconciliation. The *multi-Domain-Gene-Species (mDGS)* reconciliation model defines what constitutes a valid joint reconciliation of the given gene trees with the species tree and of the given domain trees with the gene trees. As with DGS reconciliation, mDGS reconciliation models the primary evolutionary events that shape gene family evolution in multicellular eukaryotes: *speciation, gene duplication,* and *gene loss.* Similarly, the reconciliation of a domain tree with one or more gene trees models the elementary evolutionary events that shape domain family evolution within genes: *co-divergence, domain transfer, domain duplication,* and *domain loss.* Formally:

Definition 1 (mDGS-reconciliation). *Given a collection of domain trees $\mathcal{D}$ and a collection of gene trees $\mathcal{G}$ that form a DG-group, and given a species tree S and leaf-mappings $\mathcal{L}^{\mathcal{D}}: Le(\mathcal{D}) \to Le(\mathcal{G})$ and $\mathcal{L}^{\mathcal{G}}: Le(\mathcal{G}) \to Le(S)$, an mDGS reconciliation for $\mathcal{D}, \mathcal{G},$ and S is a nine-tuple $\langle \mathcal{M}^{\mathcal{D}}, \mathcal{M}^{\mathcal{G}}, \Sigma^{\mathcal{D}}, \Sigma^{\mathcal{G}}, \Delta^{\mathcal{D}}, \Delta^{\mathcal{G}}, \Theta, \Xi, \tau \rangle$, where $\mathcal{M}^{\mathcal{D}}: V(\mathcal{D}) \to V(\mathcal{G})$ and $\mathcal{M}^{\mathcal{G}}: V(\mathcal{G}) \to V(S)$ map each node of $\mathcal{D}$ to a node from $\mathcal{G}$ and each node from $\mathcal{G}$ to a node of S, respectively, the sets $\Sigma^{\mathcal{D}}$, $\Delta^{\mathcal{D}}$, and Θ partition $I(\mathcal{D})$ into co-divergence, domain-duplication, and domain-transfer nodes, respectively, the sets $\Sigma^{\mathcal{G}}$ and $\Delta^{\mathcal{G}}$ partition $I(\mathcal{G})$ into speciation and gene-duplication nodes, respectively, Ξ is a subset of domain tree edges that represent domain-transfer events, and $\tau: \Theta \to V(\mathcal{G})$ specifies the recipient gene for each domain-transfer event, subject to:*

Gene-Species constraints:

1. *If $g \in Le(\mathcal{G})$, then $\mathcal{M}^{\mathcal{G}}(g) = \mathcal{L}^{\mathcal{G}}(g)$.*
2. *If $g \in I(\mathcal{G})$ and g' and g'' denote the children of g, then,*
   (a) *$\mathcal{M}^{\mathcal{G}}(g) \geq_S lca(\mathcal{M}^{\mathcal{G}}(g'), \mathcal{M}^{\mathcal{G}}(g''))$,*
   (b) *$g \in \Sigma^{\mathcal{G}}$ if and only if $\mathcal{M}^{\mathcal{G}}(g) = lca(\mathcal{M}^{\mathcal{G}}(g'), \mathcal{M}^{\mathcal{G}}(g''))$ and $\mathcal{M}^{\mathcal{G}}(g')$ and $\mathcal{M}^{\mathcal{G}}(g'')$ are incomparable,*
   (c) *$g \in \Delta^{\mathcal{G}}$ only if $\mathcal{M}^{\mathcal{G}}(g) \geq_S lca(\mathcal{M}^{\mathcal{G}}(g'), \mathcal{M}^{\mathcal{G}}(g''))$.*

Domain-Gene constraints:

3. *If $d \in Le(\mathcal{D})$, then $\mathcal{M}^{\mathcal{D}}(d) = \mathcal{L}^{\mathcal{D}}(d)$.*
4. *If $d \in I(\mathcal{D})$ and d' and d'' denote the children of d, then,*
   (a) *$\mathcal{M}^{\mathcal{D}}(d) \nleq_{\mathcal{G}} \mathcal{M}^{\mathcal{D}}(d')$ and $\mathcal{M}^{\mathcal{D}}(d) \nleq_{\mathcal{G}} \mathcal{M}^{\mathcal{D}}(d'')$,*
   (b) *At least one of $\mathcal{M}^{\mathcal{D}}(d')$ and $\mathcal{M}^{\mathcal{D}}(d'')$ is a descendant of $\mathcal{M}^{\mathcal{D}}(d)$ (in the same gene tree).*
5. *Given any edge $(d, d') \in E(\mathcal{D})$, $(d, d') \in \Xi$ if and only if $\mathcal{M}^{\mathcal{D}}(d)$ and $\mathcal{M}^{\mathcal{D}}(d')$ are in different gene trees or incomparable in the same gene tree.*
6. *If $d \in I(\mathcal{D})$ and d' and d'' denote the children of d, then,*
   (a) *$d \in \Sigma^{\mathcal{D}}$ if and only if $\mathcal{M}^{\mathcal{D}}(d) = lca(\mathcal{M}^{\mathcal{D}}(d'), \mathcal{M}^{\mathcal{D}}(d''))$ (in the same gene tree) and $\mathcal{M}^{\mathcal{D}}(d')$ and $\mathcal{M}^{\mathcal{D}}(d'')$ are incomparable,*
   (b) *$d \in \Delta^{\mathcal{D}}$ only if $\mathcal{M}^{\mathcal{D}}(d) \geq_{\mathcal{G}} lca(\mathcal{M}^{\mathcal{D}}(d'), \mathcal{M}^{\mathcal{D}}(d''))$ (in the same gene tree),*
   (c) *$d \in \Theta$ if and only if either $(d, d') \in \Xi$ or $(d, d'') \in \Xi$.*
   (d) *If $d \in \Theta$ and $(d, d') \in \Xi$, then $\mathcal{M}^{\mathcal{D}}(d)$ and $\tau(d)$ must either be in different gene trees or incomparable in the same gene tree, $\mathcal{M}^{\mathcal{G}}(\mathcal{M}^{\mathcal{D}}(d)) = \mathcal{M}^{\mathcal{G}}(\tau(d))$, and $\mathcal{M}^{\mathcal{D}}(d') \leq_{\mathcal{G}} \tau(d)$.*

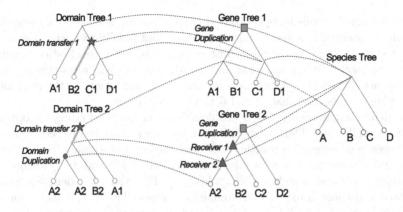

Fig. 1. The figure shows an mDGS reconciliation for two domain trees, two gene trees, and a species tree on 4 taxa. The mappings of the domain trees into the gene trees and of the gene trees into the species tree are shown by the dotted red lines. Domain-gene leaf associations are specified by shared leaf labels, and gene-species leaf associations are specified by shared letters (A, B, C, or D). In the gene-species reconciliation, a gene-duplication event (marked by the blue square) is invoked at the root of gene tree 1 while all other internal nodes of the gene trees represent speciation events. In the domain-gene reconciliation, two domain transfer events are invoked at the nodes with the orange star, one in domain tree 1 and one in domain tree 2, and duplication event is invoked at the node with the orange circle in domain tree 2. The bolded edges in the domain trees represent the domain-transfer edges; in both domain trees the domains are copied from gene tree 1 to gene tree 2, and the recipient genes for domain transfer 1 and domain transfer 2 are marked as "receiver 1" and "receiver 2", respectively. As required by the model, for both transfer events, the donor gene and recipient gene both map to the same species tree node. (Color figure online)

Constraints 1 and 2 above apply to the reconciliation of the gene trees with the species tree and are based on the classical *Duplication-Loss* model [3,9] extended to allow suboptimal gene-species reconciliations. Constraints 3, 4, 5, and 6 apply to the reconciliation of the domain tree with the gene trees. Overall, the mDGS reconciliation model is nearly identical to the DGS reconciliation model [6,7], except that we reconcile multiple domain trees instead of just one. We refer the reader to [7] for a detailed explanation of the underlying model and of each constraint. Figure 1 shows an example of a valid mDGS reconciliation.

We point out that the interdependence between domain-gene and gene-species reconciliations stems from Constraint 6d, which specifies which genes may be designated as the recipient gene for any given domain-transfer event. In the absence of horizontal gene transfer, the transfer of a domain from one gene to another can only happen within the same genome. Thus, Constraint 6d explicitly enforces that the donor gene and recipient gene for any domain transfer event must map to the same species in the species tree. It is this relationship between gene-species mappings and domain-transfer events that necessitates the computation of a *joint* reconciliation, so that one cannot simply compute optimal DGS or mDGS reconciliations by optimizing domain-gene and gene-species reconciliations independently. It is also important to note that mDGS reconciliation is not a direct generalization of the DGS problem since mDGS reconciliation requires $\mathcal{D}$ and $\mathcal{G}$ to form a DG-group. Valid input instances for DGS reconciliation may therefore not be valid input instances for mDGS reconciliation. In the remainder of this paper we assume that $\mathcal{D}$ and $\mathcal{G}$ form a DG-group.

We define a parsimony based problem formulation for finding an optimal mDGS reconciliation. Thus, each evolutionary event other than speciation and co-divergence is assigned a positive cost, and the computational objective is to find an mDGS reconciliation of minimum total cost. $P_\Delta^{\mathcal{G}}$ and $P_{loss}^{\mathcal{G}}$ denote the gene-duplication and gene-loss costs, while $P_\Delta^{\mathcal{D}}$, $P_\Theta^{\mathcal{D}}$, and $P_{loss}^{\mathcal{D}}$ denote domain-duplication, domain-transfer, and domain-loss costs. The model allows for the use of two separate costs $P_{\Theta 1}^{\mathcal{D}}$ and $P_{\Theta 2}^{\mathcal{D}}$ instead of a single $P_\Theta^{\mathcal{D}}$, so that a distinction can be made between domain transfers that remain within the same gene family from those that cross gene family boundaries.

Definition 2 (Reconciliation cost). *Given an mDGS reconciliation α, the reconciliation cost for α is the total cost of all events invoked by α.*

Note that, while domain-duplication, domain-transfer, and gene-duplication events are directly specified in the mDGS reconciliation, domain-losses and gene-losses are not. However, given an mDGS reconciliation, one can directly count the minimum number of gene-losses and domain-losses implied by the reconciliation as shown in [7].

Definition 3 (Optimal mDGS Reconciliation Problem). *Given $\mathcal{D}$, $\mathcal{G}$ and S, along with $P_\Delta^{\mathcal{G}}$, $P_{loss}^{\mathcal{G}}$, $P_\Delta^{\mathcal{D}}$, $P_{\Theta 1}^{\mathcal{D}}$, $P_{\Theta 2}^{\mathcal{D}}$, and $P_{loss}^{\mathcal{D}}$, the Optimal mDGS Reconciliation problem is to find an mDGS reconciliation for $\mathcal{D}$, $\mathcal{G}$ and S with minimum reconciliation cost.*

The NP-hardness of the optimal mDGS reconciliation problem follows from the NP-hardness proof for optimal DGS reconciliation [7]. Specifically, even though mDGS reconciliation is not a direct generalization of DGS reconciliation, the gadget used in [7] yields a valid input instance (i.e., the domain tree and gene trees form a DG-group) for the optimal mDGS reconciliation problem as well.

3 A Heuristic for Optimal mDGS Reconciliation Problem

Algorithms for DGS reconciliation cannot be used for computing mDGS reconciliations due to differences in the problem formulation and final objective. However, optimal DGS reconciliations may still serve as a good starting point for computing optimal mDGS reconciliations (we demonstrate this later in our experiments). Our proposed heuristic is based on this idea and we show how to modify an existing algorithm for DGS reconciliation to estimate optimal mDGS reconciliations.

Currently, two algorithms exist for DGS reconciliation problem: An efficient dynamic programming based heuristic algorithm from [7], and an exact integer linear programming (ILP) based algorithm from [6]. Since, problem instances for mDGS reconciliation are generally much larger (more domain trees and gene trees) than those for DGS reconciliation, the exact ILP based algorithm is not well-suited. We therefore focused on extending the efficient dynamic programming based heuristic algorithm from [7] which has also been previously shown to compute optimal DGS reconciliations (i.e., same as those computed using the exact ILP approach) in the vast majority of test cases [6]. We will refer to this dynamic programming heuristic as the *DGS-algorithm*. We refer the reader to [7] for a complete description of the DGS-algorithm; however, for the current discussion it suffices to view it as a black box that estimates optimal DGS reconciliations. The DGS-algorithm takes as input a single domain tree D, set of associated gene trees $\mathcal{G}$, and a species tree S for the species under consideration. The output of the algorithm is a domain-gene reconciliation of D with $\mathcal{G}$ and gene-species reconciliations for each $G \in \mathcal{G}$ with S (with the domain-gene reconciliation satisfying the constraints imposed on it by the gene-species reconciliations, and vice versa).

Observe that algorithms for DGS reconciliation cannot be used for computing mDGS reconciliations since reconciling each domain tree of $\mathcal{D}$ individually may lead to conflicting gene-species reconciliations for one or more gene trees. Our heuristic for mDGS reconciliation, which we will refer to as the *mDGS-algorithm*, identifies such conflicts and resolves them. In particular, it preserves the domain-gene reconciliations inferred through DGS reconciliation, but adjusts any conflicting gene-species mappings to create a single gene-species mapping for each gene tree. Before we describe the algorithm in detail, we need the following notation: Given any gene tree $G \in \mathcal{G}$, let $\mathcal{D}_G$ be the set containing those domain trees from $\mathcal{D}$ that are represented in G. Analogously, given any domain tree $D \in \mathcal{D}$, let $\mathcal{G}_D$ denote the set containing exactly those gene trees from $\mathcal{G}$ that are represented in D.

mDGS-Algorithm $(\mathcal{D}, \mathcal{G}, S, \mathcal{L}^\mathcal{D}, \mathcal{L}^\mathcal{G})$

1. For each domain tree $D \in \mathcal{D}$
   (a) Run $DGS\text{-}algorithm(D, \mathcal{G}_D, S, \mathcal{L}^D, \mathcal{L}^{\mathcal{G}_D})$. This yields a gene-species mapping for each $G \in \mathcal{G}_D$.
2. For each gene tree $G \in \mathcal{G}$
   (a) Consider the (up to) $|\mathcal{D}_G|$ different gene-species mapping for G generated above. Let these mapping be denoted by $\mathcal{M}_1^G, \ldots, \mathcal{M}_{|\mathcal{D}_G|}^G$.
   (b) For each $g \in I(G)$ in post order, let $\mathcal{M}^G(g) = lca(\mathcal{M}^G(g'), \mathcal{M}^G(g''), \mathcal{M}_1^G(g), \ldots, \mathcal{M}_{|\mathcal{D}_G|}^G(g))$, where g' and g'' denote the two children of $g \in V(G)$.
3. For each domain tree $D \in \mathcal{D}$
   (a) For each transfer event d in a post-order traversal of D
      i. Let g and g' denote the donor and recipient gene nodes for the transfer event at d, and let G and G' denote the gene trees containing g and g', respectively.

    ii. If $\mathcal{M}^G(g) \neq \mathcal{M}^{G'}(g')$ then $\mathcal{M}^G(g) = \mathcal{M}^{G'}(g') = lca(\mathcal{M}^G(g), \mathcal{M}^{G'}(g'))$.

4. Repeat Steps 2 and 3 above until no further changes are made to $\mathcal{M}^{\mathcal{G}}$.

5. Return the domain-gene reconciliation for each $D \in \mathcal{D}$ as computed in Step 1, and the gene-species reconciliation $\mathcal{M}^G$ for each $G \in \mathcal{G}$ as computed above.

It is easy to see that this heuristic is guaranteed to yield a valid mDGS reconciliation. It is also not difficult to show that, after the initial runs of DGS-algorithm in Step 1, the heuristic above requires no more than $O((m \times n \times |Le(S)|))$ time, where m is the total number of leaves in all domain trees of $\mathcal{D}$ and n is the total number of leaves in all gene trees of $\mathcal{G}$. We found the heuristic to be very efficient in practice, requiring less than an hour to run on our entire dataset of 3847 domain trees and 7165 gene trees from 12 species (described in the next section) using a single core on a desktop computer.

Empirical Justification. Observe that the mDGS-algorithm resolves conflicts by simply taking their least common ancestor in case of conflicting mappings for the same gene node. Despite its simplicity, this algorithm is expected to work well if (i) the number of gene nodes that are assigned conflicting mappings under different domain trees is small, and/or (ii) for the gene nodes that do have conflicting mappings, those conflicting mappings are close together on the species tree. This is exactly what we find in our empirical data analysis. Specifically, we find that different domain trees are remarkably consistent in their gene-species mappings under DGS reconciliation and only a very small fraction of gene nodes had conflicting mappings that had to be resolved by the mDGS-algorithm. These results appear in the next section.

4   Experiments and Results

Dataset. To experimentally study the impact of using mDGS reconciliation instead of DGS reconciliation, we used a biological dataset of 3847 rooted domain trees and 7165 rooted gene trees from 12 fly species. This dataset was first created and used in [7] to evaluate the performance of the heuristic algorithm for DGS reconciliation and was subsequently also used in [6]. The domain trees and gene trees in this dataset were constructed and error-corrected using state-of-the-art methods [7,14], and each gene tree contains at least one domain present in the domain trees. On average, each gene in the dataset contains 1.4 domains, each gene family contains 1.68 domain families, and each domain family is associated with 2.93 gene families.

Structure of DG-Groups. We first computed all DG-groups on our dataset and studied their structural properties. We found that the 3847 domain trees and 7165 gene trees could be partitioned into 2010 DG-groups. Among these, 1241 DG-groups consist of a single domain tree and single gene tree, and 386 DG-groups have a single domain tree but at least two gene trees. Note that, for these two types of DG-groups, using mDGS reconciliation is the same as using

DGS reconciliation. The remaining 383 DG-groups each had multiple domain trees and we refer to these as *complex DG-groups*. Among the 383 complex DG-groups, 149 had a single gene tree and 234 had multiple gene trees. One of the complex DG-groups is extremely large and contains 1205 domain trees and 2394 gene trees, constituting almost one-third of the entire dataset. For the remaining 382 complex DG-groups the average number of domain and gene trees is 2.74 and 2.85, respectively, with the largest DG-group having 15 domain trees and 23 gene trees. Among the 2220 domain trees in the 383 complex DG-groups, 1032 evolve inside only one gene tree and the others in multiple gene trees, including 239 that evolve inside more than five. Likewise, among the 3418 gene trees in these DG-groups, 1823 are associated with only one domain tree, 1061 with two, and only 61 gene trees are associated with more than 5 domain trees.

Impact of mDGS Reconciliation. We applied our mDGS-algorithm on the 383 complex DG-groups and compared the resulting gene-species reconciliations with those inferred through DGS reconciliation. We observed that gene-species mappings inferred through DGS reconciliation are highly consistent in general, but that there are several gene nodes for which different domain trees imply conflicting gene-species mappings. Overall, we found there were 12201 internal gene tree nodes that were assigned gene-species mappings by at least two domain trees, and among these gene nodes 148 were assigned conflicting mappings. Thus, only a small fraction of the total of 66854 internal gene tree nodes present in the 383 complex DG-groups was assigned conflicting mappings. This shows that, in the vast majority of cases, optimal mDGS reconciliations are composed of optimal DGS reconciliations.

We also found that the mDGS-algorithm rectified these conflicts without significantly increasing the total gene-species reconciliation cost or significantly affecting other conflict-free gene-species mappings. Specifically, in the largest DG-group the total gene-species reconciliation cost for the 2394 gene trees increased by only 4.6% compared to DGS reconciliation, and total number of gene nodes that deviate from their LCA (least common ancestor) mapping increased by only 294 (increased from 501 to 795) among a total of 46693 total internal gene nodes. These are very small numbers considering that there are 6,577 domain transfer events in the largest DG-group. Similarly, in the remaining 382 DG-groups, the total gene-species reconciliation cost for the 1024 gene trees increased by only 3.47% compared to DGS reconciliation, and total number of gene nodes that deviate from their LCA mapping increased by only 51 (increased from 106 to 157) among a total of 20161 total internal gene nodes. The total number of domain transfers in these DG-groups was 1786.

Splitting Large DG-Groups into Smaller Communities. As seen in our dataset, DG-groups can become extremely large, comprising of thousands of domain trees and gene trees. Upon closer inspection of the largest complex DG-group in our dataset (with 1205 domain trees and 2394 gene trees), we found that it is composed of many small well-connected communities of domain and gene trees, with different communities connected to each other through small numbers of shared gene trees. We refer to these communities within a larger DG-group as *DG-communities*, and gene trees that "connect" different DG-communities as *connecting gene trees*. Figure 2 illustrates how a larger DG-group can be split into smaller DG-communities connected through connecting gene trees.

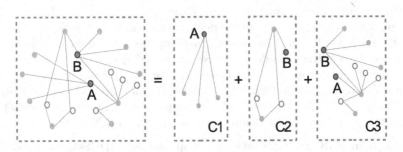

Fig. 2. This figure shows how the DG-group on the left can be split into three smaller DG-communities. Blue dots represent domain trees. Solid orange circles (labeled A and B) represent connecting gene trees, and hollow orange circles represent other gene trees. As shown, each DG-community is connected to at least one other DG-community through connecting gene trees. (Color figure online)

To systematically identify DG-communities within large DG-groups and study their relevance, we devised a simple algorithm for identifying clusters in bipartite graphs. While many clustering algorithms exist for bipartite graphs, we found that these could not be directly applied for identifying DG-communities since most clustering algorithms seek to partition the set of nodes into distinct clusters (effectively by deleting edges). This would not work in the current setting since we wish to retain all domain-gene edges and some gene trees must therefore appear in multiple DG-communities.

Our new clustering algorithm is specifically designed for identifying DG-communities. It partitions all domain trees into different DG-communities, but allows some gene trees to appear in multiple DG-communities. The algorithm makes use of a similarity measure between domain trees to do the clustering and we define this similarity in a manner that is meaningful for detecting DG-communities. Specifically, given domain trees D_1 and D_2, we define the similarity between them, denoted $sim(D_1, D_2)$, as follows:

$$sim(D_1, D_2) = \frac{|\mathcal{G}_{D_1} \cap \mathcal{G}_{D_2}|}{|\mathcal{G}_{D_1}|} + \frac{|\mathcal{G}_{D_1} \cap \mathcal{G}_{D_2}|}{|\mathcal{G}_{D_2}|}. \tag{1}$$

A high-level description of the proposed clustering algorithm follows. In addition to $\mathcal{D}$ and $\mathcal{G}$, the algorithm takes as input a clustering parameter ρ.

Find-Communities $(\mathcal{D}, \mathcal{G}, \mathcal{L}^{\mathcal{D}}, \rho)$

1. Compute $sim(D_1, D_2)$ for each pair of domain trees $D_1, D_2 \in \mathcal{D}$.
2. Initialize the set *pool* to include all domain trees in $\mathcal{D}$.
3. While $|pool| \geq 2$ and $\max_{D_1, D_2 \in pool} sim(D_1, D_2) \geq \rho$.

(a) Choose a pair of domain trees from *pool* with greatest similarity and create a new community with that pair. Add all gene trees associated with the two domain trees to this community.

(b) Repeatedly choose one domain tree from *pool* that has maximal average similarity to the domain trees in the current community and add this domain tree to the current community. Add all gene trees associated with this new domain tree to the community. Repeat this step until the maximal average similarity falls below ρ.

4. Add all remaining domain trees in *pool* to their own single-domain communities, along with their associated gene trees.

There are several crucial reasons for decomposing large DG-groups into smaller DG-communities. First, DG-communities are expected to represent clusters of domains and genes that are closely related and biologically meaningful, whereas the domains and genes in a large DG-group are likely to be only weakly associated. Second, DG-communities reveal the underlying structure of DG-groups and help identify connecting gene families. And third, each DG-community can be viewed as a smaller DG-group for the purposes of mDGS reconciliation and it may be more appropriate to use these smaller DG-communities than larger weakly connected DG-groups.

Analyzing DG-Communities. We applied our clustering algorithm to the largest complex group in our dataset with clustering parameter $\rho = 1.0$. This resulted in the identification of 532 DG-communities, of which 304 DG-communities contain only one domain tree and the remaining 228 DG-communities together contain 901 domain trees. Among the 2394 gene trees in the largest complex DG-group, 647 (or 1.22 per DG-community on average) were identified as connecting gene trees. We found that these connecting gene trees were often larger in size and contained more domains, on average, than the other gene trees. More precisely, the 647 connecting gene trees contained 2.8 domains each, on average, compared to 1.81 domains over all gene trees within the DG-group. Similarly, connecting gene trees each contained 29.2 leaf nodes on average, compared to 20.0 leaf nodes for all gene trees in the DG-group. This is not entirely surprising since any connecting gene tree must necessarily contain domains from at least two different domain trees while no such constraint applies to other gene trees.

Next, we applied our mDGS reconciliation heuristic to each DG-community separately and compared the resulting gene-species reconciliations against those obtained by applying the heuristic to the entire DG-group. Recall that, when mDGS reconciliation was applied to the entire DG-group, the total number of gene nodes that deviate from their LCA mapping increased to 795 from the 501 observed for the base DGS-algorithm. In contract, when the mDGS-algorithm is applied separately to each DG-community in this DG-group, the total number of gene nodes that deviate from their LCA mapping increases to only 567. In other words, to make the underlying DGS reconciliations consistent in their gene-species mappings, mDGS reconciliation on the entire DG-group required 294 additional gene tree nodes to deviate from their LCA mappings, while this number falls dramatically to only 66 gene nodes when mDGS reconciliation is applied to all DG-communities in that DG-group. Thus, the vast majority of gene nodes that deviate from their LCA mappings appear on connecting gene trees. One possible explanation for this surprising result is that nodes in connecting gene trees are more likely to be assigned conflicting mappings by their associated domain trees; however, we observed that this was not the case. In fact, we found that conflicting gene trees had only 31 gene tree nodes with conflicting mapping assignments compared to 115 for all 2394 gene trees in the DG-group. This implies that the abundance of gene nodes on conflicting gene trees that deviate from their LCA mappings is caused by greater disagreement between the conflicting mappings (i.e., the conflicting mappings may be far apart on the species tree), causing the mapping of such nodes to be moved higher up towards the root than for other nodes with conflicting mappings.

We performed further analysis to assess if the sizes or other features of connecting gene trees may explain this overabundance of gene nodes deviating from their LCA mappings. We found that even though connecting gene trees are larger, on average, than other gene trees, they together contained less than 40% of the total number of gene tree nodes in this DG-group. We also evaluated if the larger number of domain families, on average, represented in connecting gene trees may explain the overabundance, but found that connecting gene trees constitute only 61% of all gene trees with at least two domain families and that these gene trees contain the same average number of domain families as connecting gene trees. Thus, the overabundance of gene nodes deviating from their LCA mappings on connecting gene trees is adequately explained neither by their size nor by their domain content.

One possible explanation for this surprising result is a higher error rate for connecting gene trees. Such error in gene trees could be caused by domain chaining, discordance in domain evolutionary histories, and other reasons. Thus, the identification of DG-communities within DG-groups may not only lead to more accurate mDGS reconciliations but also help identify erroneous multi-domain gene trees.

5   Conclusion

In this work, we extended the existing DGS reconciliation framework to address the problem of inconsistent gene-species mappings. We introduced the mDGS reconciliation framework and provided an efficient heuristic for estimating optimal mDGS reconciliations. Using an extensive experimental study on real biological data, we demonstrated the importance of mDGS reconciliation and showed that the proposed heuristic is effective at estimating optimal mDGS reconciliations. We also developed a technique to further improve the accuracy of mDGS reconciliation by introducing the notion of a DG-community, which is a subset of the domain and gene trees under consideration, and providing a clustering algorithm to find such DG-communities.

Several important research questions remain to be explored. First, our heuristic for mDGS reconciliation is rather simplistic, changing only the gene-species mappings to achieve consistency and preserving the domain-gene mappings computed using DGS reconciliation. Simultaneous correction of both the domain-gene and gene-species mappings may lead to more optimal reconciliations. Second, a thorough simulation study is needed to systematically assess the impact of using mDGS reconciliation instead of DGS reconciliation and to properly assess the effectiveness of the proposed heuristic. The recent development of a probabilistic simulation framework for gene and subgene evolution [5] will facilitate such studies. And third, it would be interesting to further study the connecting gene families identified by our algorithm for finding DG-communities. It is possible that some of them represent cases of domain chaining.

References

1. Behzadi, B., Vingron, M.: Reconstructing domain compositions of ancestral multi-domain proteins. In: Bourque, G., El-Mabrouk, N. (eds.) RCG 2006. LNCS, vol. 4205, pp. 1–10. Springer, Heidelberg (2006). https://doi.org/10.1007/11864127_1
2. Ekman, D., Björklund, Å.K., Frey-Skött, J., Elofsson, A.: Multi-domain proteins in the three kingdoms of life: orphan domains and other unassigned regions. J. Mol. Biol. **348**(1), 231–243 (2005)
3. Goodman, M., Czelusniak, J., Moore, G.W., Romero-Herrera, A.E., Matsuda, G.: Fitting the gene lineage into its species lineage. A parsimony strategy illustrated by cladograms constructed from globin sequences. Syst. Zool. **28**, 132–163 (1979)
4. Han, J.-H., Batey, S., Nickson, A.A., Teichmann, S.A., Clarke, J.: The folding and evolution of multidomain proteins. Nat. Rev. Mol. Cell Biol. **8**, 319–330 (2007)
5. Kundu, S., Bansal, M.S.: SaGePhy: an improved phylogenetic simulation framework for gene and subgene evolution. Bioinformatics (2019, in press)
6. Li, L., Bansal, M.S.: An integer linear programming solution for the domain-gene-species reconciliation problem. In: Proceedings of the 2018 ACM International Conference on Bioinformatics, Computational Biology, and Health Informatics, BCB 2018, pp. 386–397. ACM, New York (2018)
7. Li, L., Bansal, M.S.: An integrated reconciliation framework for domain, gene, and species level evolution. IEEE/ACM Trans. Comput. Biol. Bioinform. **16**(1), 63–76 (2019)
8. Moore, A.D., Bjorklund, A.K., Ekman, D., Bornberg-Bauer, E., Elofsson, A.: Arrangements in the modular evolution of proteins. Trends Biochem. Sci. **33**, 444–451 (2008)
9. Page, R.D.M.: Maps between trees and cladistic analysis of historical associations among genes, organisms, and areas. Syst. Biol. **43**(1), 58–77 (1994)
10. Stolzer, M., Siewert, K., Lai, H., Xu, M., Durand, D.: Event inference in multidomain families with phylogenetic reconciliation. BMC Bioinform. **16**(14), S8 (2015)
11. Tordai, H., Nagy, A., Farkas, K., Banyai, L., Patthy, L.: Modules, multidomain proteins and organismic complexity. FEBS J. **272**(19), 5064–5078 (2005)
12. Vogel, C., Bashton, M., Kerrison, N.D., Chothia, C., Teichmann, S.A.: Structure, function and evolution of multidomain proteins. Curr. Opin. Struct. Biol. **14**(2), 208–216 (2004)
13. Wiedenhoeft, J., Krause, R., Eulenstein, O.: The plexus model for the inference of ancestral multidomain proteins. IEEE/ACM Trans. Comput. Biol. Bioinform. **8**(4), 890–901 (2011)

14. Wu, Y.-C., Bansal, M.S., Rasmussen, M.D., Herrero, J., Kellis, M.: Phylogenetic identification and functional characterization of orthologs and paralogs across human, mouse, fly, and worm. bioRxiv (2014)
15. Wu, Y.-C., Rasmussen, M.D., Kellis, M.: Evolution at the subgene level: domain rearrangements in the drosophila phylogeny. Mol. Biol. Evol. 29(2), 689–705 (2012)

Systems Biology

IDNDDI: An Integrated Drug Similarity Network Method for Predicting Drug-Drug Interactions

Cheng Yan[1,2], Guihua Duan[1], Yayan Zhang[1], Fang-Xiang Wu[3], Yi Pan[4], and Jianxin Wang[1(✉)]

[1] School of Computer Science and Engineering, Central South University, Changsha 410083, China
jxwang@mail.csu.edu.cn
[2] School of Computer and Information, Qiannan Normal University for Nationalities, Duyun 558000, Guizhou, China
[3] Division of Biomedical Engineering and Department of Mechanical Engineering, University of Saskatchewan, Saskatoon, SK S7N5A9, Canada
[4] Department of Computer Science, Georgia State University, Atlanta, GA 30302, USA

Abstract. A drug-drug interaction(DDI) was defined as the pharmacological effect(s) of a drug influenced by another drug. The positive DDIs can improve the therapeutic effect of patients. However, the negative DDIs can lead serious results, such as drug withdrawal from market and even patient death. Currently, multiple pharmaceutical drugs have widely been used to treat complex diseases, such as cancer. The traditional biomedical experiments are very time-consuming and very costly to validate new DDIs. Therefore, it is appealing to develop computational methods to discover potential DDIs. In this study, we propose a new computational method (called IDNDDI) to predict novel DDIs. Based on the binary vector of drug chemical, biological and phenotype data, IDNDDI computes the integrated drug feature similarity by the cosine similarity method. In addition, the node-based drug network diffusion method is used to calculate the relational initial scores for new drugs. To systematically evaluate the prediction performance of IDNDDI and compare it with other prediction methods, we conduct the 5-fold cross validation and de novo drug validation. In terms of the AUC (area under the ROC curve)value, IDNDDI achieves the better prediction performance in the 5-fold cross validation, specifically, the AUC value is 0.9691, which is larger than the state-of-the-art L1E (L1 Classifier ensemble method) results of 0.9570. In addition, IDNDDI also obtains the best prediction result in the de novo drug validation and the AUC reaches 0.9292. The prediction ability in application of our method is also illustrated by case studies. IDNDDI is an effective DDI prediction method which can help to reduce adverse drug reactions and improve the efficiency of drug development progress.

© Springer Nature Switzerland AG 2019
Z. Cai et al. (Eds.): ISBRA 2019, LNBI 11490, pp. 89–99, 2019.
https://doi.org/10.1007/978-3-030-20242-2_8

1 Introduction

A drug-drug interaction (DDI) is defined as a drug's pharmacological effect influenced by another drug, which usually appears when two or more drugs are administered simultaneously for a patient. The positive DDIs can provide more effective treatment to patients. However, the undesirable DDIs are the major cause of adverse reaction events. In serious cases, these drugs may be withdrawn from market and lead the death of patients [1]. In fact, we usually use multi-drugs to treat diseases, such as cancer [2]. The main object of multi-drug therapy is to improve the life quality of patients and increase the overall survival rate. However, the adverse DDIs must be avoided as much as possible. Therefore, in order to systematically understand DDIs, reduce the cost of drug development and improve the treatment effect, discovering the novel DDIs has become an important issue of bioinformatics research. With the development of medical technologies, it is necessary to predict potential DDIs by computational methods which have time and cost advantage.

Many computational approaches have been developed to infer new DDIs in recent years. A method has been developed to discover novel DDIs by Tatonetti *et al.*, which uses the drug adverse event profiles [3]. Based on two types of drug interactions: potential CYP (Cytochrome P450)-related DDIs, and non-CYP-related DDIs (NCRDs), INDI (INferring Drug Interactions) method has been developed to infer hidden DDIs, it computes the drug chemical similarities, side effect similarities, protein-protein interactions similarities and target sequence similarities, respectively [4]. Based on machine learning classifiers, Cheng *et al.* propose a DDI prediction method which also uses the drug phenotypic, therapeutic, chemical and genomic properties. Based on the 2D and 3D molecular structure, interaction profile, target and side-effect data, Vilar *et al.* develop a method to infer novel DDIs, which is a protocol applicable on large scale datasets [5]. The label propagation method is proposed to discover novel DDIs by using drug chemical structures, drug side effects and off side effects [6]. The multitask dyadic regression model has been used to predict DDI types based on known DDIs and their types [7]. Drug interaction profile fingerprints (IPFs) have been used to predict new DDIs [8]. Three ensemble methods, including weight average ensemble (WAE), L1 classifier ensemble (L1E) and L2 classifier ensemble (L2E) have also been developed to predict novel DDIs [9]. They integrate drug chemical, biological, phenotype and known DDIs network information.

Although some effective results for predicting hidden DDIs have been achieved by these computational methods, there are still some limits that desire to be improved. The features of drugs and known DDIs should be more effectively integrated. Furthermore, the DDIs prediction of new drugs also should be paid more attention. Therefore, in order to improve the treatment effects, we should develop more effective computational methods to predict potential DDIs.

In this study, based on chemical, biological and phenotype information of drugs, we develop a method (called IDNDDI) to predict potential DDIs. A binary vector is constructed by combing all these chemical, biological and phenotype

Table 1. The description of dataset

Data type	Data	Database	Dimensionality
Chemical	Chemical substructures	PubChem	881
Biological	Drug-targets	DrugBank	780
	Drug transporters	DrugBank	18
	Drug enzymes	DrugBank	129
	Drug pathways	KEGG	253
Phenotypic	Drug indications	SIDER	4897
	Drug side effects	SIDER	4897
	Drug off side effects	OFFSIDES	9496
Interaction	Drug-drug interactions	TWOSIDES	DDIs:48,584

information of drugs, then IDNDDI computes the feature similarity between drugs by the cosine similarity method. IDNDDI also computes the Gaussian Interaction Profile (GIP) similarity of drugs via known DDIs. In order to predict DDIs for new drugs, the node-based drug network diffusion method is used to calculate the relational initial scores for new drugs. IDNDDI uses the regularized least squares (RLS) classifier to infer the novel DDIs. We conduct the 5-fold cross validation and de novo validation to assess the prediction performance and compare it with other methods. AUC (area under the ROC curve) value is metric to evaluate prediction ability. In 5-fold cross validation, IDNDDI obtains the best prediction performance, the AUC value reaches 0.9691, which is larger than the state-of-the-art L1E results of 0.9599. In de novo validation, its AUC value reaches 0.9292 and is larger than the best result of WAE whose AUC value is 0.9073. The case studies illustrate the prediction ability of IDNDDI in practical applications.

2   Materials

In this study, the benchmark dataset of known DDIs is obtained from the TWO-SIDES database [10]. We also extract the chemical substructure information of drugs from PubChem Compound database. The number of dimensions with Pub-Chem substructure is 881. Drug-indications, drug-side effects and drug-off side effects are three types of phenotypic data of drugs. The former two are downloaded from SIDER database that consists of 1430 drugs and 5580 side effect terms. The latter is extracted from OFF-SIDES database.

Furthermore, biological features of drugs are constructed by drug-target interactions, drug enzymes, drug transports and drug pathways. DrugBank database provides drug-target interactions, drug enzymes and drug transports [11]. Drug pathways are extareracted from KEGG database [12].

A DDI benchmark dataset is constructed after projecting drugs in TWO-SIDES to PubChem, DrugBank, KEGG, SIDER and OFFSIDER, which contains 548 drugs and 48,584 DDIs. Tabel 1 shows the basic description about data types, data sources and dimensions of these dataset. In addition, the previous literature also provide the address to download it [9].

3   Methods

3.1   Similarity with Chemical, Biological and Phenotypic Data

In order to compute their feature similarity, we construct a binary vector by the chemical, biological and phenotypic data of drugs. The dimensionality of these binary vectors is the sum of all dimensionality of all chemical, biological and phenotypic data. There are 21,351 types of drug features, and the values of 1 and 0 in these vectors indicate the presence or absence of chemical substructures, targets, transporters, enzymes, pathways, indications, side effects and off side effects. We calculate the feature similarity of drugs by the cosine method. For example, for drugs d_i and d_j, the similarity value computation method is defined as follows:

$$Sim_d(d_i, d_j) = \frac{\sum_{l=1}^{M} d_i(l) d_j(l)}{\sqrt{\sum_{l=1}^{M} d_i^2(l)} \sqrt{\sum_{l=1}^{M} d_j^2(l)}} \tag{1}$$

where $d_i(l)$ and $d_j(l)$ are the lth element of feature vector of drugs d_i and d_j, respectively, and M is the dimensionality of drug features (21,351). The feature similarity values all range from 0 to 1.

3.2   Gaussian Interaction Profile Kernels Similarity

We also compute the Gaussian Interaction Profile (GIP) similarity of drug pairs by known DDIs. Let $D = \{d_1, d_2, ..., d_N\}$ be the set of N drugs. The matrix $Y \in N * N$ represents the adjacency matrix of the DDI network. The value of y_{ij} is 1 when drug d_i interacts with d_j, otherwise is 0. Specifically, for drugs d_i and d_j, the GIP kernel similarity $K_{GIP,d}$ can be computed as follows:

$$K_{GIP,d}(d_i, d_j) = exp(-\gamma_d ||yd_i - yd_j||^2), \tag{2}$$

$$\gamma_d = \gamma'_d / (\frac{1}{N} \sum_{i=1}^{N} ||yd_i||^2), \tag{3}$$

where $yd_i = \{y_{i1}, y_{i2}, ..., y_{iN}\}$ and $yd_j = \{y_{j1}, y_{j2}, ..., y_{jN}\}$ are the association profiles of drugs d_i and d_j, respectively. N is the number of drugs. γ_d is the regularization parameter of kernel bandwidth and γ'_d is set to 1 according to previous studies [13, 14].

3.3 Integrating Similarities of Drugs

Based on the drug feature similarity Sim_d and drug GIP similarity $K_{GIP,d}$, the final drug similarity S_d is computed by the mean as follows:

$$S_d = \frac{Sim_d + K_{GIP,d}}{2}, \tag{4}$$

in which the Sim_d, $K_{GIP,d}$ and S_d are the $N * N$ matrices.

3.4 Regularized Least Squares (RLS) Classifier

The machine learning method classifier has been widely used in other prediction issues, such as drug-target associations [13]. In this study, we also adopt RLS to predict hidden DDIs, the computation method of drug-drug pairs is defined as follows:

$$
\begin{aligned}
Y_p^T &= S_d(S_d + \sigma I)^{-1}Y \\
\hat{Y} &= \frac{Y_p + Y_p^T}{2}
\end{aligned}
\tag{5}
$$

where I is the unit matrix and Y is the symmetrical adjacency matrix of known DDIs. σ is the regularization parameter and also set to be 1 [15]. Since RLS classifier can not guarantee the predicted Y_p^T to be symmetric, we obtain a predicted symmetric matrix $\hat{Y}$ from the mean of matrix Y_p and its transpose [13].

3.5 Node-Based Drug Network Diffusion for New Drugs

We all known that some prediction models can not predict novel DDIs for new drugs which have no known DDIs. In this study, we add an initial interaction profile process before inferring new DDIs for them. Inspired by the successful applications of CSN method [16]. We also adopt node-based drug network diffusion to calculate the relational score. The drug network is constructed by using drug-target interactions, drug-indications and known DDIs. Comparing with the similarity network constructed by drug-target interactions, drug-indications and known DDIs, the node-based drug network is much sparser, especially when the number of drugs is very large. The probabilities for a seed node reaching each neighbor to be low in the diffusion process in the similarity-based drug network. The similarity-based drug network is sensitive to noises [16].

Firstly, we construct an adjacency matrix A as follows:

$$
A = \begin{bmatrix} Y & M_{dt} & M_{di} \\ M_{dt}^T & 0 & 0 \\ M_{di}^T & 0 & 0 \end{bmatrix}
\tag{6}
$$

where Y represents DDIs, which includes the new drugs but the interaction profile is 0 vector. Adjacency matrix M_{dt} represents the known drug-target interactions and adjacency matrix M_{di} also represents known drug indication interactions. Therefore, A is a $(N + N_t + N_i) * (N + N_t + N_i)$ matrix, where N, N_t and N_i are the number of drugs, targets and indications, respectively.

The node-based drug networks diffusion is a two-round resource transfer process. In the first transfer process, the scores of linked target nodes and indication nodes of new drug nodes are transferred to the other drug nodes based on the assigned transfer weight. For drug d_{new}, the initial resource transferred from its linked target node t_i to drug d_j is calculated as follows:

$$R_{st}(t_i, d_j) = \frac{A(t_i, d_j)}{\sum\limits_{l=1}^{N} A(t_i, l)} * A(t_i, d_{new}), \tag{7}$$

Then the drug d_j obtains the allocated resource by adding the contributions from all target nodes and indication nodes associated to it as follows:

$$R_{st}(d_j) = \sum\limits_{l=N+1}^{N_t + N_i} R_{st}(l, d_j), \bullet \tag{8}$$

In the second transfer process, we allocated the resource of drugs obtained in the first round to the drugs according to transfer weights from drugs to drugs. For example, the resource transfer from drug d_j to drug d_i is calculated as follows:

$$R_{nd}(d_j, d_i) = \frac{A(d_j, d_i)}{\sum\limits_{l=1}^{N} A(d_i, l)} * R_{st}(d_j), \tag{9}$$

Then the final resource allocated from drug d_{new} to drug d_i can be calculated as follows:

$$R(d_{new}, d_i) = \sum\limits_{l=1}^{N} R_{nd}(l, d_i), \tag{10}$$

Then the values of interaction score vector $R_{(d_{new}, 1 : N)}$ between drug d_{new} and other drugs are computed. When drug d_{new} is a new drug, we use these scores as the initial scores. However, considering the computed scores by node-based drug network diffusion method are too small, and the values of other known DDIs which is defined as 1, we adopt the proportion method to expend these values to higher possibility values. The expended method is defined as follows:

$$R(d_{new}, 1 : N) = R(d_{new}, 1 : N) * (\frac{\alpha}{max(R(d_{new}, 1 : N))}), \tag{11}$$

where α is the possibility score that the max value of $R(i, 1 : N)$ need to be expanded, while other values are also to be expanded according to the ratio. In this study, we set the values of α by de novo drug validation.

4 Results and Discussions

4.1 Benchmark Evaluation and Evaluation Indices

To comprehensively assess the prediction performance of IDNDDI and compare it with other DDIs prediction methods. We conduct the 5-fold cross validation and the de novo drug validation. The AUC value is used to the metric. In the 5-fold cross validation, the known DDIs are divided into 5 groups, and one group is in turn chosen as the testing set and the rest as the training set. In de novo drug validation, one drug is chosen as the test set and the other drugs as the training set at each time. Furthermore, the GIP kernel similarity is related with known DDIs, we recompute it based on training samples at each time in which the testing samples are set to be 0.

Table 2. The prediction performances of different methods in the 5-fold cross validation, the best result is in the bold face.

Method	Feature	AUC
WAE	Chemical data, biological data, phenotypic data	0.9502
L1E	Chemical data, biological data, phenotypic data	0.9570
L2E	Chemical data, biological data, phenotypic data	0.9561
LP	Drug-sub	0.9356
	Drug-label	0.9364
	Drug-off label	0.9374
IDNDDI	Chemical data, biological data, phenotypic data	**0.9691**

Table 3. The prediction performances of different methods in the de novo drug validation, the best result is in the bold face.

Method	Feature	AUC
WAE	Chemical data, biological data, phenotypic data	0.9073
LP	Drug-sub	0.8993
	Drug-label	0.8994
	Drug-off label	0.8997
IDNDDI	Chemical data, biological data, phenotypic data	**0.9292**

4.2 Comparison with Previous Methods

We compare IDNDDI with other four competing methods, including WAE, L1E, L2E and LP (label propagation method) [9]. WAE, L1E and L2E are the integrated method by using drug chemical data, biological data and phenotypic data. Furthermore, LP method used drug substructures of chemical data, drug side effects and drug off side effects to predict novel DDIs.

5-Fold Cross Validation. We calculate the AUC values of different methods over 10 repeats. The average AUC values of different methods in the 5-fold cross validation are showed in Table 2. The prediction result shows that IDNDDI is superior to other methods in terms of AUC value (IDNDDI:0.9691, WAE: 0.9502, L1E: 0.9570, L2E: 0.9561, LP (max): 0.9374).

De Novo Drug Validation. In addition, to comprehensively assess the prediction performance of computational method, we also conduct the de novo drug validation. Some methods can not predict potential DDIs for new drugs without added the initial interaction profile process. Therefore, in the de novo drug validation experiments, we compare IDNDDI with WAE method and Label propagation method. WAE used chemical data, biological data and phenotypic data and integrated the results of neighbor recommendation method and random walk method in de novo drug validation. Table 3 describes the prediction performance of IDNDDI, WAE and LP methods. The AUC value of IDNDDI is 0.9292, which is better than other methods (WAE: 0.9073, LP (max): 0.8997). By comparing the AUC values of different methods in the 5-fold cross validation and the de novo drug validation, IDNDDI method is an effective method to predict new DDIs.

4.3   Parameter Analysis for α

In this section, we analyse the parameter α which is used to expand possibility scores of the maximum value in the de novo drug validation. We choose the parameter to obtain the best prediction performance in the de novo drug validation.

Table 4. The AUC of IDNDDI under different settings of α, the best result is in the bold face.

α	0	0.1	0.2	0.3	0.4
AUC	0.7783	0.9005	0.9221	0.9275	0.9290
0.5	0.6	0.7	0.8	0.9	1.0
0.9292	0.9289	0.9285	0.9280	0.9275	0.9271

Table 4 shows the prediction performances of IDNDDI when α ranges from 0 to 1.0. We can see from Table 4 that the prediction performance of IDNDDI is worst when the value of α is set to be 0. $\alpha = 0$ means to directly predict DDIs without added the initial process. It also illustrates that the initial process via node-based network diffusion method can improve the prediction performance in the de novo drug validation. IDNDDI obtains the best prediction performance when the value of α is set to be 0.5 and its AUC value reaches 0.9292. Table 4 indicates that the expanding method and the node-based network diffusion method can effectively improve the prediction ability of IDNDDI.

Table 5. Top 10 new DDIs predicted by IDNDDI method.

Rank	Drug ID1	Drug ID2	Drug name1	Drug name2	Evidence
1	DB00448	DB01059	Lansoprazole	Norfloxacin	Unknown
2	DB00333	DB00213	Methadone	Pantoprazole	DrugBank
3	DB00991	DB00231	Oxaprozin	Temazepam	Unknown
4	DB00813	DB00535	Fentanyl	Cefdinir	Unknown
5	DB00863	DB00690	Ranitidine	Flurazepam	Unknown
6	DB00470	DB00331	Dronabinol	Metformin	Unknown
7	DB00989	DB01136	Rivastigmine	Carvedilol	DrugBank
8	DB00257	DB00230	Clotrimazole	Pregabalin	DrugBank
9	DB00869	DB00537	Dorzolamide	Ciprofloxacin	Unknown
10	DB00887	DB00207	Bumetanide	Azithromycin	Unknown

4.4 Case Studies

In order to validate the prediction ability of IDNDDI in practical application, we verify the top 10 new DDIs predicted by IDNDDI based on the benchmark dataset. The benchmark dataset is downloaded from the TWOSIDES database and is composed of 548 drugs and 48,584 known DDIs. The predicted new DDIs are evaluated by the latest version of DrugBank. We can see from Table 5 that 3 of top 10 DDIs are confirmed in DrugBank. For example, when Methadone is combined with Pantoprazole, its metabolism can be decreased [11]. Carvedilol is a non-selective beta blocker indicated in the treatment of mild to moderate congestive heart failure (CHF), which can increase the bradycardic activities of Rivastigmine [11]. In addition, other predicted DDIs that have not been validated in DrugBank database, which deserves to validate by biochemical experiments in the future.

5 Conclusion

In this study, we proposed a new computational method (IDNDDI) to infer novel DDIs. IDNDDI computes the feature similarity of drugs by the cosine similarity method. It uses the chemical, biological and phenotypic data of drugs to construct a binary vector. In order to improve the prediction ability for new drugs, we adopt the node-based drug network diffusion method to calculate the relational initial association scores for them. The RLS method is used to compute the probability scores of drug pairs. In the 5-fold cross validation and the de novo drug validation, IDNDDI achieves better prediction performances than other comparison methods.

In addition. We can also consider other more sophisticated methods to integrate the chemical, biological and phenotypic data of drugs. Deep learning method and other machine learning methods, such as matrix approximation method also can be adopted to predict new DDIs. To help with drug development and diseases treatment, we would develop a more effective method to predict new DDIs in the future by the above mentioned methods.

Acknowledgement. The authors are very grateful to the anonymous reviewers for their constructive comments which have helped significantly in revising this work. The authors would like to express their gratitude for the support from the National Natural Science Foundation of China under Grant No. 61772552, No. 61420106009, No. 61622213 and No. 61732009.

References

1. Kusuhara, H.: How far should we go? Perspective of drug-drug interaction studies in drug development. Drug Metab. Pharmacokinet. **29**(3), 227–228 (2014)
2. Chou, T.C.: Drug combination studies and their synergy quantification using the Chou-Talalay method. Cancer Res. **70**(2), 440–446 (2010)
3. Tatonetti, N.P., Fernald, G.H., Altman, R.B.: A novel signal detection algorithm for identifying hidden drug-drug interactions in adverse event reports. J. Am. Med. Inform. Assoc. **19**(1), 79–85 (2011)
4. Gottlieb, A., Stein, G.Y., Oron, Y., Ruppin, E., Sharan, R.: INDI: a computational framework for inferring drug interactions and their associated recommendations. Mol. Syst. Biol. **8**(1), 592 (2012)
5. Vilar, S., et al.: Similarity-based modeling in large-scale prediction of drug-drug interactions. Nat. Protoc. **9**(9), 2147–2163 (2014)
6. Zhang, P., Wang, F., Hu, J., Sorrentino, R.: Label propagation prediction of drug-drug interactions based on clinical side effects. Sci. Rep. **5**, 12339 (2015)
7. Jin, B., Yang, H., Xiao, C., Zhang, P., Wei, X., Wang, F.: Multitask dyadic prediction and its application in prediction of adverse drug-drug interaction. AAA I, 1367–1373 (2017)
8. Vilar, S., Uriarte, E., Santana, L., Tatonetti, N.P., Friedman, C.: Detection of drug-drug interactions by modeling interaction profile fingerprints. PLoS One **8**(3), e58321 (2013)
9. Zhang, W., Chen, Y., Liu, F., Luo, F., Tian, G., Li, X.: Predicting potential drug-drug interactions by integrating chemical, biological, phenotypic and network data. BMC Bioinform. **18**(1), 18 (2017)
10. Tatonetti, N.P., Patrick, P.Y., Daneshjou, R., Altman, R.B.: Data-driven prediction of drug effects and interactions. Sci. Transl. Med. **4**(125), 125ra31 (2012)
11. Law, V., et al.: DrugBank 4.0: shedding new light on drug metabolism. Nucl. Acids Res. **42**(D1), D1091–D1097 (2013)
12. Kanehisa, M., Goto, S., Furumichi, M., Tanabe, M., Hirakawa, M.: KEGG for representation and analysis of molecular networks involving diseases and drugs. Nucl. Acids Res. **38**(suppl. 1), D355–D360 (2009)
13. van Laarhoven, T., Nabuurs, S.B., Marchiori, E.: Gaussian interaction profile kernels for predicting drug-target interaction. Bioinformatics **27**(21), 3036–3043 (2011)
14. Yan, C., Wang, J., Ni, P., Lan, W., Wu, F.X., Pan, Y.: DNRLMF-MDA: predicting microRNA-disease associations based on similarities of microRNAs and diseases. IEEEACM Trans. Comput. Biol. Bioinform. (2017, to be published). https://doi.org/10.1109/TCBB.2017.2776101

15. Yan, C., Wang, J., Lan, W., Wu, F.X., Pan, Y.: SDTRLS: predicting drug-target interactions for complex diseases based on chemical substructures. Complexity (2017)
16. Chen, Y., Xu, R.: Context-sensitive network-based disease genetics prediction and its implications in drug discovery. Bioinformatics **33**(7), 1031–1039 (2017)

Model Revision of Boolean Regulatory Networks at Stable State

Filipe Gouveia[(⊠)], Inês Lynce, and Pedro T. Monteiro

INESC-ID/Instituto Superior Técnico, Universidade de Lisboa, Lisbon, Portugal
{filipe.gouveia,ines.lynce,pedro.tiago.monteiro}@tecnico.ulisboa.pt

Abstract. Models of biological regulatory networks are essential to understand the cellular processes. However, the definition of such models is still mostly manually performed, and consequently prone to error. Moreover, as new experimental data is acquired, models need to be revised and updated. Here, we propose a model revision tool, capable of proposing the set of minimum repairs to render a model consistent with a set of experimental observations. We consider four possible repair operations, giving preference to function repairs over topological ones. Also, we consider observations at stable state, *i.e.*, we do not consider the model dynamics. We evaluate our tool on five known logical models. We perform random changes considering several parameter configurations to assess the tool repairing capabilities. Whenever a model is repaired under the time limit, the tool successfully produces the optimal solutions to repair the model. Also, the number of repair operations required is less than or equal to the number of random changes applied to the original model.

1 Introduction

Biological regulatory networks are composed of genes, proteins and their interactions to describe complex cellular processes. Modelling such networks is particularly useful to be able to computationally reproduce existing observations, test hypotheses, and identify predictions *in silico*.

Different formalisms have been proposed to model the dynamical behavior resulting from the network' interacting components with different levels of detail (see [12] for a review). Here, we consider the logical formalism introduced by Thomas [20]. Network components are represented by discrete variables (here we consider the Boolean case), edges represent regulatory interactions (either positive or negative) and regulatory effects are represented by Boolean functions.

The definition of such models is typically a manual task performed by a domain expert, in particular for the definition of regulatory effects, where the study of the behaviors generated by the model are compared against existing

Fundação para a Ciência e a Tecnologia PhD grant SFRH/BD/130253/2017, national funds UID/CEC/50021/2019, grant SFRH/BSAB/143643/2019 and project grant PTDC/EEI-CTP/2914/2014.

© Springer Nature Switzerland AG 2019
Z. Cai et al. (Eds.): ISBRA 2019, LNBI 11490, pp. 100–112, 2019.
https://doi.org/10.1007/978-3-030-20242-2_9

data (*e.g.* literature or experimental). However, the study of the generated behaviors is hampered by the combinatorial explosion of the qualitative state space. To tackle this problem, several formal verification techniques have been proposed, such as: model-checking to automatically verify reachability properties [16], model reduction to reduce the size of the generated dynamics [17], and the identification of attractors [11], among others [18].

As the model is extended or new data is acquired, the model may become inconsistent, and in that case needs to be revised. One crucial step of the model revision process is the redefinition of the component's Boolean functions, a manual process that is not formally defined and therefore is prone to error.

Approaches to model revision have been proposed using Answer Set Programming (ASP) [6,8,15] and Boolean Satisfiability (SAT) [10]. Here, we propose an ASP-based model revision approach for the Boolean logical formalism, with four possible causes for model inconsistency and the corresponding repair operations.

The paper is organised as follows. Section 2 describes the logical formalism applied to biological regulatory networks. Section 3 describes the model revision process and the proposed repair operations. The proposed approach is presented in Sect. 4. The implemented tool is evaluated on five well-known biological models in Sect. 5. Section 6 presents the conclusion and future prospects.

2   Logical Regulatory Networks

Biological regulatory networks are usually represented by a directed graph $\mathcal{G} = (V, E)$, known as regulatory graph. In a regulatory graph, nodes represent the set of components V and the set of edges $E \subseteq \{(u, v, t) : u, v \in V; t \in \{-, +\}\}$ represent regulatory interactions. If the regulatory graph has an edge from v_i to v_j, then v_i is said to be a regulator of v_j. We can associate a sign t to each edge representing a positive interaction (activation) or a negative interaction (inhibition). Such regulatory graphs define the topology/structure of the network, lacking information on the components regulatory rules.

2.1   Logical Model

A logical model of a regulatory network is defined by a tuple (V, K), where $V = \{v_1, v_2, \ldots, v_n\}$ is the set of n regulatory components of the network, where each v_i is associated with an integer value in $D_i = \{0, \ldots, max_i\}$, representing the component concentration level. A state of the network is thus defined as a vector $s \in S = \prod_{v_i \in V} D_i$. Then $K = \{K_1, K_2, \ldots, K_n\}$ is the set of n regulatory functions where K_i is the regulatory function of v_i and $K_i : S \to D_i$.

In this work, we consider only Boolean logical models with $\forall_i \ max_i = 1$, *i.e.*, each components of the network is represented by a Boolean value, meaning that the component is either present (active) or absent (inactive).

2.2   Boolean Functions

Let $\mathcal{B}$ be the set $\{0,1\}$ and $\mathcal{B}^n$ be the n-dimensional cartesian product of the set $\mathcal{B}$. Given $(x_1, \ldots, x_n) \in \mathcal{B}^n$, a Boolean function $f : \mathcal{B}^n \to \mathcal{B}$ is *positive* (resp. *negative*) in x_i if $f|_{x_i=0} \leq f|_{x_i=1}$ (resp. x_i if $f|_{x_i=0} \geq f|_{x_i=1}$). Function f is *monotone* if it is either *positive* or *negative* for every x_i [2].

A monotone Boolean function f can be represented in Disjunctive Normal Form (DNF) [2], where a DNF formula is a disjunction of terms where each term is a conjunction of literals [1], and each variable x_i appears always as a positive literal (x_i) if f is positive in x_i, or it always appears negated $(\neg x_i)$ otherwise. In other words, each component regulating another component either has a positive (resp. negative) interaction always appearing as a positive (resp. negated) literal in the DNF of the regulatory function.

A nondegenerate Boolean function is a function that depends on all of its variables, *i.e.*, all variables have an impact on its result. More formally, a function f is nondegenerate if all of its variables are *essential*. A variable x_i is *essential* for f if $f|_{x_i=0}(X) \neq f|_{x_i=1}(X)$ for some $X \in \mathcal{B}^{n-1}$, and is *inessential* otherwise [21].

In this work, we restrict the domain of the regulatory functions to the set of monotone nondegenerate Boolean functions.

2.3   Dynamics

From a given initial state of the network, the value of a component can be updated following its regulatory function, and every component can potentially change its value at any given time. The generation of successors of each state can follow: a synchronous update policy, where every component is called to update their value simultaneously, yielding a single state successor; an asynchronous update policy, where the state has a distinct successor for each component changing its value; among other update policies (see [5] for details).

The generated dynamics is represented by a State Transition Graph (STG), where each node corresponds to a state of the network, and each edge represents a possible transition between states. A key property of interest is the identification of *attractors* in the STG, which typically denote subsets of states of biological interest [11]. There are two types of attractors: complex and point attractors. Complex attractors are sets of mutually reachable states defined as terminal Strongly Connected Components (SCC). If an SCC has a single state, it is denoted a point attractor or stable state, *i.e.* a state without a transition to any other state in the STG.

In this work, we focus on the set of point attractors (stable states) of the Boolean logical model.

3   Model Revision

Models of biological regulatory networks are not always in line with existing experimental observations, *i.e.*, the model cannot explain some experimental

Table 1. Causes of inconsistency and corresponding repair operations. Class F stands for function repair and class T stands for topology repair.

Type	Cause	Repair operation	Class
1	Wrong regulatory function	Function change	F
2	Wrong interaction type	Edge sign flip	T
3	Wrong regulator	Edge removal	T
4	Missing regulator	Edge addition	T

observations. In this case, we say that the model is inconsistent and must be revised and updated.

We consider a model to be *consistent*, if all its nodes are consistent. A given node is consistent if the value given by its regulatory function is the same as the one given by the experimental observation (if available). Otherwise, it is inconsistent. In the following, we consider all the model stable states as experimental observations.

Given a Boolean logical model, we define four possible causes for inconsistency and the corresponding repair operations as shown in Table 1. Repair operations can be classified as function repairs (class F) thus changing the regulatory functions or as topology repairs (class T) thus changing the topology of the regulatory graph.

In the following we describe in detail the repair operations defined as well as the complete model revision procedure, defining a preference order on the proposed repairs of Table 1. In particular, we assume that the domain expert has a higher level of confidence in the correctness of the network topology than in the regulatory functions of the model. We therefore give preference to the use of class F rather than class T repairs.

3.1 Function Repair

The definition of logical models still relies on domain experts to choose the best functions for every network component. However, given a component with k regulators, there are 2^{2^k} possible Boolean functions to choose from, rendering this manual process prone to error. In this work, we restrict the function space to the set of monotone nondegenerate Boolean functions, which still yields a large number of functions to choose from, as a function of k.

Let f and f' be two monotone nondegenerate Boolean functions in $\mathcal{B}^n \to \mathcal{B}$, with the relation $\preceq$ being defined as:

$$f \preceq f' \iff f(X) \Rightarrow f'(X). \tag{1}$$

A partial order set (POset) is then defined by the set of all monotone nondegenerate Boolean functions in $\mathcal{B}^n \to \mathcal{B}$ and the relation $\preceq$ [2,3], and is represented by an Hasse diagram (see Fig. 3). Considering the partial order relation

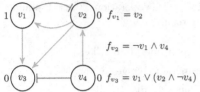

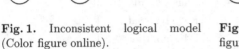

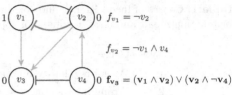

Fig. 1. Inconsistent logical model (Color figure online).

Fig. 2. Repaired logical model (Color figure online).

between functions, we rely on the work proposed by Cury *et al.* [3] to compute the functions with minimal impact to the original one, *i.e.*, their immediate neighbours. In Cury *et al.* [3] a set of rules is proposed to compute the *father/children* of a given function f without the need to compute the whole function space. Given two functions f and f', f' is a *father* of f if and only if $f \preceq f'$ and $\nexists f''$ such that $f \preceq f''$ and $f'' \preceq f'$. In this case f is said to be a *child* of f'.

When a function is inconsistent with the experimental observations, we first determine whether it is necessary to generalize the function (go up in the Hasse diagram), or specify the function (go down in the Hasse Diagram). If it is necessary to generalize (resp. specify) the function, we compute the set of *fathers* (resp. *children*) of the function. We continue to go up (resp. down) the diagram if none of the *fathers* (resp. *children*) is consistent.

3.2 Topology Repair

Changing regulatory functions may not suffice to make a model consistent. In this case, it may be necessary to (also) change the topology of the network. Here, we consider three topology-changing repair operations: flip the sign of an edge; remove an edge; and add an edge.

Flipping the sign of an edge changes the role of a single regulator. Since we consider the set of monotone nondegenerate Boolean functions, there are no dual regulators, *i.e.* regulators acting both as activators and as inhibitors. Thus, a negative regulator (inhibitor) becomes a positive regulator (activator) and vice-versa. By adding (resp. removing) an edge, we are adding (resp. removing) regulators from the Boolean function, which effectively changes its dimension which will likely have a greater impact than a function repair.

3.3 Repairing a Model

A model is considered inconsistent and deemed to be repaired if there is at least one inconsistent node. A node is inconsistent with some observational data if the expected value of the node differs from the node value evaluated by the corresponding regulatory function. Here, we consider the model's stable states as the observational data.

Figure 1 shows an example of an inconsistent model. The experimental observation is next to each node in the graph. This model with the corresponding observations has two inconsistent nodes: v_1 and v_3. For example, node v_1 is only positively regulated by v_2 (which has an observed value of 0), and therefore, the function of v_1 evaluates to 0, but the experimental observation of v_1 is 1.

There might be many reasons for a node to be inconsistent. To render a node consistent, one tries first to repair the function before considering any topological change in the network. This allows to search for possible repairs in the dimension of the current function before expanding the search space considering different function dimensions. The ordering is therefore as follows:

1. Repair the function;
2. Flip the sign of an edge;
3. Add/remove an edge.

In Fig. 1, node v_3 is inconsistent and different (function and topological) repair operations can be applied. There are 9 monotone nondegenerate Boolean functions with 3 regulators (see Fig. 3). Each of the 3 incoming edges can flip their sign. Also, we can remove any of the 3 incoming edges or add the missing edge from node v_3 to itself. If we consider the addition and removal of edges, which changes the search space of the Boolean functions, and that we can apply any combination of repair operations, we obtain a set of 2087 possible combinations of repair operations. Since models of regulatory networks typically have more than 4 nodes, the number of repair operations clearly explodes.

We start by first determining the minimum number of inconsistent nodes, to determine the solution with the minimum number of topology repairs. Therefore, first we try to repair a node by changing the regulatory function. If changing the function is not sufficient to make the model consistent we then proceed with topological repairs incrementally. We start by considering applying one topological repair operation, flipping the sign of an edge, for each possible edge. Then we consider applying two topology repair operations, and so on, until no more operations are possible. Whenever a topology repair operation is applied, the regulatory function must be verified for inconsistency again, and a function repair operation is most likely necessary.

Figure 2 shows the repaired model of Fig. 1, where the edge from v_2 to v_1 flipped the sign, making node v_1 consistent, and the regulatory function of v_3 changed, making node v_3 consistent. This is an optimal model repair with minimum topology changes. Figure 3 shows the Hasse diagram for the set of monotone nondegenerate Boolean functions of node v_3. Marked in blue (dark grey) is the original function of v_3 (Fig. 1) and marked in green (light grey) is the regulatory function of v_3 after the repair (Fig. 2).

4   Approach

In this section, we describe the approach for the revision of Boolean logical models considering the set of repair operations proposed in Sect. 3.

As previously mentioned, one must first determine if a model contains inconsistencies in order to repair it. In previous work, we proposed an Answer Set

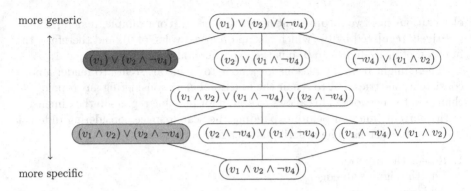

Fig. 3. Hasse diagram for monotone nondegenerate Boolean functions of three regulators (arguments v_1, v_2, and $\neg v_4$).

Programming (ASP) [7] program to verify the consistency of a model given a set of experimental observations [9]. As previously stated, in this work, we consider the model corresponding stable states, *i.e.*, we do not consider the model dynamics between specific states.

Given a model, a set of experimental observations, and a set of inconsistencies, we developed a procedure to try to repair the model using the repair operations in Table 1.

We start by verifying the consistency of the model, using the previously developed ASP program [9]. If the model is consistent with the data, no revision is necessary. In case of an inconsistent model, our ASP program returns the minimum number of nodes that are inconsistent, *i.e.*, the minimum number of nodes to which no value can be assigned. We call these nodes *inconsistent nodes*. Moreover, since we give preference to the repair of regulatory functions, as described in Sect. 3, we retrieve additional information from the ASP program regarding the inconsistent nodes. We define two possible reasons of inconsistency: the regulatory function of an inconsistent node needs to be either more specific or more generic. For example, if a regulatory function for a given node produces a 0 (resp. 1) but the value of the node should be 1 (resp. 0) in order for the node to be consistent, it is likely that a more generic (resp. specific) regulatory function is needed. These reasons for inconsistent do not imply that we can repair a model by only changing the regulatory function, but give us a direction to search for possible repairs.

The proposed procedure determines the minimum repair operations necessary to make the model consistent, using a lexicographic optimization criterion with the following order:

1. Minimize the number of add/remove edge operations;
2. Minimize the number of flip sign of an edge operations;
3. Minimize the number of change regulatory function operations.

This order of optimization gives preference to applying changes to the regulatory functions over any topological change.

As an example, let us consider an inconsistent model with a single inconsistent node. To determine the optimum solutions (*i.e.*, minimal repair operations to be applied), we start by trying to change the regulatory function of the inconsistent node, by replacing it either by a more generic or a more specific one, according to the corresponding reason of inconsistency. This is performed using an ASP program to compute the immediate neighbours (*fathers* or *children*) of a given monotone nondegenerate Boolean function, considering the set of rules proposed by Cury *et al.* [3].

Using this ASP program, our procedure computes all the fathers (resp. children) of the regulatory function. If none of the fathers (resp. children) is consistent, it is computed the corresponding fathers (resp. children), until it finds either a consistent function or there are no more functions. This approach guarantees that, if a function repair operation can be applied, then the original function is modified to (one of) the nearest function(s) in the Hasse diagram that is consistent with the experimental data, *i.e.* having less differences in the truth table. If no function is found, it proceeds to change the regulatory graph topology. In the first stage, it tries to change only the sign of the incoming edges of the inconsistent node. For each attempt of flipping the sign of an edge operation, it calculates all the possible function changes again. Then, if flipping the sign of a single edge is not enough to make a model consistent, it considers combinations of two edges, and so on until all the edges are considered. If no solution is found, the procedure advance to the next state of topological changes, by repeating this process considering adding or removing one edge. If the model is still not consistent, then considers the same process with two edges, and so on until no more edges can be added or removed.

This process is applied to every inconsistent node. Remember that we consider only the model stable states as experimental data, and thus repairing the consistency of one node does not impact the consistency of other nodes.

Even though we do not have to compute the complete Hasse diagram of all possible functions for a given n, the computation of the fathers/children of a given function still greatly increases with n. Therefore, we limit all regulatory functions to a maximum of 12 regulators, since most models have regulatory functions with an average of 3 regulators and no more than 8 (see Table 2).

Also, we can only repair a model if an inconsistent node has a single reason for its inconsistency: it either needs to be more specific or more generic, but not both simultaneously. If a function would need to be more specific in one experimental observation and more generic in another observation, in order to repair the model we would have to consider the set of all monotone nondegenerate functions. Here, we only consider the set of monotone nondegenerate functions that are *comparable* with the original function. Having this limitation allows for a reduction of the search space when repairing a function.

Finally, we developed a tool in C++[1], with the behaviour illustrated in Fig. 4. Given a logical model and a set of experimental observations, the tool decides whether the input is satisfiable. The input is said to be satisfiable either if it is consistent or if a repair can be found. Otherwise it is unsatisfiable.

---

[1] Tool available at http://sat.inesc-id.pt/~joaofrg/ISBRA2019/model_revision.zip.

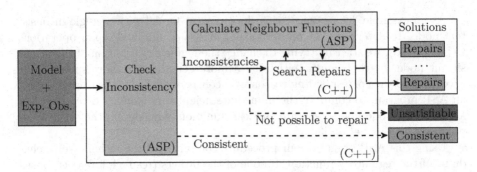

Fig. 4. Diagram of the developed tool. Model and experimental observations are the input of the tool. The tool produces as output all the optimal sets of repair operations. Marked in dashed arrows are alternative flows, where the model is consistent (no need of repair), or it is not possible to repair (no repairs produced). In yellow are represented the ASP components of the tool, and in white are the C++ components (Color figure online).

Table 2. Boolean logical models considered for evaluation with corresponding: used abbreviation (Abbr.), number of nodes (#Nodes), number of edges (#Edges), number of stable states (#SS), average number of regulators per node (Avg.Reg.), maximum number of regulators (M.Reg.), and bibliographic reference (Ref.).

Abbr.	Model	#Nodes	#Edges	#SS	Avg.Reg.	M.Reg.	Ref.
FY	Fission yeast	10	27	12	3	5	[4]
SP	Segment polarity (1 cell)	19	57	7	3	8	[19]
TCR	TCR signalisation	40	57	7	1,425	5	[13]
MCC	Mammalian cell cycle	10	35	1	3,5	6	[5]
Th	Th cell differentiation	23	35	3	1,842	5	[14]

Additionally, we provide a few options to the user. First, we allow the user to prevent some repair operations. Second, the user can define nodes as fixed nodes, preventing them from being considered inconsistent. We also allow the user to define some edges as fixed, preventing solutions with repair operations that change the sign or remove fixed edges.

5   Evaluation

In order to evaluate the proposed approach, we considered a set of five known logical models representative of different processes and organisms (see Table 2): the cell-cycle regulatory network of fission yeast by Davidich and Bornholdt [4]; the segment polarity network which plays a role in the fly embryo segmentation by Sanchèz et al. [19]; the T-Cell Receptor (TCR) signaling network by Klamt et al. [13]; the core network controlling the mammalian cell cycle by Fauré et al. [5]; and the regulatory network controlling T-helper cell differentiation by Mendoza and Xenarios [14].

Table 3. Results for FY, SP, TCR, MCC and Th. F%, E%, R%, and A% are the probabilistic parameters used to change the original model. Time (T), in seconds, is the median of time for the solved instances. #TO is the number of timeouts. 10 instances were considered per configuration per model.

(%)				FY		SP		TCR		MCC		Th	
F	E	R	A	T (s)	#TO	T (s)	#TO	T (s)	#TO	T (s)	#TO	T (s)	#TO
5	0	0	0	0,034	0	0,036	0	0,047	0	0,021	0	0,028	0
25	0	0	0	0,059	0	4,734	0	0,063	0	0,021	0	0,061	0
50	0	0	0	0,060	0	14,003	2	0,097	0	0,033	0	0,677	0
100	0	0	0	0,072	0	18,937	2	0,129	0	0,046	0	0,751	0
0	5	0	0	0,070	0	0,105	0	0,050	0	0,033	0	0,061	1
0	10	0	0	0,070	0	1,566	1	0,050	0	0,101	0	0,044	0
0	15	0	0	0,035	1	0,168	3	0,050	0	0,039	0	0,051	1
0	20	0	0	0,071	2	0,284	4	0,050	0	0,136	0	0,062	1
0	0	1	0	0,034	0	0,635	0	0,045	0	0,020	0	0,025	1
0	0	5	0	0,069	0	5,021	1	0,046	0	0,020	0	0,026	1
0	0	10	0	0,095	2	24,481	4	0,060	0	0,019	0	0,589	2
0	0	15	0	0,083	2	32,896	3	7,106	0	0,029	0	1,613	2
0	0	0	1	0,874	0	0,130	2	0,152	0	0,020	0	0,028	3
0	0	0	5	0,096	0	42,684	7	2,518	3	0,219	0	0,497	8
0	0	0	10	0,842	1	-	10	-	10	0,234	1	-	10
0	0	0	15	6,003	4	-	10	-	10	0,622	0	258,022	9
25	5	0	0	0,062	0	5,358	0	0,063	0	0,032	0	0,108	0
50	25	0	0	0,127	2	13,989	4	0,187	0	0,570	0	0,724	1
5	25	5	5	0,453	4	-	10	3,979	8	0,549	1	0,781	9
10	10	5	5	0,601	2	24,637	8	50,662	6	0,142	1	0,745	7

We developed a tool to make a set of random changes to a logical model according to four given probabilistic parameters. Our goal was to change a logical model, and then assess the repairing capabilities of the proposed tool to make the model consistent again. The set of four probabilistic parameters were considered: changing a function (F%); changing the sign of an edge (E%); removing an existing edge (R%); and adding a missing edge (A%).

We changed each model with several parameters configurations, and considered 10 instances per configuration per model (see Table 3). We considered generating instances where only the functions were changed, simulating cases where the topology of the network is correct. We also considered instances where only the sign of the edges was changed, instances where we only removed edges, and instances where only were added new edges. We generated instances with more functions changes than topology changes, since we assume greater confidence in the correctness of the topology than of the regulatory functions.

Due to the limitations of our tool, we only considered instances that have regulatory functions with less than 12 regulators, and instances with a single reason for inconsistency per node (as described in Sect. 4). We also only considered instances different from the original models, *i.e.*, instances where at least one change was applied. All experiments were run on an AMD Opteron(TM) 1.4 GHz 32-core Linux machine, with a time limit of 600 s.

Table 3 presents the results of our tool applied to different models (FY, SP, TCR, MCC, and Th) and different changes. The time is presented in seconds and corresponds to the median of times of solved instances. Whenever the model repair was possible under the time limit, the tool successfully repaired the model with a less or equal number of repair operations than the number of changes applied to the original model.

Table 3 shows that most of the repaired models can be repaired under 60 s. It is also possible to verify that changing the topology of the network has a bigger impact, increasing the number of timeouts as the number of changes increases. Perturbing the model with the addition of new edges has a bigger impact in the model revision process than the removal of edges. This is due to the dimension increase of the regulatory function, which greatly increases the search space for possible function repairs.

Comparing the results for the SP and TCR models we can observe that, although the TCR model is composed of more nodes (40) than SP model (19), repairing the SP model takes more time in general, for the same number of edges (57). This means that the latter has a more interconnected network, with regulatory functions depending on a higher number of regulators. The dimension of the regulatory function greatly impacts the performance of our tool, since the number of monotone nondegenerate Boolean functions to be considered increases with a double exponential with the number of regulators.

6   Conclusion and Future Work

In this work we propose a logical model revision tool without considering dynamics. We use ASP to verify the consistency of a model and retrieve useful information in case of inconsistency, and compute possible regulatory functions replacement candidates. And C++ to search the set of repair operations. Four repair operations are proposed: regulatory function change; edge sign flipping; edge addition; and edge removal. Our tool receives as an input a logical model and a set of experimental observations, and produces sets of repair operations to render the model consistent, under an optimization criteria (Sect. 4).

The tool was successfully tested using several well-known biological models, being able to repair most of the instances under 60 s. We were able to conclude that the dimension of the regulatory functions has the biggest impact on the tool performance, since the number of monotone nondegenerate Boolean functions increases.

As a future work, time-series data could be used to also consider the model dynamics. Also, the possibility to repair models with inconsistent nodes with multiple reasons for inconsistency (see Sect. 4 for more details). The proposed tool produces all the optimal solutions to repair a model. Heuristics could be used to reduced the number of solutions produced (see [15] for details).

References

1. Biere, A., Heule, M., van Maaren, H.: Handbook of Satisfiability, vol. 185. IOS Press, Amsterdam (2009)
2. Crama, Y., Hammer, P.L.: Boolean Functions: Theory, Algorithms, and Applications. Cambridge University Press, Cambridge (2011)
3. Cury, J.E., Monteiro, P.T., Chaouiya, C.: Partial Order on the set of Boolean Regulatory Functions. arXiv preprint arXiv:1901.07623 (2019)
4. Davidich, M.I., Bornholdt, S.: Boolean network model predicts cell cycle sequence of fission yeast. PLoS ONE **3**(2), e1672 (2008)
5. Fauré, A., Naldi, A., Chaouiya, C., Thieffry, D.: Dynamical analysis of a generic Boolean model for the control of the mammalian cell cycle. Bioinformatics **22**(14), e124–e131 (2006)
6. Gebser, M., et al.: Repair and prediction (under inconsistency) in large biological networks with answer set programming. In: KR (2010)
7. Gebser, M., Kaminski, R., Kaufmann, B., Schaub, T.: Answer set solving in practice. Synth. Lect. Artif. Intell. Mach. Learn. **6**(3), 1–238 (2012)
8. Gebser, M., Schaub, T., Thiele, S., Veber, P.: Detecting inconsistencies in large biological networks with answer set programming. TPLP **11**(2–3), 323–360 (2011)
9. Gouveia, F., Lynce, I., Monteiro, P.T.: Model revision of logical regulatory networks using logic-based tools. In: ICLP 2018 (Technical Communications). Schloss Dagstuhl-Leibniz-Zentrum fuer Informatik (2018)
10. Guerra, J., Lynce, I.: Reasoning over biological networks using maximum satisfiability. In: Milano, M. (ed.) CP 2012. LNCS, pp. 941–956. Springer, Heidelberg (2012). https://doi.org/10.1007/978-3-642-33558-7_67
11. Hopfensitz, M., Müssel, C., Maucher, M., Kestler, H.A.: Attractors in Boolean networks: a tutorial. Comput. Stat. **28**(1), 19–36 (2012)
12. Karlebach, G., Shamir, R.: Modelling and analysis of gene regulatory networks. Nat. Rev. Mol. Cell Biol. **9**(10), 770 (2008)
13. Klamt, S., Saez-Rodriguez, J., Lindquist, J.A., Simeoni, L., Gilles, E.D.: A methodology for the structural and functional analysis of signaling and regulatory networks. BMC Bioinform. **7**(1), 56 (2006)
14. Mendoza, L., Xenarios, I.: A method for the generation of standardized qualitative dynamical systems of regulatory networks. Theor. Biol. Med. Model. **3**(1), 13 (2006)
15. Merhej, E., Schockaert, S., De Cock, M.: Repairing inconsistent answer set programs using rules of thumb: a gene regulatory networks case study. Int. J. Approx. Reason. **83**, 243–264 (2017)
16. Monteiro, P.T., Ropers, D., Mateescu, R., Freitas, A.T., De Jong, H.: Temporal logic patterns for querying dynamic models of cellular interaction networks. Bioinformatics **24**(16), i227–i233 (2008)
17. Naldi, A., Remy, E., Thieffry, D., Chaouiya, C.: Dynamically consistent reduction of logical regulatory graphs. Theor. Comput. Sci. **412**(21), 2207–2218 (2011)
18. Paulevé, L.: Reduction of qualitative models of biological networks for transient dynamics analysis. IEEE/ACM Trans. Comput. Biol. Bioinform. **15**, 1167–1179 (2017)
19. Sánchez, L., Chaouiya, C., Thieffry, D.: Segmenting the fly embryo: logical analysis of the role of the segment polarity cross-regulatory module. Int. J. Dev. Biol. **52**(8), 1059–1075 (2002)

20. Thomas, R.: Boolean formalization of genetic control circuits. J. Theor. Biol. **42**(3), 563–585 (1973)
21. Wegner, I.: The critical complexity of all (monotone) Boolean functions and monotone graph properties. Inf. Control. **67**(1–3), 212–222 (1985)

Gene- and Pathway-Based Deep Neural Network for Multi-omics Data Integration to Predict Cancer Survival Outcomes

Jie Hao[1], Mohammad Masum[1], Jung Hun Oh[2],
and Mingon Kang[1,3]($\boxtimes$)

[1] Analytics and Data Science Institute, Kennesaw State University,
Kennesaw, GA, USA
[2] Department of Medical Physics, Memorial Sloan Kettering Cancer Center,
New York, NY, USA
[3] Department of Computer Science, Kennesaw State University,
Marietta, GA, USA
mkang9@kennesaw.edu

Abstract. Data integration of multi-platform based omics data from biospecimen holds promise of improving survival prediction and personalized therapies in cancer. Multi-omics data provide comprehensive descriptions of human genomes regulated by complex interactions of multiple biological processes such as genetic, epigenetic, and transcriptional regulation. Therefore, the integration of multi-omics data is essential to decipher complex mechanisms of human diseases and to enhance treatments based on genetic understanding of each patient in precision medicine. In this paper, we propose a gene- and pathway-based deep neural network for multi-omics data integration (MiNet) to predict cancer survival outcomes. MiNet introduces a multi-omics layer that represents multi-layered biological processes of genetic, epigenetic, and transcriptional regulation, in the gene- and pathway-based neural network. MiNet captures nonlinear effects of multi-omics data to survival outcomes via a neural network framework, while allowing one to biologically interpret the model. In the extensive experiments with multi-omics data of Gliblastoma multiforme (GBM) patients, MiNet outperformed the current cutting-edge methods including SurvivalNet and Cox-nnet. Moreover, MiNet's model showed the capability to interpret a multi-layered biological system. A number of biological literature in GBM supported the biological interpretation of MiNet. The open-source software of MiNet in PyTorch is publicly available at https://github.com/DataX-JieHao/MiNet.

Keywords: Multi-omics data · Deep neural network · Glioblastoma · Survival analysis

© Springer Nature Switzerland AG 2019
Z. Cai et al. (Eds.): ISBRA 2019, LNBI 11490, pp. 113–124, 2019.
https://doi.org/10.1007/978-3-030-20242-2_10

1   Introduction

Data integration of multi-platform based omics data (e.g., genomics, proteomics, and metabolomics) from biospecimens holds promise of improving survival prediction and personalized therapies in cancer [11,12]. The importance of integrative studies has been increasingly emphasized along with the rapid development of various types of high-throughput multi-omics data. A large scale of multi-omics data sets have been generated in various cancer projects, such as The Cancer Genome Atlas (TCGA) and The Cancer Genome Project in Wellcome Trust Sanger Institute. In particular, TCGA provides various types of omics data of more than 33 cancers, including tissue exome sequencing, gene expression, Copy Number Alternation (CNA), DNA variation, DNA methylation, and microRNA, as well as clinical data such as race, tumor stage, and survival status and months of cancer patients.

Multi-omics data provide comprehensive descriptions of human genomes regulated by complex interactions of multiple biological processes such as genetic, epigenetic, and transcriptional regulation [16]. Thus, the integration of multi-omics data can be leveraged to decipher complex mechanisms of human diseases and to enhance cancer treatments based on genetic understanding of each patient in precision medicine. Specifically, genes are activated by sequential interactions of DNA variations, CNA, histone modifications, transcription factors, DNA methylation, and other genes in relevant pathways [1,23]. CNA, which is a modified gene structure, often alters downstream pathways or regulatory networks, and DNA methylation often reduces gene expression in a nearby gene when the methyl groups are added to the DNA. Hence, monozygotic twins discordance in disease is often caused due to different CNA, although they have nearly identical genetic variants [3,17].

Recently, multi-omics data have been widely incorporated in an increasing number of research projects in survival analysis, rather than using a single type of genomic data that most genomic research traditionally has analyzed. Multi-omics data such as CNA, DNA methylation, and gene expression were integrated to identify knowledge-driven genomic interactions with clinical outcomes of interest in ovarian carcinoma [15]. The meta-dimensional models, which incorporate biological pathways with multi-omics data, enhanced the model interpretability in the biological pathway level. A multi-block bipartite graph was proposed not only to identify intra- and inter-block interaction effects of multi-omics data, but also to predict quantitative traits such as gene expression and survival time [14]. SurvivalNet integrated multi-omics data such as DNA mutation, CNA, protein, and mRNA along with clinical information into a deep neural network to improve survival prediction of patients in cancers [24]. Feature selection techniques were applied to each omics dataset separately, and selected features of the multi-omics data and clinical data were combined into a large augmented matrix in SurvivalNet. Another deep learning-based model integrated RNA-Seq, miRNA-Seq, and DNA methylation data to differentiate survival groups in hepatocellular carcinoma [5]. Furthermore, the differential subgroups identified several significant multi-omics features.

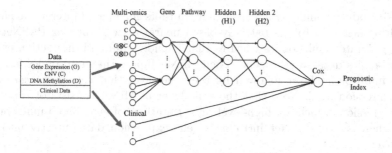

Fig. 1. The architecture of MiNet

In this study, we propose a novel approach, called MiNet, to integrate multi-omics data and clinical data using a pathway-based deep neural network for survival analysis. Our previously published model, Cox-PASNet, which is a pathway-based deep neural network for predicting survival outcome, has considered only gene expression data as well as clinical data [9]. The main contributions of MiNet are as follows: (1) to introduce a multi-omics layer that represents gene-based interaction effects of multi-omics data and (2) to interpret the model in a biological pathway level.

The rest of the paper is organized as follows. In Sect. 2, our proposed model is elaborated in detail. The experimental setting and results are demonstrated in Sect. 3. Section 4 discusses the model interpretation with biological findings, while Sect. 5 concludes the discussion.

2 Methods

We propose a gene- and pathway-based multi-omics integrative deep neural network (MiNet) to predict cancer survival outcomes. MiNet introduces a gene-based multi-omics layer to integrate multi-omics data, leveraging the advantages of the pathway-based neural network framework in Cox-PASNet [9]. The neural network structure of MiNet follows a biological system, which is multi-layered with multi-omics data and their interactions along with clinical features, by utilizing prior knowledge of biological pathways. The biologically inspired neural network architecture provides a rich interpretation of a biological system.

2.1 Multi-omics Integration

Most studies have integrated multi-omics data by combining all types of omics data to a single matrix and performed analysis, e.g., survival analysis. However, the consideration of the augmented multi-omics data as independent features lacks of representing interaction effects of genomic and epigenomic data with gene expressions. Note that CNA and DNA methylation often regulate transcriptional mechanisms of genes, so some genes may be down- or up-regulated caused by interaction effects of other omics data.

We introduce a multi-omics layer that transfers gene-based interaction effects of multi-omics data to the pathway-based neural network of Cox-PASNet [9]. MiNet generates multi-omics features that include main and interaction effects of multi-omics data on each gene. Then, MiNet inputs the multi-omics features to the multi-omics layer followed by the gene layer that represents *canonical* gene expression level. Note that the gene layer of MiNet consists of *canonical* gene expressions which are high-level representations of gene-based multi-omics data, whereas Cox-PASNet introduces gene expression data directly into the gene layer.

We consider cis-regulatory interaction effects of CNA and DNA methylation to a nearest gene. Multi-omics feature vectors $\mathbf{x}_i$ are generated as:

$$
\mathbf{x}_i = \begin{bmatrix} \mathbf{g}_i \\ \mathbf{c}_i \\ \mathbf{d}_i \\ \mathbf{g}_i \otimes \mathbf{c}_i \\ \mathbf{g}_i \otimes \mathbf{d}_i \end{bmatrix}^\top
\begin{array}{l}
\text{// Main effect of gene expression} \\
\text{// Main effect of CNA} \\
\text{// Main effect of DNA methlyation} \\
\text{// Interaction effect with CNA} \\
\text{// Interaction effect with DNA methlyation}
\end{array}
, \quad (1)
$$

where $\mathbf{g}_i$, $\mathbf{c}_i$, and $\mathbf{d}_i$ are sample vectors of gene expression, CNA, and DNA methylation for the i-th gene, respectively. Note that we consider the genes that have at least a gene expression feature. Then, *canonical* gene expression ($\tilde{\boldsymbol{g}}_i$) for the i-th gene is expressed by:

$$
\tilde{\boldsymbol{g}}_i = \sigma(\mathbf{x}_i \mathbf{w}_i), \quad (2)
$$

where $\mathbf{w}_i$ is a weight vector, $\sigma(\cdot)$ is an activation function, and $\otimes$ is element-by-element multiplication. The main or interaction effects are ignored if there is no CNA or DNA methylation associated to the i-th gene, so genes may have different numbers of multi-omics features.

2.2   The Architecture of MiNet

The architecture of MiNet is composed of a multi-omics layer, a gene layer, a pathway layer, multiple hidden layers, a clinical layer, and a Cox layer, as shown in Fig. 1. The multi-omics layer is an input layer, which introduces multi-omics features (see Eq. 1) from genomics (CNV), epigenomics (DNA methylation), and transcriptomics (gene expression) data into MiNet. The multi-omics layer contains multi-omics features of all genes, and the connections between multi-omics features and genes are implemented by a boolean mask matrix. Note that the associations of multi-omics features are determined with the nearest gene. Most databases often provide genes that CNV and DNA methlayion are mapped to. At the end, every multi-omics features are connected to only a node in the gene layer.

The gene layer represents *canonical* gene features computed by Eq. 2, where each node indicates a gene in a biological system. Since a set of genes are involved in biological pathways, genes in the gene layer transfer to corresponding pathway nodes in the pathway layer. Note that the connections between genes and pathways are given by pathway databases, so the number of nodes in the pathway layer is identical with the number of known biological pathways. Hidden layers show hierarchical representations of multiple pathways. A hidden node contains the interaction effect of a set of pathways. More hidden layers may capture more complex interactions of biological pathways.

The clinical layer is an additional input layer for clinical features (e.g., sex, age, and tumor stage). The clinical data are introduced to the output layer as additional features of the last hidden layer, rather than concatenating with the multi-omics layer. The independent clinical layer prevents a few input features from dominating others and makes the model interpretation effective in genomic level. Clinical features, such as age, have often been shown as significant covariates in several cancer studies. The effects of genomic features may be suppressed by clinical features or vice versa. Moreover, genomic data and clinical data should be separated for the model interpretation.

The output layer with one node is named as a Cox layer. A linear activation function without bias is applied to this layer to adopt Cox regression. The final outcome of MiNet is Prognostic Index (PI) which is a linear combination of covariates, and PI is introduced to the hazard function for the Cox proportional hazards model as:

$$\lambda(t|\mathbf{x}) = \lambda_0(t) \exp(\text{PI}), \tag{3}$$

where PI is an outcome of the Cox layer in MiNet.

2.3   Training MiNet with Sparse Coding

MiNet minimizes the average negative log partial likelihood with L^2 regularization. MiNet adapts the training strategy introduced in Cox-PASNet for effective training with high-dimensional, low-sample-size data, where small sub-networks are randomly selected and trained with sparse coding. For the parameter initialization, all layers are fully-connected with He's initialization strategy [10].

The connections between the multi-omics layer and the gene layer are masked by the given boolean mask matrix during the entire training process, similarly in the connections between the gene layer and the pathway layer. Note that the connections between the multi-omics layer and the pathway layer are defined by prior biological knowledge. Sparse coding is applied to the hidden layers following the pathway layer.

We apply sparse coding (L^1 regularization) individually on each layer pair, instead of entire weight matrix. Inspired by LASSO, a soft-thresholding strategy is applied to the connections on each layer pair. Thus, weight matrix is further optimized on each layer pair by:

$$\mathbf{W}^\star \leftarrow S(\mathbf{W}, Q_s), \tag{4}$$

where $S(\mathbf{W}, s) = sign(\mathbf{W})(|\mathbf{W}| - Q_s)_+$ is the soft-thresholding function and $sign(\mathbf{W})$ returns a sign of $\mathbf{W}$. $(|\mathbf{W}| - Q_s)_+$ returns $|\mathbf{W}| - Q_s$ if $|\mathbf{W}| - Q_s > 0$, otherwise, $(|\mathbf{W}| - Q_s)_+ = 0$. Q_s is the optimal threshold with respect to the optimal sparsity level s. The optimal sparsity level s is estimated with the strategy proposed in Cox-PASNet [9].

Table 1. Performance comparison of MiNet with the benchmark methods using C-index in over 20 experiments

Model	C-index ($\mu \pm \sigma$)
Cox-EN [20]	0.5163 ± 0.0359
SurvivalNet [24]	0.5567 ± 0.0312
Cox-nnet [7]	0.5655 ± 0.0287
MiNet (proposed)	**0.6214 ± 0.0352**

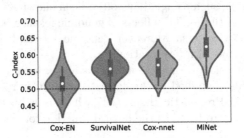

Fig. 2. Distribution of C-index with 20 experiments

Table 2. Statistical assessment

	Wilcoxon rank-sum test
MiNet vs. Cox-EN	1e-4*
MiNet vs. Cox-nnet	2e-4*
MiNet vs. SurvivalNet	2e-4*

*shows the statistical significance with significance level = 0.05.

3   Experimental Results

In this paper, we conducted experiments with multi-omics data and clinical data in Glioblastoma Multiforme (GBM), which is the most invasive brain tumor. We downloaded multi-omics data including gene expressions, CNAs, and DNA methylations, and clinical data of GBM patients from The Cancer Genome Atlas (TCGA)[1]. We retrieved age, survival status (living or deceased), and survival months of the GBM patients. Age was used as a clinical feature, and both survival status and survival months were used for response variables. The other clinical features were not considered because of large missing values. We filtered out samples with missing values in survival information.

For pathway-based analysis, we downloaded KEGG and Reactome pathway databases from the Molecular Signatures Database (MSigDB) [21]. The pathway databases consist of gene sets of well-known biological pathways, which have

---

[1] https://cancergenome.nih.gov.

molecular interactions in a cell that simultaneously lead to a certain biological process. Small pathways with less than 25 genes were excluded to avoid large redundancy with other pathways [18].

For the experiments, we considered genes that belong to at least one pathway. In particular, 5,481 genes were associated with 507 pathways in the dataset. We included CNAs and DNA methylations associated to the 5,481 genes. Missing values in CNV and DNA methylation features were imputed by 1-Nearest Neighbor (1-NN). Finally, we used 24,803 multi-omics features including interactions and one clinical feature (i.e. age) from 523 samples. The dataset for benchmark models has 14,142 multi-omics features and age from 523 samples, where interactions were excluded. Note that the benchmark methods considered much less numbers of input features than our model.

We compared the performance of MiNet with the current cutting-edge methods: Cox regression with elastic net regularization (Cox-EN) [20], Survival-Net [24], and Cox-nnet [7]. Concordance index (C-index) was measured to evaluate the performance of the methods. C-index is commonly used to measure the predictive performance in survival analysis. We randomly split the entire data into three subsets of training (64%), validation (16%), and test data (20%) by stratified sampling with survival status, so that each subset preserves the same proportion of censored samples as the entire data. Then, all features were normalized to zero mean with variance of one. Validation and test data were normalized with the mean and variance obtained from training data. Validation data were used to perform early stopping and grid search for finding the optimal hyper-parameters. We repeated the experiments 20 times to show the reproducibility of the performance.

Our proposed method MiNet was implemented by PyTorch 1.0 with CUDA 10. We used ReLU for the activation function, and dropout and L^2 regularization were applied to avoid overfitting problems. Adaptive Moment Estimation (Adam) optimizer was performed to take advantage of a fast convergence and a reduced oscillation. The structure of MiNet was constructed with two hidden layers following multi-omics, gene, and pathway layers, as empirically showing better performance than with a single hidden layer. We considered 22 and 5 nodes in the two hidden layers (H1 and H2) respectively, following the rule of thumb that the number of hidden nodes is the square root of the number of input nodes. Dropout rates were empirically set as 0.7 and 0.5 for pathway layer and hidden layer, respectively. The optimal initial learning rate (η) and L^2 regularization (λ) were determined by grid search that maximizes C-index in validation data on each experiment. All experiments were performed with four NVIDIA Tesla M40 (12GB memory) Graphics Processing Units (GPU). The source code of MiNet is publicly available in GitHub[2].

Experiments of SurvivalNet [24] and Cox-nnet [7] were performed by the Python packages published on GitHub[3,4]. Bayesian optimization [19] was

---

[2] https://github.com/DataX-JieHao/MiNet.
[3] https://github.com/CancerDataScience/SurvivalNet.
[4] https://github.com/lanagarmire/cox-nnet.

employed in SurvivalNet for the optimal neural network structure and hyper-parameters, such as number of layers, number of nodes, dropout rate, L^1 regularization, and L^2 regularization. For Cox-nnet, grid search strategy was applied for optimal regularization parameter (L^2). Cox-EN was implemented by the package *Glmnet Vignette* in Python [20]. The tunning hyperparamter λ and the elastic-net penalty term α ($\alpha \in [0, 1]$) were optimized by grid search. Kaplan-Meier analysis and log-rank test were performed by using the Python package *lifelines*.

C-index scores obtained from Cox-EN, SurvivalNet, Cox-nnet, and MiNet over 20 experiments with GBM data are shown in Table 1. Our proposed method, MiNet, produced the highest C-index of 0.6214 ± 0.0352 among the benchmark methods, whereas Cox-EN, SurvivalNet, and Cox-nnet showed 0.5163 ± 0.0359, 0.5567 ± 0.0312, and 0.5655 ± 0.0287, respectively. Figure 2 depicts the distribution of C-index of the experiments. Moreover, we performed Wilcoxon rank-sum tests to assess the statistical significance of the model improvement. As shown at Table 2, the outperformance of MiNet against the other benchmarks was statistically assessed, i.e., p-values < 0.05.

4   Model Interpretation with GBM

For the model interpretation of MiNet with GBM data, we trained the model with the entire data again using the optimal hyper-parameters that have been selected most frequently over 20 experiments (i.e., $\lambda = 0.02$ and $\eta = 0.005$). In consequence, the C-index of the re-trained model was 0.91, which was not overfitted to the input data.

We first examined the six covariates, which are the input nodes to the Cox layer. Five covariates are in the last hidden layer (H2), and one covariate (age) is from the clinical layer. Figure 3a illustrates the H2 and age node values, where the nodes are ranked by the partial derivatives with respect to the H2 layer and the clinical layer. Overall, the node values show high correlation with PI. Specifically, Node 2 in H2 (the first column in Fig. 3a) appeared as the most important covariate for predicting survival time in MiNet with GBM data. For evaluating each covariate, we separated the samples into two groups of high-risk and low-risk by the median of PI. Then, p-values were computed with logrank test. The p-values are shown in the upper plot in Fig. 3a, where all covariates were statistically significant (i.e., p-values < 0.05). Kaplan-Meier plots are depicted in Fig. 3b and c with the two top-ranked covariates, which demonstrates significantly distinct survival curves. Moreover, the six nodes are visualized by t-SNE in Fig. 4, which shows a highly linear correlation between the six covariates and the survival outcomes.

Table 3 shows five top-ranked pathways by MiNet, where pathway nodes are ranked by the partial derivatives with respect to the pathway layer. It was discovered that GnRH receptor is expressed in GBM [8]. Interestingly, GnRH signaling pathway was not identified with single omics data, but significantly enriched with multi-omics data [13]. MiNet accordingly ranked GnRH signaling

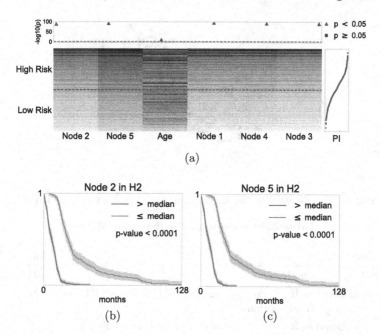

Fig. 3. Graphical interpretation of the last hidden layer (H2) and the clinical layer. (a) Heatmap of the H2 and age node values. The horizontal dashed line separates high-risk and low-risk groups, which were separated by the median of PI. The upper dot plot shows -log10(p-values) from the logrank test between high-risk and low-risk groups for every single node. The right curve shows the distribution of PI with the corresponding samples on the heatmap. (b) – (c) Kaplan-Meier plots for the two top-ranked covariates.

pathway as a significant factor with multi-omics data. Furthermore, the other four pathways have been also recognized in GBM with several biological literature. The references are listed in Table 3.

Two genes of NRAS and PRKACA are identified as significant in GnRH signaling pathway (see Table 4). Then, we traced back to the multi-omics layer of the genes. Somatic mutation of NRAS in GBM and its critical role in PI3K-AKT pathway were reported [2]. For NRAS, the main effects of gene expression was the most important factor, followed by the interaction effects of gene expression and CNA and the main effect of DNA methylation. The numbers in parenthesis show partial derivatives with respect to the input nodes, and the higher values indicate the more important factors. For PRKACA, the main effect of CNA and gene expression were highly ranked as the most important multi-omics factors and followed by the interaction effect of CNA, so CNA may play an important role in regulating PRKACA in GBM.

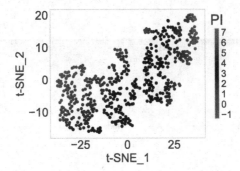

Fig. 4. Visualization of the H2 and age nodes in MiNet using t-SNE.

Table 3. Five top-ranked pathways by MiNet

Pathways	Size	Ref.
GnRH signaling pathway	101	[13]
Genes involved in RNA Polymerase I, RNA Polymerase III, and Mitochondrial Transcription	122	–
Genes involved in Response to elevated platelet cytosolic Ca2+	89	[4]
Melanogenesis	102	[6]
Genes involved in Extracellular matrix organization	87	[22]

Table 4. Two top-ranked genes in GnRH signaling pathway

Genes	Multi-omics	Ref.
NRAS	**G** (0.001829), **G** ⊗ **C** (0.000888), **D** (0.000791), **C** (0.000319), **G** ⊗ **M** (0.000037)	[2]
PRKACA	**C** (0.000774), **G** (0.000738), **G** ⊗ **C** (0.000698)	-

5   Conclusion

In this paper, we propose a gene- and pathway-based deep neural network for multi-omics data integration, named MiNet, to predict cancer survival outcomes. In MiNet, gene-based multi-omics features are generated by considering main and interaction effects of multi-omics data in the multi-omics layer. The multi-omics features produce *canonical* gene expression in the gene layer. The hierarchical representations of biological processes of multi-omics, genes, and pathways are captured in MiNet. MiNet showed the outstanding performance to predict cancer survival outcomes with GBM patients. More importantly, MiNet provides the capability to interpret a multi-layered biological system. A large number of biological literature supported our biological findings from MiNet.

The multi-omics layer of MiNet is designed as a neural network module for the integration of multi-omics data, and is compatible to the pathway-based neural network, Cox-PASNet. The high flexibility and expandability of the model architecture would allow one to take an advantage of utilizing the well-established pathway-based framework.

References

1. Aure, M.R., et al.: Individual and combined effects of DNA methylation and copy number alterations on miRNA expression in breast tumors. Genome Biol. **14**(11), R126 (2013). https://doi.org/10.1186/gb-2013-14-11-r126
2. Bleeker, F.E., et al.: Mutational profiling of kinases in glioblastoma. BMC Cancer **14**(1), 718 (2014). https://doi.org/10.1186/1471-2407-14-718
3. Bruder, C.E., et al.: Phenotypically concordant and discordant monozygotic twins display different DNA copy-number-variation profiles. Am. J. Hum. Genet. **82**(3), 763–771 (2008). https://doi.org/10.1016/j.ajhg.2007.12.011
4. Catacuzzeno, L., Franciolini, F.: Role of KCa3.1 channels in modulating Ca^{2+} oscillations during glioblastoma cell migration and invasion. Int. J. Mol. Sci. **19**(10), 2970 (2018). https://doi.org/10.3390/ijms19102970
5. Chaudhary, K., Poirion, O.B., Lu, L., Garmire, L.X.: Deep learning-based multi-omics integration robustly predicts survival in liver cancer. Clin. Cancer Res. **24**(6), 1248–1259 (2018). https://doi.org/10.1158/1078-0432.CCR-17-0853
6. Chi, D.D.J., et al.: Molecular detection of tumor-associated antigens shared by human cutaneous melanomas and gliomas. Am. J. Pathol. **150**(6), 2143–2152 (1997). https://www.ncbi.nlm.nih.gov/pubmed/9176405
7. Ching, T., Zhu, X., Garmire, L.X.: Cox-nnet: an artificial neural network method for prognosis prediction of high-throughput omics data. PLOS Comput. Biol. **14**(4), 1–18 (2018). https://doi.org/10.1371/journal.pcbi.1006076
8. Gründker, C., Emons, G.: The role of gonadotropin-releasing hormone in cancer cell proliferation and metastasis. Front. Endocrinol. **8**, 187 (2017). https://doi.org/10.3389/fendo.2017.00187
9. Hao, J., Kim, Y., Mallavarapu, T., Oh, J.H., Kang, M.: Cox-PASNet: pathway-based sparse deep neural network for survival analysis. In: 2018 IEEE International Conference on Bioinformatics and Biomedicine (BIBM), pp. 381–386 (2018). https://doi.org/10.1109/BIBM.2018.8621345
10. He, K., Zhang, X., Ren, S., Sun, J.: Delving deep into rectifiers: surpassing human-level performance on imagenet classification. In: 2015 IEEE International Conference on Computer Vision (ICCV), pp. 1026–1034 (2015). https://doi.org/10.1109/ICCV.2015.123
11. Higdon, R., et al.: The promise of multi-omics and clinical data integration to identify and target personalized healthcare approaches in autism spectrum disorders. OMICS: J. Integr. Biol. **19**(4), 197–208 (2015). https://doi.org/10.1089/omi.2015.0020
12. Huang, S., Chaudhary, K., Garmire, L.X.: More is better: recent progress in multi-omics data integration methods. Front. Genet. **8**, 84 (2017). https://doi.org/10.3389/fgene.2017.00084
13. Jayaram, S., Gupta, M.K., Raju, R., Gautam, P., Sirdeshmukh, R.: Multi-omics data integration and mapping of altered kinases to pathways reveal gonadotropin hormone signaling in glioblastoma. OMICS: J. Integr. Biol. **20**(12), 736–746 (2016). https://doi.org/10.1089/omi.2016.0142

14. Kang, M., et al.: Multi-block bipartite graph for integrative genomic analysis. IEEE/ACM Trans. Comput. Biol. Bioinform. **14**(6), 1350–1358 (2017). https://doi.org/10.1109/TCBB.2016.2591521

15. Kim, D., et al.: Using knowledge-driven genomic interactions for multi-omics data analysis: metadimensional models for predicting clinical outcomes in ovarian carcinoma. J. Am. Med. Inform. Assoc. **24**(3), 577–587 (2017). https://doi.org/10.1093/jamia/ocw165

16. Kristensen, V.N., et al.: Principles and methods of integrative genomic analyses in cancer. Nat. Rev. Cancer **14**, 299–313 (2014). https://doi.org/10.1038/nrc3721

17. Lyu, G., et al.: Genome and epigenome analysis of monozygotic twins discordant for congenital heart disease. BMC Genomics **19**(1), 428 (2018). https://doi.org/10.1186/s12864-018-4814-7

18. Reimand, J., et al.: Pathway enrichment analysis and visualization of omics data using g: Profiler, GSEA, Cytoscape and EnrichmentMap. Nat. Protoc. **14**(2), 482–517 (2019). https://doi.org/10.1038/s41596-018-0103-9

19. Ruben, M.C.: BayesOpt: a Bayesian optimization library for nonlinear optimization, experimental design and bandits. J. Mach. Learn. Res. **15**, 3915–3919 (2014). http://jmlr.org/papers/v15/martinezcantin14a.html

20. Simon, N., Friedman, J., Hastie, T., Tibshirani, R.: Regularization paths for cox's proportional hazards model via coordinate descent. J. Stat. Softw. **39**(5), 1–13 (2011). https://doi.org/10.18637/jss.v039.i05

21. Subramanian, A., et al.: Gene set enrichment analysis: a knowledge-based approach for interpreting genome-wide expression profiles. Proc. Natl. Acad. Sci. **102**(43), 15545–15550 (2005). https://doi.org/10.1073/pnas.0506580102

22. Ulrich, T.A., de Juan Pardo, E.M., Kumar, S.: The mechanical rigidity of the extracellular matrix regulates the structure, motility, and proliferation of glioma cells. Cancer Res. **69**(10), 4167–4174 (2009). https://doi.org/10.1158/0008-5472.CAN-08-4859

23. Wagner, J.R., et al.: The relationship between DNA methylation, genetic and expression inter-individual variation in untransformed human fibroblasts. Genome Biol. **15**(2), R37 (2014). https://doi.org/10.1186/gb-2014-15-2-r37

24. Yousefi, S., et al.: Predicting clinical outcomes from large scale cancer genomic profiles with deep survival models. Sci. Rep. **7**(1), 11707 (2017). https://doi.org/10.1038/s41598-017-11817-6

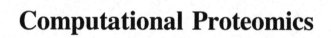

Computational Proteomics

Identifying Human Essential Genes by Network Embedding Protein-Protein Interaction Network

Wei Dai[1], Qi Chang[1], Wei Peng[1,2(✉)], Jiancheng Zhong[3],
and Yongjiang Li[2]

[1] Faculty of Information Engineering and Automation,
Kunming University of Science and Technology, Kunming 650050, China
dw@cnlab.net, changqil994@163.com
[2] Computer Technology Application Key Lab of Yunnan Province,
Kunming University of Science and Technology, Kunming 650050, China
weipeng1980@gmail.com, yongjiang318@gmail.com
[3] College of Engineering and Design, Hunan Normal University,
Changsha 410081, China
jczhongcs@gmail.com

Abstract. Essential genes play an indispensable role in cell viability and fertility. Identifying human essential genes helps us to study the functions of human genes, but also provides a way for finding potential targets for cancer and other diseases. Recently, with the publishing of human essential gene data and the availability of a large amount of biological data, some computational methods have been proposed to predict human essential genes based on genes' DNA sequence or their topological properties in the protein-protein interaction (PPI) network. However, there is still some room to improve the prediction accuracy. In this work, we propose a novel supervised method to predict human essential genes by network embedding protein-protein interaction network. Our method extracts the features of the genes in network by mapping them to a latent space of features that maximally preserves the relationships between the genes and their network neighborhoods. After that, the features are input into a SVM classifier to predict human essential genes. Two human PPI networks are employed to evaluate the effectiveness of our method. The prediction results show that our method outperforms the method that only uses genes' sequence information, but also is obviously superior to the method utilizing genes' centrality properties in the network as input features.

Keywords: Human essential genes · Protein-protein interaction network · Network embedding · Feature representation

1 Introduction

A gene can be defined as essential gene if it plays indispensable role in cell viability and fertility [1]. Studying essential genes of prokaryotic or simple eukaryotic organisms closely associates with the emerging science of synthetic biology, which aims to create a cell with minimal genome. Detecting essential genes of bacterial cell also

© Springer Nature Switzerland AG 2019
Z. Cai et al. (Eds.): ISBRA 2019, LNBI 11490, pp. 127–137, 2019.
https://doi.org/10.1007/978-3-030-20242-2_11

points the way to new approaches for finding potential drug targets for new antibiotics [2]. Recently, some studies have pointed out that the essential genes have closed association with human diseases [3]. The researchers are eager to identify essential genes of human to seek guidance for treating diseases. Some systematic approaches to identifying the essential genes of an eukaryote cell, such as single-gene knockouts [4], conditional knockouts [5], and RNA interference [6] are pioneered in yeast and worm. However, those systematic genetic screens are hard to be applied on human cells. Recently, with advances in gene editing enabled by the CRISPR-Cas system, three research teams have identified about 2000 essential genes in human cancer cell lines by using CRISPER screen and gene trap technology [7–9]. Their studies give a clear definition of human essential genes that are required for the viability of individual human cell types. These data not only present an opportunity to further comprehensively study the functions of human genes, but also provides potential targets for cancer and other diseases.

Designing an effective computational method to identify essential genes from nonessential genes has always been one of the hottest fields in bioinformatics, which aims to provide guidance and proof for biological experiments. In the past few years, some computational methods have been successfully applied on some simple eukaryotic organisms, such as S.cerevisiae [10], C. elegans [11] and A.thaliana [12], Generally, these computational methods can be classified into two categories, unsupervised and supervised methods [13].

Unsupervised approaches make use of genes' features to assign a ranking score for each gene, and the genes ranked in top list are regarded as candidate essential genes. One of the most important features related to the essentiality of genes are their topological properties in biological networks [14]. It has been found that in some species, such as S.cerevisiae (yeast), C. elegans (worm), and D.melanogaster (fruit fly), the proteins (genes) with larger number of interactive partners in protein-protein interaction (PPI) network are tended to be essential proteins (genes), because removing them from the networks will cause the breakdown of the network. Therefore some centrality methods have been proposed for identifying essential genes, such as Betweenness Centrality (BC) [15], Closeness Centrality (CC) [16], Degree Centrality (DC) [17], Eigenvector Centrality (EC) [18], Information Centrality (IC) [19], Edge Clustering Coefficient Centrality (NC) [20] and Subgraph Centrality (SC) [21]. However, the performance of these centrality-based methods is limited to the incomplete and errorprone PPI data currently available. Therefore, recently, some researchers have tried to improve the prediction accuracy of essential genes by integrating PPI networks with other biological information, such as gene functional annotation data [22], gene expression data (PeC [23], WDC [24]), subcellular information [25], protein domain information [25], orthologous information [26].

Supervised approaches usually identify essential genes by inputting some features to train a classifier, i.e. SVM, Random forest, and then employ the classifier to predict essential genes of the same organism or the other organisms. For example, Chen et al. [10] trained a SVM classifier by the features of some known yeast essential proteins, including the combination of phyletic retention, protein evolutionary rate, paralogy, protein size, and the degree centrality in PPI networks and gene-expression networks. After that, the classifier is adopted to predict other potential yeast essential genes.

Recently, Zhong et al. [27] developed a xgboost based classifier to integrate multiple topological properties of essential genes in PPI network to predict yeast and E.coli essential genes. However, lacking gold dataset of human essential genes hinders us to apply the supervised methods on predicting essential genes in human cells. With the publishing of human essential gene data, Guo et al. [28] proposed a supervised method to predict human essential genes by inputting the genes' nucleotide sequence into a SVM classifier. Their work inspires us to design a more effective method to improve the prediction accuracy of human essential genes.

In this work, we propose a novel supervised method to predict human essential genes by network embedding protein-protein interaction network. As far as we know, genes and their protein products work together to perform their functions in cell. Their topological properties in PPI network are very powerful in essentiality prediction. To fully taking advantage of genes' properties in PPI network, we adopt a method that was originally proposed for natural language processing [29] to represent the features of genes in PPI network, and then input these features into different types of classifiers to predict human essential genes. Compared with previous supervised methods, our method extracts the features of the genes in network by mapping them to a latent space of features that maximally preserves the relationships between the nodes and their network neighborhoods. Our method was applied on predicting human essential genes. The prediction results show that our method outperforms Guo's method [28] that only uses genes' sequence information, but also is obviously superior to the methods using genes' centrality properties in the network as input features.

2 Methods

As Fig. 1 shown, our method mainly consists of two steps to predict human essential genes. The first step is to learn feature representation of every node in PPI network. The second step is to put these feature representations (also called feature vectors) into a classifier to predict human essential genes.

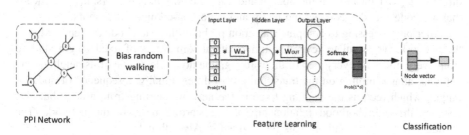

Fig. 1. An overview of framework for identifying essential genes.

2.1 Network Feature Learning

Inspired by node2vec model [29], our method takes two steps to learn the feature representation of every node in PPI network, meanwhile preserving their neighborhood relationships. At first, a bias random walk is implemented on the PPI network starting from each node in the network, and then a sequence of nodes is extracted, which represents the network context of the start node. After that, these sequences are put into a Skip-gram-based architecture to learn the feature representation of each node.

2.1.1 Bias Random Walking

Feature learning methods have been originally developed in the context of natural language process, where the input data is a sequence of words with neighborhood relationship. However, the input of this work is a network. Hence, the first step of feature learning is to use a bias random walk method to extract the linear neighborhood relationship of each node in the network.

A PPI network can be represented as an undirected graph $G = (V, E, W)$, where a node $v \in V$ represents a gene and an edge $e(u,v) \in E$ denotes an interaction between two genes v and u. $w(u,v)$ denotes the weight of edge $e(u,v)$. In this work, the weights of all edges are set to 1. At random walking, the nodes in the network travel to their neighbors at certain probability. Considering the differences in the neighbors of a node, a bias random walk is implemented. That is, given a node v, it walks to one of its neighbors x by evaluating the transition probabilities $\pi(v, x)$ on $e(v, x)$ leading from v. We set the transition probabilities $\pi(v, x) = \beta(v, x) * w(v, x)$.

$$\beta(v,x) = \begin{cases} \frac{1}{p}, & d_{vx} = 0 \\ 1, & d_{vx} = 1 \\ \frac{1}{q}, & d_{vx} = 2 \end{cases} \tag{1}$$

Where d_{vx} represents the distance from the previous node v to the next node x. The return hyper-parameter p controls the probability of jumping back to the previous node v. the in-out hyper-parameter q controls the probability of jumping to the next node x without common neighborhoods. Specially, $d_{vx} = 0$ means node v and x connect directly. $d_{vx} = 1$ means that the node v and x have common neighbors. $d_{vx} = 2$ means that the node v and x connect indirectly and have non-common neighbors.

After preprocessing to compute transition probabilities each node and each edge in the network, we simulate several numbers of random walks of fixed length starting from every node. At every step of the walking, one note selects one of its neighbors by applying alias sampling on the transition probabilities. Finally, a sequence of nodes is output, which records the walking trace in the network starting from a node but also reserves the neighborhood relationship of the starting node in the network. The pseudocode for Bias random walking is given in Algorithm 1.

Algorithm 1 Bias random walking

Input G=(V,E,W), walk_length, Num_walks, Parameters p and q;
Output node lists:walks
π =PreprocessTransitionProbabilities(G,p,q)// computing transition probabilities for each node and each edge
G'=(V,E, π)
Initialize walks to Empty
for walk_iter = 1 to Num_walks do
  for all nodes u∈ V do
  walk=[u]
  while len(walk) < walk_length:
   curr = walk[-1]
   V_{cur_nbrs} = sort(GetNeighbors(curr,G')) // getting neighbor list of current node
   next = AliasSample(V_{cur_nbrs} , π) //applying alias sampling on the transition probabilities to select a neighbor node
   Append next to walk
  Append walk to walks
return walks

2.1.2 Feature Learning Methods

In this work, a skip-gram with negative sampling (SGNS) architecture is employed to learn the feature representation for every node in the network, which was introduced by Mikolov et al. in [30] and was also called word2vec method. In the rest of this section, we provide a brief overview of the SGNS method.

Given a sequence of nodes $\{v_i^m\}$ from a finite node set $\{V = v_i^k\}$, here, m is the number of nodes in the node sequence and k is the number of nodes in the training set, let f: $v_j \in V -> R^d$ be the mapping function from node vector to feature representations. Here d is the number of dimensions of the feature representation. For every node in the input sequence of nodes, $v_j \in V$ and $v_j \in N(v_i)$ denote the neighborhood node of v_i. The aim of the SGNS is to maximize the log-probability of neighborhood nodes $N(v_i)$ surrounding a node v_i conditioned on its feature representation, given by f. Formally, the objective function is following.

$$\max_f \sum_{v_i \in V} \log P(N(v_i)|f(v_i)) \tag{2}$$

We assume that the possibility of observing a neighborhood node is independent of other neighborhood nodes given the feature representation of the source v_i.

$$P(N(v_i)|f(v_i)) = \prod_{v_j \in N(v_i)} P(v_j|f(v_i)) \tag{3}$$

Where $P(v_i|f(v_j))$ is the softmax function.

$$P(v_j|f(v_i)) = \frac{\exp(f(v_j) \cdot f(v_i))}{\sum_{v_s \in V} \exp(f(v_s) \cdot f(v_i))} \tag{4}$$

SGNS architecture is shown in Fig. 1. Firstly, the SGNS counts the number of distinct nodes in the training set. After that, the SGNS represents an input node sequence as a one-hot vector with k components. Only one position where the node appears in the vector is placed in 1 and the others are placed in 0 s. The output of the architecture is also a single vector with k components, which represents the probability that a node in the training set should be the neighbor of the input node. To get the neighborhood nodes of an input node, we slide a window along the node sequence. The goal of the training is to learn the weights of the hidden layer and to use SGD (Stochastic Gradient Descent) to optimize the weight matrix. The final feature representation of an input node is generated by multiplying the input 1 * k vector and a k * d weight matrix. Note that computing $\sum \exp (f(v_s) \cdot f(v_i))$ in Eq. 4 is too expensive for the large network during training. We alleviated the computational problem by using negative sampling for all our experiments.

2.2   Classification

After learning the feature represent of every node in the network, we put these feature vectors into a classifier to predict essential genes. These node features are randomly divided into testing samples and training samples. There are various classifiers available to finish the prediction task. In this work, we focus on proofing that the genes' features maximally preserving their neighborhood relationships in the network are helpful to predict human essential genes. We adopt one of following popular classifiers to predict human essential genes. They are support vector machine (SVM) [10], random forest (RF) [31], decision tree (DT) [32], extra tree (ET), logistic regression (LR) [33], Naive Bayes (NB) [34] and k-Nearest Neighbor (KNN) [35].

3   Results

3.1   Datasets

The human essential genes were from the supplementary files of the reference [28], which was extracted from the DEG database (http://tubic.tju.edu.cn/deg/), the updated version of which contained human gene essentiality information from three recent works [7–9]. The dataset consisted of 1516 essential genes and 10499 non-essential genes.

Two human PPI network datasets were employed to evaluate the performance of our method. The one was from a human functional interaction (FI) network, including protein-protein interactions, gene coexpression data, protein domain interaction, Gene Ontology (GO) annotations and text-mined protein interactions from Reactome database [36]. The network contained 12,277 genes and 230,243 interactions. We compared all the genes in FI network with the genes from reference [28] and only kept the overlapping genes of the two datasets. Ultimately, 6747 genes were obtained as one of our benchmark dataset (called FIs), including 1,359 essential genes and 5,388 non-essential genes. Another human network data was from InBio Map database, which was aggregated from 8 source databases and spanning 87% of reviewed human UniProt IDs [37]. The network contained 17530 genes and 625641 interactions. Similarly, we obtained 10548 genes overlapping with the genes from reference [28] as another our benchmark dataset (called InWeb_IM), including 1512 essential genes and 9036 non-essential genes.

3.2   Evaluation Metrics

To evaluate the performance of our method in predicting human essential genes, five-fold cross-validation test was performed, where the benchmark dataset was randomly divided into 5 parts, the four parts was training set and the other one was testing set. We kept the ratio of essential genes to non-essential genes as 1:1 in each fold data. Some popular statistic metrics were adopted to evaluate the prediction performance of each method, including ACC, Precision, Recall, SP, NPV, F-measure and MCC (Matthews Correlation Coefficient). To evaluation the overall performance of each method, the Area Under ROC Curve (AUC) and the Area Under Precision-Recall Curve (AP) are calculated for comparison.

3.3   Comparison with Existing Methods

To evaluate the performance of our method in predicting human essential genes, we compared it with other two existing methods. The one namely Z curve method employed a λ-interval Z curve method to extract DNA sequence features of genes and input these features into a SVM classifier to make prediction [28]. The other namely centrality-based method calculated 7 popular central indices of DC, BC, CC, SC, EC, IC and NC for each gene in human network and combined them into a 7-column vector as input to a SVM classifier. All these indices were calculated by a Cytoscape plugin CytoNCA [38]. For fair comparison, our method also chose a SVM classifier for prediction. Tables 1 and 2 show the prediction performance of all comparing methods on two different human network datasets, including FIs dataset and InWeb_IM dataset. The digits in the brackets at the first column of the tables are the ratio between essential genes and non-essential genes in the course of validation.

Table 1. Performance comparison of existing methods on FIs dataset.

Methods	Precision	Recall	SP	NPV	F-measure	MCC	Accuracy	AUC	AP
Centrality (1:1)	0.852	0.553	**0.904**	0.669	0.671	0.488	0.728	0.760	0.797
Z curve (1:1)	0.733	0.800	0.708	0.780	0.765	0.511	0.754	0.824	0.783
Our method (1:1)	**0.859**	**0.836**	0.863	**0.840**	**0.847**	**0.699**	**0.849**	**0.914**	**0.902**

Table 2. Performance comparison of existing methods on InWeb_IM dataset.

Methods	Precision	Recall	SP	NPV	F-measure	MCC	Accuracy	AUC	AP
Centrality (1:1)	0.816	0.713	0.839	0.745	0.761	0.557	0.776	0.851	0.841
Z curve (1:1)	0.730	0.777	0.713	0.762	0.753	0.491	0.745	0.827	0.801
Our method (1:1)	**0.845**	**0.850**	**0.844**	**0.849**	**0.848**	**0.694**	**0.847**	**0.914**	**0.903**

As Tables 1 and 2 shown, our method achieved the best performance among all comparing methods on the two datasets. When the ratio of the two kinds of genes in the training or testing sets was as 1:1, The F-measure, MCC, AUC and AP values of our method reached up to 0.847, 0.699, 0.914 and 0.902 on FIs dataset and reached up to 0.848, 0.694 and 0.914 and 0.903 on InWeb_IM dataset. However, the highest F-measure, MCC and AUC values of the other two methods are 0.765, 0.511 and 0.824 for Z curve method on FIs dataset, and 0.761, 0.557 and 0.851 for Centrality method on InWeb_IM dataset. Note that the Centrality method has higher AP values on the two datasets than the Z curve method, i.e. 0.797 on FIs dataset and 0.841 on InWeb_IM dataset. Hence, compared with the Z-curve method that used human sequence data and the Centrality method that used topological properties in the PPI network, our method learning the genes' latent features from PPI network can effectively improve the accuracy of predicting human essential genes.

3.4   Evaluation of Impact of Different Classifiers

In order to further test the impact of different classifiers on the prediction of our method, besides SVM, other popular classifiers, such as random forest (IR), decision tree (DT), extra tree (ET), logistic regression (LR), Naive Bayes (NB) and k-Nearest Neighbor (KNN), were selected to make prediction in our method. Tables 3 and 4 list the performance comparisons of our method on two different datasets with respect to different classifiers.

The results show that the performance of our method can be further improved by selecting other more efficient classifiers. On the whole, the extra tree algorithm (ET) based on random forest has the best performance. As maintaining the essential/non-essential gene ratio as 1:1, the F-measure, MCC, AUC and AP values of our method by using ET classifier achieved 0.853, 0.712, 0.923, 0.922 on FIs dataset and 0.856, 0.707, 0.926 and 0.920 on InWeb_IM dataset.

Table 3. Performance comparison for our method with different classifiers on FIs dataset.

Methods	Precision	Recall	SP	NPV	F-measure	MCC	Accuracy	AUC	AP
DT(1:1)	0.768	0.788	0.762	0.783	0.778	0.551	0.775	0.775	0.831
NB(1:1)	0.822	0.765	0.834	0.780	0.792	0.600	0.799	0.876	0.873
KNN(1:1)	0.837	0.805	0.844	0.812	0.821	0.649	0.824	0.895	0.906
LR(1:1)	0.839	0.827	0.841	0.829	0.833	0.668	0.834	0.910	0.907
SVM(1:1)	0.859	0.836	0.863	0.840	0.847	0.699	0.849	0.914	0.902
RF(1:1)	0.859	**0.844**	0.861	**0.846**	0.851	0.705	0.852	0.921	0.921
ET(1:1)	**0.867**	0.840	**0.872**	0.845	**0.853**	**0.712**	**0.856**	**0.923**	**0.922**

Table 4. Performance comparison for our method with different classifiers on InWeb_IM dataset.

Methods	Precision	Recall	SP	NPV	F-measure	MCC	Accuracy	AUC	AP
DT(1:1)	0.790	0.774	0.794	0.778	0.782	0.568	0.784	0.784	0.838
NB(1:1)	**0.852**	0.789	**0.863**	0.803	0.819	0.653	0.826	0.901	0.898
KNN(1:1)	0.850	0.796	0.860	0.808	0.822	0.657	0.828	0.894	0.897
LR(1:1)	0.843	0.846	0.842	0.846	0.845	0.689	0.844	0.914	0.902
SVM(1:1)	0.845	0.850	0.844	0.849	0.848	0.694	0.847	0.917	0.903
RF(1:1)	0.827	**0.884**	0.815	**0.876**	0.855	0.701	0.850	0.923	0.914
ET(1:1)	0.839	0.874	0.832	0.868	**0.856**	**0.707**	**0.853**	**0.926**	**0.920**

4 Conclusion

In this work, we proposed a novel supervised method to predict human essential gene by network embedding human PPI network. Compared with the previous method, our method firstly represented the nodes in the PPI network as latent features vectors that maximally preserved the relationships between the nodes and their network neighbors and then input the feature vectors to a classifier to predict potential human essential genes. The prediction was implemented on two different human PPI datasets. The results show that the highest AUC values of our method on the two datasets reached 0.923 and 0.926 by selecting ET classifier on the proportion of original data. Even taking the same SVM classifier as Guo's method [28], our method also obviously outperformed the methods that took DNA sequences or network topological properties as input features, which verified that learning feature vectors for the nodes from the PPI network can make a great contribution to improving the prediction of the human essential genes. In our future work, more powerful network embedding method should be designed to find the latent feature representation of the nodes in the PPI network and more effective machine learning methods should be developed to predict human essential genes based on the features of genes.

Acknowledgment. This work is supported in part by the National Natural Science Foundation of China under grant No. 31560317, No. 61502214, No. 61472133, No. 61502166, No. 61702122 and No. 81560221. Natural Science Foundation of Yunnan Province of China (No. 2016FB107).

References

1. Zhang, R., Lin, Y.: DEG 5.0, a database of essential genes in both prokaryotes and eukaryotes. Nucleic Acids Res. **37**(Database issue), D455–D458 (2009)
2. Clatworthy, A.E., Pierson, E., Hung, D.T.: Targeting virulence: a new paradigm for antimicrobial therapy. Nat. Chem. Biol. **3**(9), 541–548 (2007)
3. Furney, S., Alba, M.M., Lopez-Bigas, N.: Differences in the evolutionary history of disease genes affected by dominant or recessive mutations. BMC Genom. **7**(1), 165 (2006)

4. Giaever, G., et al.: Functional profiling of the Saccharomyces cerevisiae genome. Nature **418**, 6869 (2002)
5. Roemer, T.J.B., et al.: Large-scale essential gene identification in Candida albicans and applications to antifungal drug discovery. Mol. Microbiol. **50**(1), 167–181 (2010)
6. Cullen, L.M., Arndt, G.M.: Genome-wide screening for gene function using RNAi in mammalian cells. Immunol. Cell Biol. **83**(3), 217–223 (2005)
7. Fraser, A.: Essential human genes. Cell Syst. **1**(6), 381–382 (2015)
8. Hart, T., et al.: High-Resolution CRISPR Screens Reveal Fitness Genes and Genotype-Specific Cancer Liabilities. Cell **163**(6), 1515–1526 (2015)
9. Wang, T., et al.: Identification and characterization of essential genes in the human genome. Science **350**(6264), 1096–1101 (2015)
10. Chen, Y., Xu, D.: Understanding protein dispensability through machine-learning analysis of high-throughput data. Bioinformatics **21**(5), 575–581 (2005)
11. Yuan, Y., et al.: Predicting the lethal phenotype of the knockout mouse by integrating comprehensive genomic data. Bioinformatics **28**(9), 1246–1252 (2012)
12. Lloyd, J.P., et al.: Characteristics of plant essential genes allow for within- and between-species prediction of lethal mutant phenotypes. Plant Cell **27**(8), 2133 (2015)
13. Wang, J., Peng, W., Wu, F.X.: Computational approaches to predicting essential proteins: a survey. PROTEOMICS-Clin. Appl. **7**(1–2), 181–192 (2013)
14. Jeong, H., et al.: Lethality and centrality in protein networks. Nature **411**(6833), 41–42 (2001)
15. Joy, M.P., et al.: High-betweenness proteins in the yeast protein interaction network. J. Biomed. Biotechnol. **2005**(2), 96–103 (2005)
16. Wuchty, S., Stadler, P.F.: Centers of complex networks. J. Theor. Biol. **223**(1), 45–53 (2003)
17. Vallabhajosyula, R.R., et al.: Identifying hubs in protein interaction networks. PLoS ONE **4**(4), e5344 (2009)
18. Bonacich, P.: Power and centrality: a family of measures. Am. J. Sociol. **92**(5), 1170–1182 (1987)
19. Stephenson, K., Zelen, M.: Rethinking centrality: methods and examples. Soc. Netw. **11**(1), 1–37 (1989)
20. Wang, J., et al.: Identification of essential proteins based on edge clustering coefficient. IEEE/ACM Trans. Comput. Biol. Bioinform. **9**(4), 1070–1080 (2012)
21. Ernesto, E., Rodríguez-Velázquez, J.A.: Subgraph centrality in complex networks. Phys. Rev. E Stat. Nonlinear Soft Matter Phys. **71**(5 Pt 2), 056103 (2005)
22. Li, M., et al.: Essential proteins discovery from weighted protein interaction networks. Bioinform. Res. Appl. Proc. **6053**, 89–100 (2010)
23. Li, M., et al.: A new essential protein discovery method based on the integration of protein-protein interaction and gene expression data. BMC Syst. Biol. **6**(1), 15 (2012)
24. Tang, X., et al.: Predicting essential proteins based on weighted degree centrality. IEEE/ACM Trans. Comput. Biol. Bioinform. (TCBB) **11**(2), 407–418 (2014)
25. Peng, W., et al.: UDoNC: an algorithm for identifying essential proteins based on protein domains and protein-protein interaction networks. IEEE/ACM Trans. Comput. Biol. Bioinform. (TCBB) **12**(2), 276–288 (2015)
26. Peng, W., et al.: Iteration method for predicting essential proteins based on orthology and protein-protein interaction networks. BMC Syst. Biol. **6**(1), 87 (2012)
27. Zhong, J., et al.: XGBFEMF: an XGBoost-based framework for essential protein prediction. IEEE Trans. Nanobioscience **17**(3), 243–250 (2018)
28. Guo, F.B., et al.: Accurate prediction of human essential genes using only nucleotide composition and association information. Bioinformatics **33**(12), 1758–1764 (2017)

29. Grover, A., Leskovec, J.: node2vec: scalable feature learning for networks. In: KDD, pp. 855–864 (2016)
30. Mikolov, T., et al.: Distributed representations of words and phrases and their compositionality. In: International Conference on Neural Information Processing Systems (2013)
31. Wu, J., et al.: WDL-RF: predicting bioactivities of ligand molecules acting with G protein-coupled receptors by combining weighted deep learning and random forest. Bioinformatics **34**(13), 2271–2282 (2018)
32. Acencio, M.L., Lemke, N.: Towards the prediction of essential genes by integration of network topology, cellular localization and biological process information. BMC Bioinformatics **10**, 290 (2009)
33. Liao, J., Chin, K.: Logistic regression for disease classification using microarray data: model selection in a large p and small n case. Bioinformatics **23**(15), 1945–1951 (2007)
34. Cheng, J., et al.: Training set selection for the prediction of essential genes. PLoS ONE **9**(1), e86805 (2014)
35. Kuo-Chen, C., Hong-Bin, S.: Predicting eukaryotic protein subcellular location by fusing optimized evidence-theoretic K-nearest neighbor classifiers. J. Proteome Res. **5**(8), 1888–1897 (2006)
36. Wu, G., Feng, X., Stein, L.: A human functional protein interaction network and its application to cancer data analysis. Genome Biol. **11**(5), 1–23 (2010)
37. Li, T., et al.: A scored human protein-protein interaction network to catalyze genomic interpretation. Nat. Methods **14**(1), 61 (2016)
38. Tang, Y., et al.: CytoNCA: a cytoscape plugin for centrality analysis and evaluation of protein interaction networks. Biosystems **127**, 67–72 (2015)

Automated Hub-Protein Detection via a New Fused Similarity Measure-Based Multi-objective Clustering Framework

Sudipta Acharya[1], Laizhong Cui[1(✉)], and Yi Pan[2]

[1] College of Computer Science and Software Engineering, Shenzhen University, Shenzhen, People's Republic of China
sudiptaszu@outlook.com, cuilz@szu.edu.cn
[2] Department of Computer Science, Georgia State University, Atlanta, USA
yipan@gsu.edu

Abstract. In the field of computational biology and bioinformatics, there have been limited studies on the development of protein-protein proximity measures which blend multiple sources of biological properties of protein. In Protein-Protein Interaction Network (PPIN), hub-proteins play a central role. There are many literature with user-studied different degree cut-offs for defining hub-proteins. Therefore, there is a need for a standard method for identifying hub-proteins without manually determining the degree cut-off. In the current research article, an effort has been made towards addressing both problems. At first, we have proposed a new Fused protein-protein Similarity measure - *FuSim*, which involves biological properties of both Gene Ontology (GO) and PPIN. Later, utilizing the proposed similarity measure, a multi-objective clustering algorithm-based automated hub-protein detection framework is developed.

Keywords: Protein-protein interaction network (PPIN) ·
Gene Ontology (GO) · Multi-objective optimization · Clustering ·
Hub proteins · Protein-protein similarity measure

1 Introduction

In PPIN, hub-proteins are small number of highly connected protein nodes which play a central role. Essential proteins plays fundamental role in survival and reproduction of an organism. According to existing literature, most of the essential proteins are found to be hubs instead of non-hubs protein [8]. Thus, identifying and understanding hub proteins is an open problem in the field of bioinformatics. Different literature follows different degree-thresholds and different conventions in defining hub-proteins ([5,7]). Hence, there is a need for a standard method to identify hub proteins from PPIN with limited manual intervention. Motivated by this fact, an effort has been made in the current paper towards determining the number of hub-proteins automatically and identify them from PPIN.

© Springer Nature Switzerland AG 2019
Z. Cai et al. (Eds.): ISBRA 2019, LNBI 11490, pp. 138–145, 2019.
https://doi.org/10.1007/978-3-030-20242-2_12

Measuring functional similarity between genes/proteins is crucial as it is the building block to further analyze the biological, molecular, cellular functionalities of genes/proteins. In literature, different gene-gene/protein-protein proximity measures were developed using various biological resources [1,2,12,16]. Genes associated with similar biological, molecular or cellular functions tend to form groups (clusters/bi-clusters) [1,2] and hence their protein products also tend to interact with each other [16]. Therefore, one favourite way to determine the functional similarity between genes/proteins among researchers is through studying interaction edges between proteins in a PPIN [12,16].

PPIN stores biological information regarding different interacting proteins and confidence of interaction. Another potential database for storing knowledge about several molecular, biological and cellular processes and sub-processes for which genes/proteins are involved is Gene Ontology (GO)[1]. It is a large ontology consists of three child ontologies like biological process (BP), molecular function (MF) and cellular component (CC). Each node of the GO tree represents one particular biological or cellular or molecular process/sub-process; which are called *GO-term*. Considering the gene annotation information as well as structural properties of GO, in the past several gene-gene semantic similarity measures have been proposed [2,10,14]. But none of them takes both GO and PPIN into the account to measure functional relatedness between genes.

In this work, in the first phase, we have proposed an integrated protein-protein similarity measure exploring both GO and PPIN. In the second phase, we have employed our proposed proximity measure to identify hub proteins automatically from PPIN through a widely used pattern recognition tool - *Clustering* [1,3]. The utility of the proposed framework has been measured through proper validity indices.

2   Relevant Existing Works and Motivation

In the past, several semantic similarity measures between *GO-terms* or genes or proteins have been proposed utilizing GO. Some of them adopted information theoretic approaches like Resnik's similarity [14], Lin's similarity [10]. Authors Wang et al. [17] proposed a similarity measure, which considers topological information of the GO graph. One GO annotation set-based method to measure the similarity between genes was proposed by [13], which is named as normalized term-overlap method. Several hybrid similarity measure, for example in Shen et al. [15], authors proposed a similarity measure that takes into account both the path length between the terms as well as the information content of the ancestor terms. Recently in [2], authors have developed some multi-factored gene-gene similarity/distance measures by considering several mutually exclusive GO and *GO-term* properties to identify functionally as well as semantically related genes from a genome. In [6], authors have proposed a protein-protein semantic similarity measure by combining similarity scores of the *GO-terms* associated with the proteins. Apart from GO-based similarity measures, many works have been

---

[1] http://geneontology.org/.

done on finding co-expressed genes or genes which are functionally similar based on their expression levels [1]. Also, there are few research articles on finding functional similarity between genes/proteins based on interaction data in PPIN. In [16], authors have proposed a method to develop gene-gene functional similarity network based on both traditional GO-based similarity measures and PPIN. Article [12] proposed a diseased-gene selection technique - termed as RelSim based on information from gene expression profile and PPIN. According to past literature, several researchers have proposed different strategies to detect hub proteins from PPIN. For example, in [5], the top 95% and 50% of the high degree nodes were defined as hubs in two different contexts; in [7], nodes with degree greater than 5 were labelled as hubs. After performing a thorough literature survey, we found no existing works considering both GO and PPIN in designing gene-gene/protein-protein similarity measure [2,13,16]. Also, a limited study has been performed in the direction of detecting hub proteins automatically from a PPIN. The underlying idea of our proposed automated hub-detection technique is based on the ground principles of any clustering algorithm [1,3]. For any gene clustering technique [1], each cluster has a center gene (cluster center), which has the highest average similarity with other genes of the same cluster but the lowest average similarity with other cluster center genes. This property is applicable for proteins in PPIN too. The hub proteins are highly connected to a set of non-hub proteins but less connected with other hub proteins. It has also been shown in existing research [12] that genes having more functional similarity have more interacting proteins. Inversely, two proteins with less interaction indicate that they and their corresponding genes are less functionally similar. Therefore, we have performed a multi-objective clustering on a set of proteins utilizing our proposed *FuSim* measure as underlying proximity measure. In the resulting clusters, center protein in a cluster is most functionally and semantically similar hence more interacting to other proteins of the same cluster and least functionally similar therefore less interacting to other center proteins. So, obtained cluster centers in our approach follow the property of hub-proteins in a PPIN. As underlying optimization strategy of used clustering algorithm, AMOSA (Archived Multi Objective Simulated Annealing) [4] is utilized which has shown its superiority over several other multi-objective as well as single objective optimization techniques in existing literature [1,3,4]. One advantage of AMOSA-based clustering is, unlike K-means or hierarchical clustering, the number of clusters in a solution is determined automatically within a given range. Hence, in our proposed hub-detection framework, the number of hubs (cluster center proteins) also gets determined and hence identified automatically.

3   Proposed Fused Protein-Protein Similarity Measure: *FuSim*

Our proposed hybrid protein-protein similarity measure includes important factors from both GO and PPIN. Two factors which participate in forming the proposed measure are as follows.

1. Multi-factored protein-protein semantic similarity based on GO [2]
2. Functional similarity between proteins based on the confidence of association in PPIN [12].

For a particular organism, let p_i and p_j represents two proteins. A_i and A_j represent sets of annotated *GO-terms* by p_i and p_j respectively. According to definition of multi-factored semantic measure [2], the multi-factored semantic similarity between two *GO-terms* gt_i and gt_j is as follows.

$$Multi\text{-}sim(gt_i, gt_j) = \frac{arctan[sim_{Lin}(gt_i, gt_j) + sim_{Shen}(gt_i, gt_j) + sim_{norm-struct_{depth}}(gt_i, gt_j)]}{\pi/2} \quad (1)$$

Where, $sim_{Lin}(gt_i, gt_j)$, $sim_{Shen}(gt_i, gt_j)$ and $sim_{norm-struct_{depth}}(gt_i, gt_j)$ are Lin's semantic similarity measure [2], Shen's similarity measure [15] and normalized structure-based semantic similarity [2].

Utilizing the above equation, the multi-factored semantic similarity between protein p_i and p_j is as follows.

$$Multi\text{-}SIM(p_i, p_j) = \frac{\frac{1}{m \times n}\sum_{gt_k \in A_i, gt_p \in A_j} Multi\text{-}sim(gt_k, gt_p) + sim_{NTO}(p_i, p_j)}{2} \quad (2)$$

where, $sim_{NTO}(p_i, p_j)$ is normalized term overlap-based similarity measure [13]. $m = |A_i|$ and $n = |A_j|$. The value of $Multi\text{-}SIM(p_i, p_j) \in [0, 1]$.

Again let, N_i is the set of interactive proteins of protein p_i in corresponding PPIN. w_{ij} is the confidence score or weight value of the interacting edge between protein $p_j \in N_i$ and p_i. Let N_{ij} is the set of proteins which are interactive neighbours of both protein p_i and p_j i.e. $N_{ij} = N_i \cap N_j$. $\tilde{N} = N_i \backslash N_j$, indicates set of proteins which are interactive neighbours of protein p_i but not of protein p_j. The functional similarity between two proteins p_i and p_j based on confidence (here weight) of association in PPIN [12] is defined as follows.

$$PPI\text{-}SIM(p_i, p_j) = \frac{\sum_{p_k \in N_{ij}} min\{w_{ik}, w_{jk}\}}{\sum_{p_k \in \tilde{N}_i} w_{ik} + \sum_{p_k \in N_{ij}} max\{w_{ik}, w_{jk}\} + \sum_{p_k \in \tilde{N}_j} w_{jk}} \quad (3)$$

Value of $PPI\text{-}SIM(p_i, p_j) \in [0, 1]$. Combining Eqs. 2 and 3, our proposed hybrid similarity measure $FuSim(p_i, p_j)$ is defined as follows.

$$FuSim(p_i, p_j) = \frac{Multi\text{-}SIM(p_i, p_j) + PPI\text{-}SIM(p_i, p_j)}{2} \quad (4)$$

where, $FuSim(p_i, p_j) \in [0, 1]$.

4   The Working Methodology of Proposed Automated Hub Detection Technique

Our overall proposed framework can majorly be divided in two phases as follows.

Phase 1: Choose dataset and generate protein-*GO-term* annotation dataset and protein-protein similarity matrix
Different steps followed in this phase are described as follows.

Step 1: Fetching protein annotation information from GO: For the experiment purpose, we have chosen Homo sapience PPI database from HitPredict [11]. The database contains 18,484 unique protein IDs and 2,60,277 number of interactions with confidence score (weight). We obtained annotation information of 18,484 proteins using GO tool - GO Consortium (See footnote 1). Out of 18,484 proteins, 16,196 proteins were mapped to one or more *GO-terms*, and 2289 proteins were unmapped. For further analysis, we have considered mapped proteins only. For our experiment, the total number of significant *GO-terms* obtained is 358 (out of which 243 for BP, 50 for MF and 65 for CC). Also, we have obtained the full GO tree[2] for calculating our proposed similarity measure.

Step 2: Generating protein-*GO-term* annotation matrix and protein-protein similarity matrix: Once the protein annotation information is obtained from the GO tool the corresponding protein-*GO-term* binary annotation matrix [2] is prepared for our chosen dataset. The dimension of obtained annotation matrix is 16,196 × 358, where 16,196 = # of mapped proteins and 358 = # of significant *GO-terms*. To form this matrix we have followed the same strategy as followed in [2]. This matrix is used as an input matrix to our applied multi-objective clustering algorithm.

Another input to our performed clustering algorithm is protein-protein similarity matrix based on which groups of functionally similar proteins are formed. First, we have calculated similarity value between each pair of proteins from the set of 16,196 proteins according to our proposed measure *FuSim* in Eq. 4. Apart from our proposed measure, for comparative analysis purpose, we have also chosen six other GO/PPIN-based similarity measures viz. Lin's measure [10], Shen's measure [15], Mistry's measure [13], $Struct_{depth}$-based measure [2], Multi-factored similarity measure [2] - which are GO-based measures, and confidence (weight) of association-based measure [12]: which is PPIN-based measure. Utilizing all these six existing measures, corresponding protein-protein similarity matrices are developed too. A dataset having p number of proteins have $p \times p$ dimensional protein-protein similarity matrix for each of the proposed and chosen similarity measures.

Phase 2: Identifying hub proteins automatically through multi-objective clustering on protein-*GO-term* annotation dataset
This is the phase of the proposed framework where hub protein detection takes place automatically. Two objective functions are simultaneously optimized in our utilized AMOSA-based clustering viz. Xie-Beni (XB) index [3] (minimized) and PBM index [3] (maximized). The AMOSA-based clustering has been applied here on proteins of prepared protein-*GO-term* annotation matrix from phase 1 utilizing our proposed as well as six of chosen existing similarity measures as mentioned

---

[2] http://purl.obolibrary.org/obo/go/go-basic.obo.

in step 2 of phase 1. Some basic concepts regarding AMOSA [4] to be noted before going through in detail of proposed algorithm are given below.

- 'Archive' in AMOSA $\approx$ 'Population' in genetic algorithm
- 'Archive_element of 'Archive' $\approx$ 'Chromosome' in 'Population' $\approx$ Complete clustering solution.
- Suppose p = # of proteins, GOT = # of $GO\text{-}terms$.
- **Input:** Protein-$GO\text{-}term$ annotation dataset of dimension $p \times GOT$ and protein-protein similarity matrix of dimension $p \times p$.
- **Output:** Set of hub proteins. Let us denote this as P_h.

The Pseudo-code of proposed hub-selection algorithm is shown in Fig. 1.

```
Set T_max, T_min , HL, SL, iter, α, tmp=T_max
/* T_max = Maximum temperature, T_min = Minimum temperature
HL= Hard limit, SL= Soft limit */
Initialization of Archive.
current-sol = random(Archive). /* solution chosen randomly from Archive*/
while (tmp > T_min)
    for (i=0; i< iter; i++)
        new-sol=perturb(current-sol)
        Checking domination status of new-sol and current-sol. current-sol gets updated
    End for
tmp = α * tmp.
End while
if Archive-size > SL
    Cluster Archive to HL number of clusters.
V= max(Sil_1, Sil_2...Sil_l) /* Sil_i = Silhouette index for i^th Archive_element*/
    Sol_best = V. Archive-element
/* V.Archive-element = Archive_element with maximum Silhouette index */
P_h ← Cluster-centers (Sol_best). /* P_h = set of hub proteins */
Validate P_h using GO consortium.
```

Fig. 1. Pseudo-code of proposed automated hub detection algorithm

5 Experimental Results and Discussion

To compare the accuracy of the obtained hub and non-hub sets of proteins we need a reference set which can be used as gold standard/true labels. We have prepared the reference set according to literature [9] where top 10% of highly connected proteins are considered as hub proteins after consulting definitions of hub-proteins in some literature. As high connectivity is the fundamental property of any hub-protein, therefore, this approach to create the reference hub/non-hub protein set seems convincing to us.

Table 1 reports the Silhouette index values of best clustering solution corresponding to each of seven similarity measures (out of which one is *FuSim* and rest six measures are existing). From the reported values of Table 1, we can see that the quality of best-obtained clustering solution by our proposed hybrid similarity measure *FuSim* is better than best clustering solutions obtained by other chosen similarity measures according to Silhouette measure. This result also validates the logic that both GO and PPIN are unavoidable sources to measure the functional similarity between proteins. Below to each Silhouette value, the number of clusters in the corresponding solution is indicated. The number of hub proteins for each measure-based approach is equal to the number of clusters shown in this table, and each center protein of each cluster is treated as hub-protein in this work.

Once the sets of hub/non-hub proteins are identified by our developed framework the next step is to validate obtained set through biological significance test through GO Consortium. We conducted the biological significance test for hub protein set obtained by *FuSim*-based approach (1,494 number of hub proteins) as well as six other similarity measure-based approaches. We found that on average all hub-proteins were annotated with a large number of *GO-terms* compared to non-hub proteins. This is another essential property of hub-protein, i.e. involvement in a large number of biological activities. After performing the biological significance test, we have compared obtained hub/non-hub protein set with reference hub/non-hub protein set developed before with respect to two external validity measures, i.e. ARI and %CA. The obtained ARI and %CA values are reported in Table 2. If we further analyse Table 2, we can see that for *FuSim*-based automated hub detection approach the accuracy of identified protein hub/non-hub sets has improved with respect to both indices compared to other similarity measures. Reported result experimentally supports our argument on utility of proposed GO and PPIN-based measure - *FuSim* over other existing measures which are build upon single biological source like either GO or PPIN.

Table 1. The Silhouette index value corresponding to best obtained clustering solution for AMOSA-based clustering for all of seven similarity measures. The number of clusters K corresponding to the clustering solution is indicated too.

Similarity measure	*FuSim*	Multi-factored [2]	Conf. of asso. [12]	Shen	Lin	Mistry	Struct$_{depth}$
Silhouette value	**0.623**	0.57	0.55	0.553	0.547	0.53	0.51
# clusters (K)	1,494	1,265	1,370	993	997	1005	910

Table 2. Comparison between obtained hub/non-hub protein sets by seven similarity measure-based approaches with reference hub/non-hub protein set with respect to Adjusted Rand Index (ARI) and Classification Accuracy (%CA) values.

Similarity measure	*FuSim*	Multi-factored [2]	Conf. of asso. [12]	Shen	Lin	Mistry	Struct$_{depth}$
ARI	**0.71**	0.67	0.65	0.62	0.623	0.59	0.57
%CA	**95.6%**	92.3%	90.4%	82.3%	81.7%	78.1%	77.6%

In future, one or more mutually exclusive biological factors can be fused with *FuSim* measure to configure it into more refined protein-protein similarity measure. Also, to prove that the obtained results are statistically and biologically significant a thorough statistical/biological significance tests can be performed. Authors are working in that direction.

References

1. Acharya, S., Saha, S.: Importance of proximity measures in clustering of cancer and mirna datasets: proposal of an automated framework. Mol. BioSyst. **12**(11), 3478–3501 (2016)
2. Acharya, S., Saha, S., Pradhan, P.: Multi-factored gene-gene proximity measures exploiting biological knowledge extracted from gene ontology: application in gene clustering. IEEE/ACM Trans. Comput. Biol. Bioinform. (2018). https://doi.org/10.1109/TCBB.2018.2849362
3. Bandyopadhyay, S., Saha, S.: Unsupervised Classification: Similarity Measures, Classical and Metaheuristic Approaches, and Applications. Springer, Heidelberg (2012). https://doi.org/10.1007/978-3-642-32451-2
4. Bandyopadhyay, S., Saha, S., Maulik, U., Deb, K.: A simulated annealing-based multiobjective optimization algorithm: amosa. IEEE Trans. Evol. Comput. **12**(3), 269–283 (2008)
5. Batada, N.N., et al.: Stratus not altocumulus: a new view of the yeast protein interaction network. PLoS Biol. **4**(10), e317 (2006)
6. Dutta, P., Basu, S., Kundu, M.: Assessment of semantic similarity between proteins using information content and topological properties of the gene ontology graph. IEEE/ACM Trans. Comput. Biol. Bioinform. **15**(3), 839–849 (2018)
7. Han, J.D.J., et al.: Evidence for dynamically organized modularity in the yeast protein-protein interaction network. Nature **430**(6995), 88 (2004)
8. He, X., Zhang, J.: Why do hubs tend to be essential in protein networks? PLoS Genet. **2**(6), e88 (2006)
9. Hsing, M., Byler, K.G., Cherkasov, A.: The use of gene ontology terms for predicting highly-connected'hub'nodes in protein-protein interaction networks. BMC Syst. Biol. **2**(1), 80 (2008)
10. Lin, D., et al.: An information-theoretic definition of similarity. In: ICML, vol. 98, pp. 296–304. Citeseer (1998)
11. López, Y., Nakai, K., Patil, A.: Hitpredict version 4: comprehensive reliability scoring of physical protein–protein interactions from more than 100 species. Database (2015)
12. Maji, P., Shah, E., Paul, S.: Relsim: an integrated method to identify disease genes using gene expression profiles and PPIN based similarity measure. Inf. Sci. **384**, 110–125 (2017)
13. Mistry, M., Pavlidis, P.: Gene ontology term overlap as a measure of gene functional similarity. BMC Bioinform. **9**(1), 327 (2008)
14. Resnik, P.: Using information content to evaluate semantic similarity in a taxonomy. arXiv preprint cmp-lg/9511007 (1995)
15. Shen, Y., Zhang, S., Wong, H.S.: A new method for measuring the semantic similarity on gene ontology. In: 2010 IEEE International Conference on Bioinformatics and Biomedicine (BIBM), pp. 533–538. IEEE (2010)
16. Tian, Z., Guo, M., Wang, C., Liu, X., Wang, S.: Refine gene functional similarity network based on interaction networks. BMC Bioinform. **18**(16), 550 (2017)
17. Wang, J.Z., Du, Z., Payattakool, R., Yu, P.S., Chen, C.F.: A new method to measure the semantic similarity of GO terms. Bioinformatics **23**(10), 1274–1281 (2007)

Improving Identification of Essential Proteins by a Novel Ensemble Method

Wei Dai[1], Xia Li[1], Wei Peng[1,2(✉)], Jurong Song[2], Jiancheng Zhong[3], and Jianxin Wang[4]

[1] Faculty of Information Engineering and Automation,
Kunming University of Science and Technology, Kunming 650050, China
dw@cnlab.net, lixialxl995@gmail.com
[2] Computer Technology Application Key Lab of Yunnan Province,
Kunming University of Science and Technology, Kunming 650050, China
weipengl980@gmail.com, sjunrong@gmail.com
[3] College of Engineering and Design, Hunan Normal University,
Changsha 410081, China
jczhongcs@gmail.com
[4] Computer Science, Central South University, Changsha 410081, China
jxwang@mail.csu.edu.cn

Abstract. Essential proteins are indispensable for cell survival, and the identification of essential proteins plays a critical role in biological and pharmaceutical design research. Recently, some machine learning methods have been proposed by introducing effective protein features or by employing powerful classifiers. Seldom of them focused on improving the prediction accuracy by designing efficient strategies to ensemble different classifiers. In this work, a novel ensemble learning framework called by Tri-ensemble was proposed to integrate different classifiers, which selected three weak classifiers and trained these classifiers by continually adding the samples that are predicted to have abnormally high or abnormally low properties by the other two classifiers. We applied Tri-ensemble on predicting the essential protein of Yeast and E.coli. The results show that our approach achieves better performance than both individual classifiers and the other ensemble learning methods.

Keywords: Essential proteins · Ensemble learning · Machine learning · Tri-ensemble

1 Introduction

Essential proteins are indispensable for cells to survive and play a crucial role in the cellular function of each organism [1]. Lacking essential proteins will lead to the function loss of generating relevant protein complexes and even to cell death. Identifying essential proteins can help us better understand the minimal requirements for cell life, and it is also critical for biological and pharmaceutical design research.

In recent years, many methods have been proposed to identify essential proteins based on their topological features in biological network, protein sequence features and some other biological features, such as protein domain and protein orthology

Z. Cai et al. (Eds.): ISBRA 2019, LNBI 11490, pp. 146–155, 2019.
https://doi.org/10.1007/978-3-030-20242-2_13

properties. Previous studies found that essential proteins tend to be the center of protein-protein interaction (PPI) network, because removing them from the networks will cause the lethality and break down of the networks [2]. Therefore, many centrality methods have been proposed to identify essential proteins from PPI network, such as Betweenness Centrality (BC) [3, 4], Closeness Centrality (CC) [5], Degree Centrality (DC) [6], Eigenvector Centrality (EC) [7], Information Centrality (IC) [8], Edge Clustering Coefficient Centrality (NC) [9] and Subgraph Centrality (SC) [10]. However, these centrality methods overly depended on the topological features extracted from the PPI network and ignored their other biological features. Recently, new methods that combine the topological features with the biological ones have been developed. Li et al. [11] have proposed a new method to predict essential proteins called PeC and Tang et al. [12] have developed another one, WDC, which integrates network topology with gene expression profiles. Considering the fact that essential proteins are more conserved than non-essential ones [13] and they frequently connect to each other [14], Peng et al. [15] have proposed an iterative method to predict essential proteins based on the orthology and PPI networks.

Meanwhile, many machine learning algorithms were also applied to the identification of essential proteins. Gustafson et al. [16] extracted the topological features such as degree centrality (DC), and biological features such as paralogs, open reading frame (ORF) length, then put these features into a Naive Bayes classifier for essential protein prediction. Hwang et al. [17] combined different kinds of PPI network topological features (DC, BC, CC, etc.) and some biological features including ORF length, phyletic retention (PHY) and strand to predict essential proteins by using SVM method. There are also some methods by integrating different classifiers. Zhong et al. [18] combined topological features and biological features by using GEP-based method to predict essential protein. Acencio et al. [19] combined multiple decision tree classifiers by a voting strategy. They used local effects of subcellular localization, biological features and network topological features as the input of the classifiers. Deng et al. [20] have also combined multi-model including Naive Bayes classifier, C4.5 decision tree, CN2 rule and logistical regression model to predict essential proteins. Chen and Xu et al. [21] combined support vector machines (SVM) and Artificial Neural Networks (ANN) to predict essential proteins. However, the aforementioned methods took very simple strategies, such as voting strategy, to integrate different classifiers.

In order to better identify the essential proteins, in this work, a novel ensemble learning framework called by Tri-ensemble was proposed to integrate different classifiers. The main premise of ensemble learning is that by combining multiple models, the errors of a single one will likely be compensated by others, and as a result, the overall prediction performance of the ensemble would be improved. Ensemble learning consists of two parts, the one is how to generate different individual classifiers, and the other is how to integrate them. At present, there are two kinds of approaches for ensemble learning to generate individual classifiers. The one is bagging [22] that generates multiple classifiers through different training samples and uses them to get an aggregated predictor. There is no strong dependency between individual classifiers for bagging approaches. The other is boosting. There is a strong dependency between individual classifiers. Freund and Schapire [23] have developed a boosting approach named Adaboost, which gave weights to samples and classifiers and updated the weights based on error rate. Chen et al. [24] have proposed a scalable tree boosting system named Xgboost, which made prediction by weighted summing the prediction scores of different regression trees. The next step for ensemble learning is to integrate individual classifiers. The current combination strategies include average method, voting method and stacking [25].

Compared with previous ensemble learning method, Tri-ensemble took different strategies to generate and integrate classifiers. It selected three weak classifiers and the training set of one classifier was not fixed but changed with respect to the prediction results of the other two weak classifiers. One of the three classifiers was continually trained through increasing the samples that were misclassified or abnormally predicted by the other two classifiers. Finally, the prediction of the three classifiers was combined by a logistic regression model. The basic idea of our method using three weak classifiers comes from Tri-training for Semi-supervised learning algorithm proposed by Zhou et al. [26], which put forward the idea of multi-view and selected three classifiers and learned the differences between them by voting. We applied Tri-ensemble to predict the essential proteins of Yeast and E.coli. The results show that our approach achieves better results than individual classifiers and the other ensemble learning methods.

2   Methods

Figure 1 shows the workflow of Tri-ensemble for predicting essential proteins. It firstly divided data into a training set and a testing set. Every one of the three weak classifiers was trained by continually adding the samples that are predicted to have abnormally high or abnormally low properties by the other two classifiers. After generating the three classifiers, a logistic regression model was adopted to integrate the output of the three classifiers to make a final prediction.

2.1   Partitioning Data

The original data is randomly divided into five parts, four parts taken as the training set, and the remaining one as the test set. And then the training set is further randomly divided into P and R, where P is a quarter of the training set and R is the remaining part of the training set. After that, bootstrap sampling was done on P to generate three data sets with the same size, denoted by P_1, P_2, P_3. The size was set to 1000 in the Yeast data experiment and set to 530 in the E.coli data experiment. Meanwhile, set R was divided into n mutually exclusive subsets, denoted by $R_1, R_2 \ldots R_n$. Their relationship can be formally expressed as follows. $R = R_1 \cup R_2 \cup \ldots \cup R_n, \text{ where } R_1 \cap R_2 \cap \ldots \cap R_n = \emptyset.$ $P_1 \subset P, P_2 \subset P, P_3 \subset P, \text{ where } P_1 \cap P_2 \cap P_n \neq \emptyset.$

2.2   Training Weak Classifiers

Three weak classifiers h_1, h_2, h_3, were initially trained by the training samples in P_1, P_2, P_3, respectively. After that, every one of the three weak classifiers made a prediction on the samples in R_j, where $R_j \in \{R_1, R_2 \ldots R_n\}$. Then, every one of the three weak

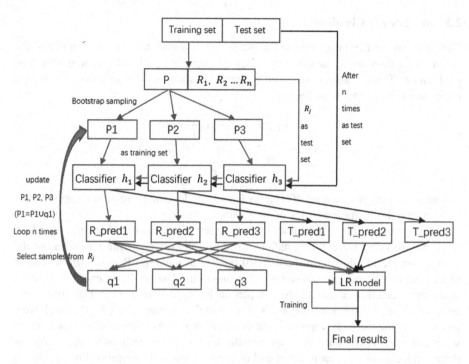

Fig. 1. An overview of Tri-ensemble. R_pred1, R_pred2, R_pred3 are the predicted probability values of three weak classifiers on the samples in R_j and T_pred1, T_pred2, T_pred3 are the predicted probability values of three weak classifiers on the test set

classifiers attached part of samples in R_j to its corresponding training sample set and were implemented training again. The newly attached samples were selected if they were misclassified or abnormally predicted by the other two weak classifiers. The samples were considered to be misclassified or abnormally predicted by a classifier if their prediction values satisfied Eq. 1.

$$|h_i(x) - E_i| > \xi * \sigma_i^2, \quad i \in [1, 3] \tag{1}$$

Where $h_i(x)$ denotes the output positive probability of sample x by the weak classifier h_i. E_i and σ_i^2 are the mean and variance of the output positive probability of the weak classifier h_i over all samples. ξ is a custom coefficient, it was set to 3 in the Yeast data experiment and set to 1 in the E.coli data experiment. For example, the weak classifier h_1 will attach some samples in R_j to its initial training set P_1 if the positive probabilities of these samples predicted by the weak classifier h_2 and h_3 deviate from the corresponding mean values of the two weak classifiers.

The process was repeated n times until all sample in $\{R_1, R_2 ... R_n\}$ were predicted by the three weak classifiers. Meanwhile, the three classifiers were well trained with the increase of training samples.

2.3 Integration Classifiers

The final step of the Tri-ensemble for predicting essential proteins is to integrate the output of the three weak classifiers by a logistic regression (LR) model and get the final predictions. This is also a stacking integration strategy. The equation of logistic regression (LR) model is as follows.

$$x = w_0 + w_1 x_1 + w_2 x_2 + w_3 x_3 \qquad (2)$$

$$f(x) = \frac{1}{1 + e^{-x}} \qquad (3)$$

Where w_0, w_1, w_2, w_3 are a set of weights learned by logistic regression algorithms and x_1, x_2, x_3 are the 3 characteristics of each sample.

The parameters in the LR model were learned in the course of training the three weak classifiers by using their prediction results and corresponding real labels of the training samples. Following is the detailed process of training LR model. In the course of training the three weak classifiers, when inputting data set R_1 to every weak classifier, the prediction results of the samples in R_1 by the three weak classifiers were collected. The process was repeated n times until all samples in $\{R_1, R_2 \ldots R_n\}$ were predicted by the three classifiers and the corresponding prediction results at each repeat were collected. After that, we combined the n of result collections predicted by the three weak classifiers at each repeat and input them and their corresponding real labels into LR model to learn its parameters. When the LR model was well trained, the test data was input and the final prediction results were generated.

3 Results

3.1 Datasets

Saccharomyces cerevisiae (Yeast) has the most complete and reliable essential protein set and PPI network data among all species, so we use Yeast data to test our method. Meanwhile, to further prove the high performance and to reflect the generalization ability of our method, we also applied it on E. coli data.

For Yeast data, we collected essential proteins from the MIPS database [27], the SGD database [28], the DEG database [1], and the SGDP (Saccharomyces Genome Deletion Project) database [29]. The yeast PPI network data was from the DIP database [30]. There were 5093 proteins and 24743 edges in the PPI network after removing self-referenced and duplicated edges. Among the 5093 proteins, there were 1167 essential proteins and the remaining 3926 proteins were regarded as non-essential proteins. Therefore, the ratio of essential proteins to non-essential proteins reached 1:3.36.

The essential proteins of E.coli were collected in the DEG database. The PPI network data of E. coli was downloaded from the DIP database, including 2727 proteins and 11803 edges. Among the 2727 proteins, there were 254 essential proteins and 2473 non-essential proteins, and the ratio reached 1:9.74.

The inputting features of the classifiers for Yeast data consisted of 10 topological features and 16 subcellular localization features. The topological features included 7 centrality methods, such as Betweenness Centrality (BC) [3, 4], Closeness Centrality (CC) [5], Degree Centrality (DC) [6], Eigenvector Centrality (EC) [7], Information Centrality (IC) [8], Edge Clustering Coefficient Centrality (NC) [9] and Subgraph Centrality (SC) [10], which were calculated by a Cytoscape plugin CytoNCA [31]. And three composited features (PeC [11], WDC [12], and ION [15]) that integrate the topological features with other biological features. ION integrated orthology data with PPI network to predict essential proteins. PeC and WDC predicted essential proteins by combing PPI network with gene expression profiles. The orthology data used in ION came from InParanoid database [32] and the gene expression data used by PeC and WDC came from Tu's literature [33]. The 16 subcellular localization features involving in Vacuole, Vesicles, Lysosome, Membrane, Mitochondrion, Peroxisome, Secretory pathway, Cell wall, Cytoskeleton, Endoplasmic reticulum, Golgi, Transmembrane, Cytoplasm, Nucleus, Endosome and Extracellular were download from the eSLDB database [34]. The inputting features of the classifiers for E.coli data included BC, CC, DC, EC, IC, NC, SC and ION.

3.2   Evaluation Metrics

Both Yeast dataset and E. coli dataset are unbalanced datasets, where the ratio of essential proteins to non-essential proteins is 1:3.36 in Yeast dataset and 1:9.74 in E.coli dataset. We proportional separated the samples to 5 folds, according to the original ratio of essential to non-essential proteins. One of the five folds was selected as a test set, while the remaining four folds were used as training set. The process was repeated 5 times until each one of the five folds was used for testing.

To evaluate the performance of our method, some popular statistic evaluation metrics are adopted, including SP, SN, FPR, PPV, NPV, F-measure, ACC and MCC metrics. Additionally, to evaluate the overall performance of each method, the ROC curves are drawn with different thresholds and the Area Under ROC Curve (AUC) is calculated for comparison.

3.3   Comparing with Existing Methods

To test the effectiveness of our method, three different classifiers were selected as the weak classifiers of our method and the prediction performance of our method was compared with that of any individual classifiers and two state-of-the-art ensemble learning methods, i.e. Adaboost and Xgboost. Softmax refers to single layer neural network plus softmax regression classifier. We carried out two experiments by using our method to combine different three weak classifiers, namely, Tri-ensemble[1], Tri-ensemble[2]. Tri-ensemble[1] integrated single layer neural network plus softmax regression classifier (NN + Softmax), Adaboost and Random Forest. Tri-ensemble[2] integrated single layer neural network plus softmax regression classifier (NN + Softmax), Xgboost and Random Forest. We also compared with Logistic Regression (LR) method because our method used it to integrate the output of the weak classifiers. Adaboost and Xgboost are two boosting-based ensemble learning methods, whose default weak classifiers are decision trees and regression trees respectively. Tables 1 and 2 show the performance comparisons of our method and individual classifiers on Yeast and E.coli dataset.

Table 1. The comparisons of SN, SP, FPR, PPV, NPV, F-MEASURE, ACC, MCC and AUC for Tri-ensemble and individual classifiers on Yeast dataset.

Methods	SN	SP	FPR	PPV	NPV	F-measure	ACC	MCC	AUC
Softmax	0.5278	0.8597	0.1403	0.5278	0.8597	0.5278	0.7836	0.3875	0.7771
Random Forest	0.5073	0.8535	0.1465	0.5073	0.8535	0.5073	0.7742	0.3608	0.7536
LR	0.5321	0.8609	0.1391	0.5321	0.8609	0.5321	0.7856	0.3931	0.7762
Adaboost	0.5373	0.8625	0.1375	0.5373	0.8625	0.5373	0.7879	0.3997	0.7695
Xgboost	0.5278	0.8597	0.1403	0.5278	0.8597	0.5278	0.7836	0.3837	0.7707
Tri-ensemble[1]	0.5458	0.865	0.135	0.5458	0.865	0.5484	0.7919	0.4108	0.7833
Tri-ensemble[2]	**0.5467**	**0.8653**	**0.1347**	**0.5467**	**0.8653**	**0.5467**	**0.7927**	**0.412**	**0.7847**

The AUC values are calculated by averaging the area under the ROC curves of all the testing samples in the 5-fold cross-validation, which illustrated the overall performance of each method. In addition, we ranked all Yeast proteins or E.coli proteins in descending order according to their score calculated by each method and selected the top of 1167 proteins as predicted essential proteins in Yeast dataset and the top of 254 proteins as predicted essential proteins in E.coli dataset. After that, the SN, SP, FPR, PPV, NPV, F-Measure, ACC and MCC values were calculated for each method based on the predicted essential proteins. Note that there are exactly 1167 real essential proteins in Yeast dataset and 254 real essential proteins in E.coli dataset.

As can be seen from Tables 1 and 2 on Yeast and E.coli dataset, all of our methods that integrated three different weak classifiers outperformed the corresponding individual classifiers and the other ensemble learning methods, i.e. Adaboost and Xgboost. Among all of our methods, Tri-ensemble[2] that integrated single layer neural network plus softmax regression classifier(NN + Softmax), Xgboost and Random Forest had better performance than that of Tri-ensemble[1]. Consequently, we used the results of Tri-ensemble[2] for following comparison.

3.4   Comparing with Other Machine Learning-Based Methods

To further evaluate our methods, we also compared it with the other machine learning methods, such as GEP, SVM, SMO, NaiveBayes, Bays Network and NaiveBayes Tree. GEP is a newly proposed method and has excellent performance for essential protein prediction. All the other machine learning methods except GEP were implemented by WEKA software. The parameters of these methods were set to their default values. Table 3 shows the AUC values comparison between Tri-ensemble and the other machine learning methods on Yeast and E.coli dataset.

Table 2. The comparison of SN, SP, FPR, PPV, NPV, F-MEASURE, ACC, MCC and AUC for Tri-ensemble and individual classifiers on E.coli dataset.

Methods	SN	SP	FPR	PPV	NPV	F-measure	ACC	MCC	AUC
Softmax	0.3268	0.9309	0.0691	0.3268	0.9309	0.3268	0.8746	0.2576	0.7611
Random Forest	0.3504	0.9333	0.0667	0.3504	0.9333	0.3504	0.879	0.2837	0.7612
LR	0.3228	0.9305	0.0696	0.3228	0.9305	0.3228	0.8739	0.2533	0.7617
Adaboost	0.3425	0.9325	0.0675	0.3425	0.9325	0.3425	0.8775	0.275	0.7606
Xgboost	0.3701	0.9353	0.0647	0.3701	0.9353	0.3701	0.8827	0.3054	0.7674
Tri-ensemble[1]	0.3819	0.9365	0.0635	0.3819	0.9365	0.3819	0.8849	0.3184	0.7787
Tri-ensemble[2]	**0.3898**	**0.9373**	**0.0627**	**0.3898**	**0.9373**	**0.3898**	**0.8863**	**0.3271**	**0.7828**

Table 3. The AUC values comparison between Tri-ensemble and other machine learning methods on Yeast and E.coli dataset.

Methods	AUC of Yeast data	AUC of E.coli data
Tri-ensemble[2]	0.7847	0.7828
GEP	0.773	0.779
SVM	0.577	0.5
SMO	0.608	0.5
NaiveBayes	0.744	0.7437
Bayes Network	0.731	0.7258
NaiveBayes Tree	0.746	0.7204

As can be seen from Table 3, the values of AUC of Tri-ensemble[2] were higher than that of the other methods on Yeast and E.coli dataset, which suggests that applying our ensemble learning method to integrate suitable weak classifiers, such as single layer neural network plus softmax regression classifier(NN + Softmax), Xgboost and Random Forest, can achieve better performance on essential protein prediction than GEP that has excellent prediction accuracy.

4   Conclusions

This paper proposed a novel ensemble framework named by Tri-ensemble to predict essential proteins, which improved the prediction accuracy by integrating different weak classifiers. Tri-ensemble firstly partitioned the data into training set and testing set and further divided the training set into two parts. And then three weak classifiers were selected and were initially trained by a small part of training samples. After that, the three weak classifiers were trained through continually attaching remaining training samples that were misclassified or abnormally predicted by the other two classifiers. Finally, a stacking strategy was adopted to integrate the output of the three weak classifiers by a logistic regression model. Compared with previous ensemble learning method, the Tri-ensemble selected three weak classifiers and the training sets of one classifier was not fixed but changed with respect to the prediction results of the other two classifiers. We carried out experiments on Yeast data and E.coli data and the results show that our approach can achieve better prediction performance than both individual classifiers and the other state-of-the-art ensemble learning methods.

Acknowledgment. This work is supported in part by the National Natural Science Foundation of China under grant No. 31560317, No. 61502214, No. 61502166, No. 61702122 and No. 81560221. Natural Science Foundation of Yunnan Province of China (No. 2016FB107).

References

1. Ren, Z., Yan, L.: DEG 50, a database of essential genes in both prokaryotes and eukaryotes. Nucleic Acids Res. **37**(Database issue), D455 (2009)
2. Jeong, H., Mason, S.P., Barabasi, A.L., Oltvai, Z.N.: Lethality and centrality in protein networks. Nature **411**(6833), 41–42 (2001)
3. Freeman, L.C.: A set of measures of centrality based on betweenness. Sociometry **40**(1), 35–41 (1977)
4. Joy, M.P., Brock, A., Ingber, D.E., Huang, S.: High-betweenness proteins in the yeast protein interaction network. J. Biomed. Biotechnol. **2005**(2), 96 (2014)
5. Stefan, W., Stadler, P.F.: Centers of complex networks. J. Theor. Biol. **223**(1), 45–53 (2003)
6. Vallabhajosyula, R.R., Deboki, C., Samina, L., Animesh, R., Alpan, R.: Identifying hubs in protein interaction networks. PLoS ONE **4**(4), e5344 (2009)
7. Bonacich, P.: Power and centrality: a family of measures. Am. J. Sociol. **92**(5), 1170–1182 (1987)
8. Stephenson, K., Zelen, M.: Rethinking centrality: methods and examples ☆. Soc. Netw. **11**(1), 1–37 (1989)
9. Wang, J., Li, M., Wang, H., Pan, Y.: Identification of essential proteins based on edge clustering coefficient. IEEE/ACM Trans. Comput. Biol. Bioinf. **9**(4), 1070–1080 (2012)
10. Ernesto, E., Rodríguez-Velázquez, J.A.: Subgraph centrality in complex networks. Phys. Rev. E Stat. Nonlinear Soft Matter Phys. **71**(5 Pt 2), 056103 (2005)
11. Li, M., Zhang, H., Fei, Y.: Essential protein discovery method based on integration of PPI and gene expression data. J. Cent. South Univ. **44**(3), 1024–1029 (2013)
12. Tang, X., Wang, J., Yi, P.: Identifying essential proteins via integration of protein interaction and gene expression data (2012)
13. Jordan, I.K., Rogozin, I.B., Wolf, Y.I., Koonin, E.V.: Essential genes are more evolutionarily conserved than are nonessential genes in bacteria. Genome Res. **12**(6), 962 (2002)
14. Hart, G.T., Lee, I., Marcotte, E.M.: A high-accuracy consensus map of yeast protein complexes reveals modular nature of gene essentiality. BMC Bioinform. **8**(1), 1–11 (2007)
15. Peng, W., Wang, J., Wang, W., Liu, Q., Wu, F.X., Pan, Y.: Iteration method for predicting essential proteins based on orthology and protein-protein interaction networks. BMC Syst. Biol. **6**(1), 1–17 (2012)
16. Gustafson, A.M., Snitkin, E.S., Parker, S.C., Delisi, C., Kasif, S.: Towards the identification of essential genes using targeted genome sequencing and comparative analysis. BMC Genom. **7**(1), 265 (2006)
17. Hwang, Y.C., Lin, C.C., Chang, J.Y., Mori, H., Juan, H.F., Huang, H.C.: Predicting essential genes based on network and sequence analysis. Mol. BioSyst. **5**(12), 1672–1678 (2009)

18. Zhong, J., Wang, J., Peng, W., Zhang, Z., Pan, Y.: Prediction of essential proteins based on gene expression programming. BMC Genom. **14**(S4), S7 (2013)
19. Acencio, M.L., Lemke, N.: Towards the prediction of essential genes by integration of network topology, cellular localization and biological process information. BMC Bioinform. **10**(1), 290 (2009)
20. Deng, J., et al.: Investigating the predictability of essential genes across distantly related organisms using an integrative approach. Nucleic Acids Res. **39**(3), 795–807 (2011)
21. Chen, Y., Xu, D.: Understanding protein dispensability through machine-learning analysis of high-throughput data. Bioinformatics **21**(5), 575–581 (2005)
22. Breiman, L.: Bagging predictors. Mach. Learn. **24**(2), 123–140 (1996)
23. Schapire, R.E., Singer, Y., Singhal, A.: Boosting and Rocchio applied to text filtering. In: SIGIR Proceedings of Annual International Conference on Research & Development in Information Retrieval, pp. 215–223 (1998)
24. Chen, T., Guestrin, C.: XGBoost: a scalable tree boosting system (2016)
25. Breiman, L.: Stacked regressions. Mach. Learn. **24**(1), 49–64 (1996)
26. Li, M., Zhou, Z.-H.: Tri-training exploiting unlabeled data using three classifiers. IEEE Trans. Knowl. Data Eng. **17**(11), 1529–1541 (2005)
27. Mewes, F.D., et al.: MIPS: analysis and annotation of proteins from whole genomes in 2005. Nucleic Acids Res. **34**(Database issue), 169–172 (2004)
28. Cherry, J.M., et al.: SGD: saccharomyces genome database. Nucleic Acids Res. **26**(1), 73–79 (1998)
29. Saccharomyces Genome Deletion Project. http://www-sequence.stanford.edu/group/yeast_deletion_project/deletions3.html
30. Xenarios, I., Salwinski, L., Duan, X.J., Higney, P., Kim, S.M., Eisenberg, D.: DIP, the database of interacting proteins: a research tool for studying cellular networks of protein interactions. Nucleic Acids Res. **30**(1), 303 (2002)
31. Tang, Y., Li, M., Wang, J., Pan, Y., Wu, F.X.: CytoNCA: a cytoscape plugin for centrality analysis and evaluation of protein interaction networks. Biosystems **127**, 67–72 (2015)
32. Gabriel, O., et al.: InParanoid 7: new algorithms and tools for eukaryotic orthology analysis. Nucleic Acids Res. **38**(Database issue), D196 (2010)
33. Tu, B.P., Andrzej, K., Maga, R., Mcknight, S.L.: Logic of the yeast metabolic cycle: temporal compartmentalization of cellular processes. Science **310**(5751), 1152 (2005)
34. Andea, P., Pier Luigi, M., Piero, F., Rita, C.: eSLDB: eukaryotic subcellular localization database. Nucl. Acids Res. **35**(Database issue), 208–212 (2007)

Machine and Deep Learning

Deep Learning and Random Forest-Based Augmentation of sRNA Expression Profiles

Jelena Fiosina[1], Maksims Fiosins[2,3,4(✉)], and Stefan Bonn[2,3]

[1] Clausthal University of Technology, Clausthal-Zellerfeld, Germany
jelena.fiosina@gmail.com
[2] German Center for Neurodegenerative Diseases, Tübingen, Germany
maksims.fiosins@gmail.com
[3] Institute for Medical Systems Biology, Center for Molecular Neurobiology,
University Medical Center Hamburg-Eppendorf, Hamburg, Germany
sbonn@uke.de
[4] Genevention GmbH, Göttingen, Germany

Abstract. The lack of well-structured annotations in a growing amount of RNA expression data complicates data interoperability and reusability. Commonly used text mining methods extract annotations from existing unstructured data descriptions and often provide inaccurate output that requires manual curation. Automatic data-based augmentation (generation of annotations on the base of expression data) can considerably improve the annotation quality and has not been well-studied. We formulate an automatic augmentation of small RNA-seq expression data as a classification problem and investigate deep learning (DL) and random forest (RF) approaches to solve it. We generate tissue and sex annotations from small RNA-seq expression data for tissues and cell lines of *homo sapiens*. We validate our approach on 4243 annotated small RNA-seq samples from the Small RNA Expression Atlas (SEA) database. The average prediction accuracy for tissue groups is 98% (DL), for tissues - 96.5% (DL), and for sex - 77% (DL). The "one dataset out" average accuracy for tissue group prediction is 83% (DL) and 59% (RF). On average, DL provides better results as compared to RF, and considerably improves classification performance for 'unseen' datasets.

Keywords: Augmentation · Deep learning · Random forest · Ontology · Small RNA · Expression counts · Contamination

1 Background

Qualitative and standardized annotations (tissue, disease, age, sex, cell line, etc.) of expression data is a key aspect to enable data interoperability and reusability. Data should be findable, accessible, interoperable, and reusable (FAIR), which ultimately facilitate knowledge discovery [16]. Annotations are essential part of

© Springer Nature Switzerland AG 2019
Z. Cai et al. (Eds.): ISBRA 2019, LNBI 11490, pp. 159–170, 2019.
https://doi.org/10.1007/978-3-030-20242-2_14

semantic data integration systems [9]. In various databases, data annotations are available in different often-unstructured text formats and many times important information on e.g. age, sex, and sometimes even tissue of sample origin is missing (i.e GEO [3]). This leads to missing and/or inaccurate annotations, and requires revision and correction by an expert [6]. While state-of-the-art expression databases such as the small RNA Expression Atlas (SEA, http://sea.ims. bio) [10] provide well-structured, ontology-based annotations of publicly available small RNA-seq (sRNA-seq) data, this is achieved by curation of annotations, and missing information is still a problem in many experimental databases.

A fundamental hypothesis is that augmentation from the source (here, expression counts) data can annotate missing information with high accuracy, allowing for the subsequent analysis of the (meta) data. We suppose that data with similar expression profiles should have similar annotations. Several publications highlight the possibility to use machine learning (ML) approaches to augment expression information, for small RNAs (sRNAs) as well as messenger RNAs (mRNAs). In [4,6], the sex in different micro RNA (miRNA) tissues was defined. In [6], the authors used the DESeq package and analysis of variance (ANOVA) to detect sex differences in several tissue in miRNAs. In [2], age, sex, and tissue were predicted in mRNA sequencing (mRNA-seq) expressions. In [12], the sex of mRNAs was predicted, and the most important mRNAs were selected. Random Forest (RF) classifier is being widely used for classification of expression data, especially in disease diagnostics [13]. RF also enables explanation of classification by supplying variable importances.

Deep learning (DL) is making major advances in solving problems that have resisted the best attempts of the artificial intelligence community for many years [7]. DL is able to deal with big data and is robust even for massive amount of noisy labeled training data [17]. On the downside, DL requires large amount of training data [8], is prone to overfit on small training sets, and are notoriously hard to biologically interpret (extraction of feature importances) [15].

In this study we investigate whether the DL-based data augmentation could be superior to classical ML approaches, such as RF. The main hypothesis is that DL classifier trained on sufficiently large data sets would generalize more efficiently to yet unseen datasets. Whereas single unseen samples might be easy to learn, datasets usually contain a distinct experimental bias that the model has not learnt a priori. We apply DL and RF models on human sRNA-seq datasets from SEA, which contains 4243 sRNA-seq samples. Every sample is semantically annotated and analyzed with the same workflow (OASIS [11], https://oasis.dzne. de), increasing data interoperability while reducing analysis bias.

We use this data to predict tissue and sex annotations. DL performs slightly better than RF for cross-validation experiments and significantly outperforms it for "one dataset out" experiments. These results strongly suggest that DL-based expression data augmentation could significantly outperform classical ML approaches, given enough training data.

2   Methods

2.1   Data and Meta-data Acquisition

We augment sRNA-seq data with missing annotations. We use SEA sRNA-seq data integration platform that contains 4243 samples and annotations in 350 datasets. The relatively large number of high-quality samples allows us to use DL for data augmentation purposes.

We selected 128 *homo sapiens* datasets with available annotations for tissue or cell line. We avoided small datasets and samples with rare types of tissues. We used 2806 samples for tissue prediction, including 641 cell line samples with known tissue. For sex classification, we used samples with available sex (only real tissue samples, 1591 samples in 41 datasets). The female and male proportion was 42% and 58%, respectively. We constructed separate classification models for each outcome variable prediction (tissue, sex).

There are two kinds of expression data available: sRNA expression and the reads not mapped to sRNAs, but mapped to contamination organisms. We use both expression profiles, separately and together. The expression counts from SEA are normalized inside each sample using reads per million (RPM).

Available tissues are annotated in SEA as specifically as possible. For example, parts of the brain are annotated as "neocortex" or "prefrontal cortex" if this information is available from the experiment. However, using all those tissues in classification leads to a large number of small classes. To avoid this, we joined the available tissues according to used BTO ontology (Table 1). We added also the cell lines to the corresponding groups. We used a hierarchical classification approach: first, we predicted the tissue group and then the single tissue.

Table 1. Tissue and cell line grouping according to ontologies.

Tissue group	Containing tissues
blood_group	blood, blood plasma, blood serum, peripheral blood, umbilical cord blood, serum, buffy coat, immortal human B cell, liver, lymphoblastoid cell
brain_group	brain, cingulate gyrus, motor cortex, prefrontal cortex, neocortex
epithelium_group	skin, dermis, epidermis, breast, oral mucosa, larynx
gland_group	prostate gland, testis, kidney, bladder, uterine endometrium, tonsil, lymph node
intestine_group	intestine, colon, ileal mucosa

2.2 Data Scaling and Filtering

Data Scaling (DL Only): We scaled counts of each sRNA independently. We compared two alternative scalers. A MinMax scaler scales the data in the range (0,1). A standard scaler standardizes features by removing the mean and scaling to unit variance. The MinMax scaler showed better results.

SRNA Filtering (for both RF and DL): The number of features (sRNAs) was considerably greater than the number of available observations (samples). The initial number of factors was approximately 35000, while the number of available samples was 2200 (for tissue prediction, even fewer for sex). In addition, approximately 5600 contamination counts were available for each sample.

Most of the counts were equal to zero. The preliminary experiments showed that the maximal accuracy was obtained by excluding variables (sRNAs and contaminants) containing more than 30% of zeroes. After this, the number of sRNAs and contaminants was approximately 2500 and 2000 respectively.

Sample Filtering (for both RF and DL): Some tissues we could not group (i.e. milk, urine, heart, etc), especially if they were presented in only one dataset. This made "one dataset out" classification (s. Sect. 2.3, Validation) impossible, and we did not predict tissue in such datasets. We also excluded some tissues and cell lines that were presented in one dataset containing less than 9 samples. The cell lines located in the t-distributed stochastic neighbor embedding (t-SNE)

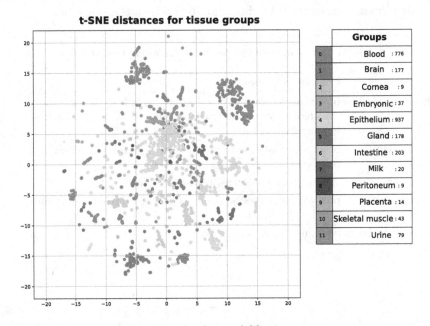

Fig. 1. t-SNE plot for available tissue types.

plot in other region as the corresponding tissue, were also excluded. The reason for this is that such cell lines are not similar to original tissue and should be predicted separately.

After this exclusion, 105 datasets are left, containing 2215 samples. The proportions of cell lines and tissue samples are 23% and 77%, respectively. Figure 1 illustrates the t-SNE plot for the tissue groups.

2.3 Models

DL Model: We used a fully connected neuronal network (NN) architecture. It has one input layer with number of inputs equal to the number of variables after the initial filtering; we tested the NN with different hyper-parameters (such as layer sizes, number of layers, and drop-out rates). Finally, we used a NN with three hidden layers containing 1000, 250, and 250 neurons, with the drop-out rates 0.5, 0.4, and 0.4. The number of neurons in the output layer was equal to the number of predicted classes.

We examined different optimizers: 'rmsprop', 'adam', 'sgd', 'adadelta'. We used the rectifier linear unit (ReLU) activation function for our initial and hidden layers. We chose the "softmax" activation for multi-class classification.

We trained the NN for 50 epochs with batch size 30.

RF Model: On both stages of the RF, the following parameters were used: mtry equal to the square root of the number of features, and down-sampling to balance the imbalanced classes (especially for tissue prediction). On the first stage (pre-classification), the RF was based on all filtered columns, and the number of trees was 100. We ordered the features according to their importance (Gini index decrease). We used the top-1000 selected features for the second stage classification with an increased number of trees (here, 500). We used RF models for obtaining variable importances

Validation: We implemented two types of cross-validation to check the accuracy of data augmentation in two different conditions. First, we used 5 fold CV as one readout, where we reported the average performance. Second, we trained a model using CV and classified a test dataset, which was not seen by the model during training. More specifically, this data came from a different experiment, which contained a different bias, but had a tissue that the model was trained on. This case is more relevant to the real situation, because for automatic augmentation, one should augment the new dataset, taking other datasets as training data. In the case of tissue prediction, such a validation technique was not available for each dataset, because some tissues are available in one dataset only. Throughout this manuscript we will refer to the 5 fold CV as 'cross validation' and the validation on unseen datasets as 'one dataset out'.

Quality Metrics: The main metric of the classification quality was accuracy. Apart from the accuracy, we used various other metrics such as: confusion matrix, precision, recall, F1 score in macro and micro versions, and Cohen's kappa, which normalizes the accuracy by the imbalance of the classes in the data. Those metrics we used internally to tune the classification models.

Software Libraries: All the scripts for DL classification are developed in R based on the "keras" library. The RF models are also implemented in R, using the "randomForest" library. We used the Python 3.5 "sklearn.manifold" t-SNE library to build the t-SNE plots.

3   Results

The main hypothesis of this study is that DL-based expression data augmentation approaches might outperform classical approaches. We therefore compared DL to RF classification to predict the target tissue and patient sex of human sRNA-seq data. A second aim was to analyse variable importance to check their biological relevance.

3.1   Robust sRNA-seq Tissue Prediction

Tissue Group Prediction:

CV Experiments. We experimented with 9 and 15 minimal number of samples per tissue class. Figure 2 (left) shows that RF is less accurate, especially for a class with a minimum of 9 samples: DL: 97%, RF: 85%. For the classes with

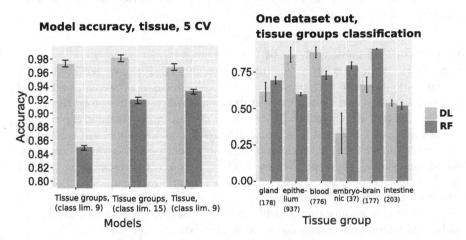

Fig. 2. CV tissue pred. accuracy (left); "One dataset out" tissue groups pred. accuracy (right)

a minimum of 15 samples, the accuracy was better: DL: 98%, RF: 92%. The DL model gave better results, in both cases, because it did not suffer from imbalanced classes, however we used an internal class balancing mechanism for the RF model.

"One Dataset Out" Experiments. After initial filtering only 6 aggregated tissues were left. The reason was that some tissues were only in one dataset and some tissues were presented in datasets containing less than 9 samples (see Sect. 2.2, Sample Filtering). In Fig. 2 (right), we present the accuracy of tissue group prediction. Notice that the datasets with the same tissue may differ from dataset to dataset because of different factor influences (e.g., library preparation methods, biological conditions of samples: cell types, diseases). This is a reason for the significantly lower model accuracy in this case. For the intestine group, which we could not detect very well, the accuracy was around 50%. We could predict most of the datasets with accuracy of 80-100%. The average accuracy is 83% (DL) and 59% (RF).

Tissue Prediction:

CV Experiments. We avoided combining any tissue or cell line. Instead, we used all the tissue and cell line classes that had more than nine samples. The DL und RF models had standard parameters, as described above. The average accuracy (Fig. 2, left) was DL: 96.5%, RF: 93%. The classes were not as imbalanced without tissue aggregation, and thus we got similar results with both models.

"One Dataset Out" Experiments. Knowing the tissue group from the previous experiments, we specified its tissue class. The dataset exclusion criteria were the same as in previous experiments with tissue groups. The resulting histogram for each dataset as a test set are presented in Fig. 3 (left). The tissue in the most datasets is predicted within the accuracy interval (0.8,1); nevertheless, tissue in some datasets is predicted with accuracy (0,0.2). In Fig. 3 (right) and then in Fig. 4, we can see the tissues and cell lines that could not be predicted well (brain, breast, colon, skin, etc.). Bad "brain" tissue prediction was caused by its identification as sub/tissues: prefrontal cortex and neocortex. It could be true, because the sub-tissue had no annotation in the given dataset.

"Breast" and "skin" tissues are very similar, and both were not identified correctly in many datasets. "Colon" tissue was identified as the "HCT116" cell line, i.e., colon cancer and as "ileal mucosa", which is very near to the colon.

We conclude that for tissue group prediction DL outperforms RF, especially in "one dataset out" case. For tissue prediction the difference was smaller, but DL was still better.

3.2   Robust sRNA-seq Sex Prediction

The DL and RF models had standard parameters described above. To improve the model accuracy apart from sRNA-seq expression counts, we extend the models with contamination expression counts (Fig. 5).

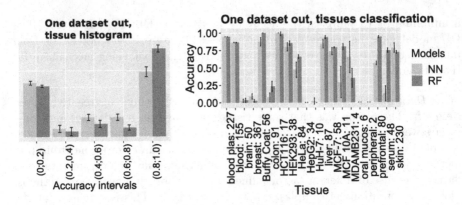

Fig. 3. "One dataset out" tissue pred. accuracy histogram; "One dataset out" tissues pred. accuracy by classes.

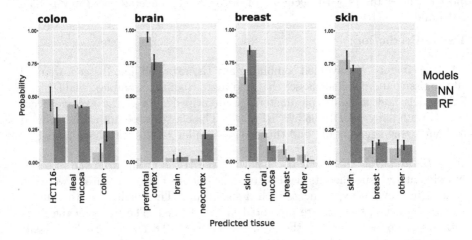

Fig. 4. "One dataset out" of tissue groups each dataset pred. accuracy.

The best models were DL and RF based on both sRNAs and contaminations, with accuracies of 77% and 76.9%. The other three models RF(RNA), DL(contaminations), and DL(RNA) gave an accuracy of approximately 76.2%. It was unexpected that the model based on contaminations only could predict the sex with an accuracy of approximately 76% for both DL and RF. So for sex prediction DL slightly outperforms RF.

4   Enrichment Tests

Given the good prediction accuracy we next investigated whether the ML models learn tissue- and/or sex-specific sRNAs. The hypothesis is that to govern accurate prediction the model has to put more emphasis on sRNAs that contain biologically relevant information, in a given context, while ignoring non-relevant information. For the tissue prediction use case, this would imply that a good classifier would put heavy emphasis on sRNAs that are tissue-specific whereas it would put little weight on house-keeping sRNA expression, which is largely invariant over tissues. The same should be true for the sex prediction.

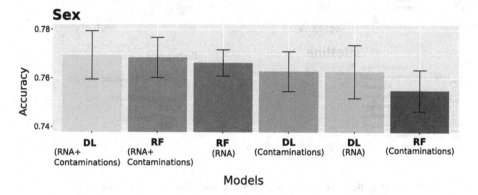

Fig. 5. CV sex prediction accuracy with different models.

We used the miRNA enrichment analysis and annotation (miEAA) tool [1] and run over-representation analysis with default settings, no reference miRNA set and checking Organs, Diseases and Age/Gender dependent miRNAs. The tool was developed for miRNAs, so we excluded other types of sRNAs from the analysis. We performed the enrichment test on prediction of tissue groups and sex. We took the top-200 miRNAs from the RF classifier (Sects. 3.1–3.2).

4.1   Tissue-Specific sRNA Enrichment

First, we investigated the enrichment of biological categories for miRNAs that are important for tissue classification. In Fig. 6 we see the enrichment of stem cells responsible for tissue-specific tissue formation, and of the cytoskeleton. Blood (including lymphocytes) and adipose tissue show some tissue-specific categories. However, the full set of top miRNAs would probably not provide a clear enrichment, because the miRNA subsets used by the classifier to detect the particular tissue groups are mixed.

Next, we clustered sRNAs by their expression. We could see specific clusters of tissue groups with highly expressed miRNAs. Next, we analyzed each specific cluster separately.

Brain: The cluster contains 43 miRNAs. 10 of them (miR-124, miR-128, miR-129, miR-137, miR-138, miR-153, miR-323, miR-708, miR-99, and miR-9) are reported as brain-specific in [5]. The enrichment test (Fig. 6) shows that most of the enriched categories are brain-specific or nervous system-specific. Therefore, this cluster is well-suited for detection of the brain group.

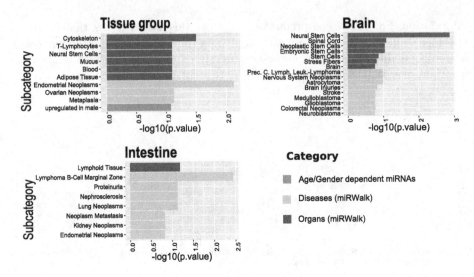

Fig. 6. Enrichment results for tissue-specific categories.

Intestine: This cluster contains 22 miRNAs. 3 of them (miR-10a, miR-196, and miR-200a) are reported as kidney-specific, and one(miR-196) as liver-specific in [5]. In addition, four of them (miR-192, miR-194 and miR-215) are reported as kidney-specific in [14]. Moreover, miR-31 is reported as brain-specific in [5]. The enrichment test (Fig. 6) shows, from organ/tissue category, that the lymphoid tissue is enriched, and may be associated with intestine. Therefore, this cluster in general suits for detection of the intestine group.

Blood: This cluster contains only six miRNAs. Four of them (miR-129, miR-9, miR-323 and miR-708) are reported as brain-specific in [5]. However, miR-129 is a candidate biomarker for heart failure, and thus is heart/blood specific. The set of six miRNAs is too small for the enrichment test. The classifier uses this cluster more for brain detection than for blood detection, as some sRNAs are highly expressed both in blood and in brain.

The results indicate that the ML learns relevant tissue-specific sRNAs, especially for the brain and intestine clusters.

4.2   Sex Specific sRNA Enrichment

We investigated the enrichment of biological categories coming from sex classification. Figure 7 (left) illustrates enrichment of sex-specific terms (upregulated in male, sex-dependent). A broad range of tissues (liver, kidney, adipose tissue, serum, skeletal muscle, bones, breast, and ovary) is enriched. This may show a sex specificity of miRNA expression in many organs.

The list of enriched diseases mostly contains cancer. Considering that cancer is more frequent in males (approx. 1.5 times), we checked whether the classifier

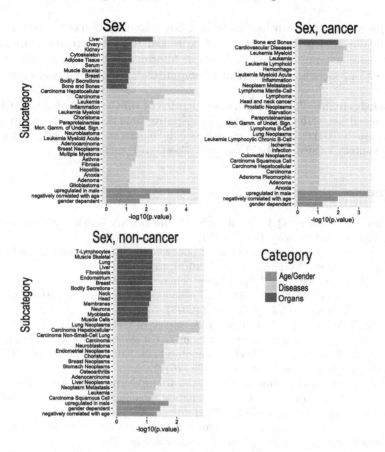

Fig. 7. Enrichment results.

had learned disease instead of learning sex. We divided all available samples into cancer-specific and non-cancer specific, and classified sex separately (Fig. 7).

For cancer-related samples, a similar list of diseases is enriched. Therefore, this classifier may be disease-based. However, for non-cancer samples, tissues and a shorter list of diseases are enriched. Therefore, this classifier is really based on organ sex specificity. This means that the classifier based on all samples causes mixed sex classification: sometimes based on disease, sometimes based on organs.

5   Conclusion and Future Work

We compared the performance of DL with classical (RF) approach for prediction of tissue and sex based of human sRNA-seq expression data. The obtained results show that DL based augmentation outperforms RF, especially in the 'one dataset out' validation. DL acts as a "black box" model, while RF allows to explain variable importance. As our future work, we are going to predict age in the same manner as sex, improve our models by stacking a combination of various models, and apply our models for other types of expression data. We are also planning to conduct more accurate variable and sample filtering, as well as more deep result interpretation, including enrichment of non-miRNA sRNAs and contaminants.

Acknowledgements. The research was supported by the German Federal Ministry of Education and Research (BMBF), project Integrative Data Semantics for Neurodegenerative research (031L0029); by German Research Foundation (DFG), project Quantitative Synaptology (SFB 1286 Z2) and by Volkswagen Foundation.

References

1. Backes, C., Khaleeq, Q.T., et al.: miEAA: microRNA enrichment analysis and annotation. Nucleic Acids Res. **44**(W1), W110–W116 (2016)
2. Ellis, S., et al.: Improving the value of public RNA-SEQ expression data by phenotype prediction. Nucleic Acids Res. **46**(9), e54 (2018)
3. Gene expression omnibus. https://www.ncbi.nlm.nih.gov/geo/
4. Guo, L., et al.: miRNA and mRNA expression analysis reveals potential sex-biased miRNA expression. Sci. Rep. **7**, 39812 (2017)
5. Guo, Z., Maki, M., et al.: Genome-wide survey of tissue-specific microRNA and transcription factor regulatory networks in 12 tissues. Sci. Rep. **4**, 5150 (2014)
6. Hadley, D., Pan, J., et al.: Precision annotation of digital samples in NCBI's gene expression omnibus. Sci. Data **4**, 170125 (2017)
7. LeCun, Y., Bengio, Y., Hinton, G.: Deep learning. Nature **521**, 436 (2015)
8. Li, Y., et al.: Deep learning in bioinformatics: introduction, application, and perspective in big data era. bioRxiv (2019)
9. Madan, S., Fiosins, M., et al.: A semantic data integration methodology for translational neurodegenerative disease research. Figshare (2018)
10. Rahman, R.U., Sattar, A., Fiosins, M., et al.: Sea: the small RNA expression atlas. bioRxiv (2017). https://www.biorxiv.org/content/early/2017/08/04/133199
11. Rahman, R.U., et al.: Oasis 2: improved online analysis of small RNA-seq data. BMC Bioinform. **19**, 54 (2018)
12. Simon, L., et al.: Human platelet microRNA-mRNA networks associated with age and gender revealed by integrated plateletomics. Blood **123**, e37–e45 (2014)
13. Statnikov, A., Wang, L., Aliferis, C.F.: A comprehensive comparison of random forests and support vector machines for microarray-based cancer classification. BMC Bioinform. **9**, 319 (2008)
14. Sun, Y., Koo, S., et al.: Development of a micro-array to detect human and mouse microRNAs and characterization of expression in human organs. Nucleic Acids Res. **32**(22), e188 (2004)
15. Webb, S.: Deep learning for biology. Nature **554**, 555–557 (2018)
16. Wilkinson, M.D., et al.: The fair guiding principles for scientific data management and stewardship. Sci. Data **3**, 160018 (2016)
17. Xiao, T., et al.: Learning from massive noisy labeled data for image classification. In: 2015 IEEE Conference on Computer Vision and Pattern Recognition (CVPR), pp. 2691–2699 (2015)

Detecting Illicit Drug Ads in Google+ Using Machine Learning

Fengpan Zhao[1]($\boxtimes$)(iD), Pavel Skums[1], Alex Zelikovsky[1], Eric L. Sevigny[2],
Monica Haavisto Swahn[3], Sheryl M. Strasser[3], and Yubao Wu[1]

[1] Department of Computer Science, Georgia State University, Atlanta, GA, USA
fzhao6@student.gsu.edu, {pskums,ywu28}@gsu.edu, alexz@cs.gsu.edu
[2] Department of Criminal Justice and Criminology, Georgia State University,
Atlanta, GA, USA
esevigny@gsu.edu
[3] School of Public Health, Georgia State University, Atlanta, GA, USA
{mswahn,sstrasser}@gsu.edu

Abstract. Opioid abuse epidemics is a major public health emergency
in the US. Social media platforms have facilitated illicit drug trading,
with significant amount of drug advertisement and selling being carried
out online. In order to understand dynamics of drug abuse epidemics
and design efficient public health interventions, it is essential to extract
and analyze data from online drug markets. In this paper, we present a
computational framework for automatic detection of illicit drug ads in
social media, with Google+ being used for a proof-of-concept. The proposed SVM- and CNN-based methods have been extensively validated
on the large dataset containing millions of posts collected using Google+
API. Experimental results demonstrate that our methods can efficiently
identify illicit drug ads with high accuracy. Both approaches have been
extensively validated using the dataset containing millions of posts collected using Google+ API. Experimental results demonstrate that both
methods allow for accurate identification of illicit drug ads.

Keywords: Illicit drug ads · Social media · Text mining ·
Deep learning

1 Introduction

The opioid abuse epidemic is a national crisis seriously affecting public health,
causing preventable harm and premature death, and devastating communities.
In 2017, 70,467 Americans died of drug overdoses that year, representing an
increase of 10 percent over the 63,938 opioid overdose deaths recorded in 2016
[1].

The drug abuse epidemic has been facilitated by modern information technology and the rise of illicit drug trading platforms. With an estimated 4.1
billion persons worldwide regularly using the Internet in 2018 [2], drug vendors can efficiently and effectively reach drug consumers via online social media

© Springer Nature Switzerland AG 2019
Z. Cai et al. (Eds.): ISBRA 2019, LNBI 11490, pp. 171–179, 2019.
https://doi.org/10.1007/978-3-030-20242-2_15

platforms. Online drug trading is both more efficient and less risky than traditional drug market exchanges, since the buyer does not need to connect with the seller in person. An open question concerns the extent to which the current opioid abuse epidemic is facilitated by the proliferation of social media. In our research, we found that most social media platforms are used extensively for illicit drug advertising. Figure 1 shows two sample posts collected from Google+. Many ads contain vendors' phone numbers, emails, Wickr IDs, and websites. Buyers can contact drug vendors using these communication methods to order drugs for delivery to a specified pickup location. Purchasing illicit drugs online seemingly has become as straightforward as making an Amazon purchase. It is therefore of paramount importance that public health and law enforcement personnel have access to efficient tools for monitoring online drug transactions using traditional epidemiological surveillance methods to inform the design of appropriate response strategies.

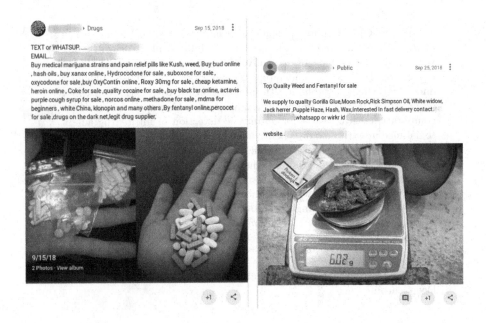

Fig. 1. Examples of illicit drug advertisements from Google+

In this paper, we develop a computational framework for detecting illicit ads in Google+, one of the largest social media platforms. We first captured relevant data posts via Google+ APIs, and then applied binary classification methods to analyze the text data in the posts. The textual analyses were used to identify illicit drug ads. We employed two methods in our approach: 1. the support vector machine (SVM) and 2. the convolutional neural network (CNN). The SVM-based method allowed for term frequency-inverse document frequency (TF-IDF) extraction of terms, which were subsequently applied to SVM for prediction [3,4]. The CNN-based methodology was applied to social media posts for text classification [5]. The first approach (SVM) required a precursory feature selection, while the second approach (CNN) automatically learns features from the text data.

2 Related Work

Illicit online drug trade has been the subject of several epidemiological and sociological studies. In particular, Mackey et al. [6] created a fictitious advertisement, offering consumers a way to buy drugs without a prescription. The advertisement was posted on four social media platforms: Facebook, Twitter, MySpace and Google+. Eventually one of these accounts was blocked due to suspicious activity, but the remaining fake illicit drug advertisements were easily accessible during the duration of the experiment. A study conducted by Stroppa et al. [7] revealed that one-fifth of collected posts advertise counterfeit and/or illicit products online. Their research emphasized that detection of illegal cyber-vendors and online tactics requires development and application of sophisticated and tailored screening/detection methods.

On the computational side, development of tools for detection of malicious and/or undesired advertisements in social media has been a subject of several studies. Hu et al. [8] provided a framework for detection of spammers on microblogging. Zheng and colleagues [9] proposed a SVM-based machine learning model to detect spammer behavior on Sina Weibo. Agrawal et al. [10] introduced an unsupervised method called Reliability-based Stochastic Approach for Link-Structure Analysis, which can be used to detect topical posts on social media. Jain et al. [11] used convolutional and long short-term memory (LSTM) neural networks to detect spam in social media, while addressing the challenges of text mining on short posts.

In contrast to the previous studies, we specifically focus on detection of illicit drug ads within social media platforms, with the aim of applying epidemiological methods to investigate online enabling structures associated with opioid abuse.

3 Methods

In this section, we describe two methods for classifying social media posts based on Support Vector Machine (SVM) and Convolutional Neural Network (CNN) approaches. For both methods, the inputs are text data extracted from Google+ posts, and the outputs are the predicted labels indicating whether each post is an illicit drug ad.

3.1 The SVM-Based Method

The proposed method pipeline consists of two stages: pre-processing and classification.

Pre-processing Steps. At this stage, text posts collected from social media are transformed into numerical feature vectors, which are further used as the inputs for the SVM classifier. It is a crucial part of traditional text mining methods because the selected features affect the performance of the classifier. Figure 2 shows the general scheme of the pre-processing stage.

Fig. 2. Pre-processing steps

Pre-processing consists of three steps. In the first step, the stop words considered noise are removed. In the second step, the root of a word is isolated by removing tenses of verbs, which is also referred to as stemming [12]. In the third step, the term frequency-inverse document frequency (TF-IDF) features are determined [13]. The TF-IDF is the product of two statistics: term-frequency and inverse document frequency. The term frequency is calculated based on the raw count of a term (word). The inverse document frequency is a measure of how much information the word provides.

Support Vector Machine (SVM) Classification. TF-IDF features computed at the pre-processing step are used to train an SVM model that can be further used to predict labels of new posts. SVM is a classical supervised learning method, which constructs a hyperplane in a multidimensional euclidean space to serve as a separator for feature vectors from two classes. We used the radial basis function (RBF) kernel SVM classifier, whose accuracy was assessed using a ten-fold cross-validation process on a labeled post text dataset manually curated by human experts.

3.2 The CNN-Based Method

This method uses the TextCNN approach [5], which first computes a word embedding and then applies the convolutional neural networks (CNN) to perform the classification. TextCNN does not require the removal of stop words or stemming.

Word Embedding. Word embedding which maps words or phrases to numerical vectors, was utilized to allow neural networks to process the text data. We used Word2vec, a commonly used word embedding model [14] that relies on the combination of skip-grams and continuous bag-of-words (CBOW) procedures [15]. CBOW generates a word based on the context, while skip-grams generates the context from a word. For example, if we treat { "Washington D.C.", "is", "the United States"} as a context, then CBOW will generate the word "capital". If given the word "capital", skip-grams will be able to predict the following words: "Washington D.C.", "is", "the United States". The numerical vectors generated by word2vec are used as the input of CNN.

Convolutional Neural Networks. TextCNN contains a single layer of neural net, which allows it to be highly scalable yet sensitive in performing text classification. Figure 3 shows the general scheme of TextCNN [16]. Let d be the dimension of word vector. Given a sentence "Buy drugs on social media without prescription" and $d = 5$, we can generate a sentence matrix in Fig. 3. Then feature maps are generated by filters operating convolutions on the sentence matrix. Here we set the region sizes to 2, 3 and 4, and each region size has two filters. A max-pooling operation is applied to the feature map to retrieve the largest number. Therefore we can take six features from six feature maps and concatenate them together to get a feature vector which will serve as the input of the softmax layer. Finally, we complete a binary classification by using this feature vector through softmax layer.

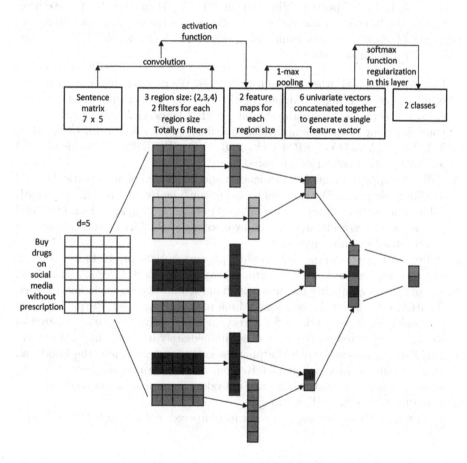

Fig. 3. Illustration of TextCNN

4   Experimental Results

In this section, we will describe the data collection and data processing, and then evaluate the performance of the SVM-based and CNN-based methods. All tools have been implemented in Python 2.7, and run on a DELL workstation with Intel Xeon E5-1603 2.80 GHz CPU, 32G memory, and Ubuntu 18.04 OS.

4.1   Data Collection

The data have been collected using Google+ API. The analyzed dataset has been formed by posts containing at least one of the following 30 keywords [17]:

opioid, alprazolam, amphetamine, antidepressant, benzodiazepine, buprenorphine, cocaine, diazepam, fentanyl, heroin, hydrocodone, meth, methadone, morphine, naloxone, narcan, opana, opiate, overdose, oxycodone, oxymorphone, percocet, suboxone, subutex, pill, rehab, sober, withdrawal, shooting up, track marks

In total, 1,162,445 posts published from 2018/01/01 to 2018/10/31 have been collected. We labeled all the posts manually. The following examples illustrate examples of illicit drug ads from the dataset. Ads 1–3 are selling illicit drugs while ad 4 is a normal post.

1. Buy pain pills and other research chemicals. We do offer discount as well to bulk buyers. Overnight Shipping with tracking numbers provided. Stay to enjoy our services. Overnight shipping with a tracking number provided for your shipment (Fast, safe and reliable delivery). We ship within USA, AUSTRALIA, CANADA, GERMANY, POLAND, SWEDEN, NEW ZEALAND and many other countries not listed here.
2. Hello we supply high quality medication and high rated pharmaceutical opioid at affordable prices. Dear buyers we bring you The Best Of real pharmaceutical product such as oxycodone, nembutal powder, fentanyl patch and fentanyl powder, subutex, adderal, demerol, hydrocodone MDMA etc, and only serious buyers should contact please.
3. Hello, I am a vendor in high quality pharmaceutical products like Xanax, Oxycodone, Fentanyl patch, Viagra, Diazapam, Percoset, Opana, Methadone, etc and also high quality medical marijuana strains like Og kush, Sativa, Kief, S hatter, Girls Scott, Lemon haze, Moon rock, Afghan kush, Purple haze etc, my packaging is very safe and discreet, also my delivery is 100% assured as we do refund or resend the same order immediately in case of any unforeseen.
4. Highlighting concerns with the pharmaceutical supply chain, the Food and Drug Administration warned McKesson, one of the nation's largest wholesalers, for failing to properly handle episodes where pharmacies received tampered medicines, including three ...
FDA scolds McKesson for naproxen in tampered oxycodone bottles -STAT-

4.2 Effectiveness Evaluation

We use precision, recall and F-score as metrics to evaluate the accuracy of the classification methods [18]. Precision is defined as the ratio of predicted and ground-truth illicit ads among all predicted illicit ads, i.e., $Prec = tp/(tp + fp)$. Recall is defined as the ratio of predicted and ground-truth illicit ads among all ground-truth illicit ads, i.e., $Recall = tp/(tp + fn)$. The F-score is the harmonic mean of precision and recall: $F\text{-}score = 2 \cdot Prec \cdot Recall/(Prec + Recall)$. We use 10-fold cross-validation procedures to evaluate the accuracy of both the SVM and CNN based methods.

In TextCNN, we set the parameters as follows: max_sequence_length 20, embedding_dim 200, validation_split 0.16, test_split 0.2 [16]. Table 1 shows the precision, recall, and F-score for SVM and TextCNN. From Table 1, we can see that TextCNN outperforms SVM in all metrics.

Table 1. Accuracy of the SVM based method and TextCNN

Methods	Pre	Recall	F-score
SVM based method	0.65	0.81	0.72
TextCNN	0.97	0.90	0.93

Table 2 shows the running time. In Table 2, the training time represents the average running times for training ten SVM or CNN models during the ten-fold cross-validation procedure. The number of posts in the input dataset for training each model is 1,046,200, which is 90% of the total of 1,162,445 posts. The testing time represents the average running time of predicting the label of a single post. In each iteration of the ten-fold cross-validation, the input number of posts is 116,244 posts. We measure the average time for each post. From Table 2, we can see that the SVM based method takes less than 1 hour while the TextCNN method takes 11 hours for training. Both methods take less than 0.05 second for prediction.

Table 2. Running time of the SVM based method and TextCNN

Methods	Training time	Testing time
SVM based method	2,469 s	0.023 s
TextCNN	3,936 s/epoch, 10 epoch	0.034 s

5 Conclusion

Social media platforms have facilitated illicit drug trading and may be an important driver of the current opioid epidemic. Thus tools for monitoring and analysis of online drug markets are needed to advance epidemiological studies and develop intervention applications. In this paper, we used the Google+ platform as a proof-of-concept to demonstrate that machine-learning-based methods allow for efficient identification of illicit drug advertisements from social media posts. Our tools could be used by health care practitioners, law enforcement officials and researchers to extract and analyze the data related to the opioid abuse epidemic, which can be examined to better understand dynamics of online drug markets, trade, and behaviors. These insights are essential in the development of tailored recommendations and public health intervention strategies that are responsive to social media and online environments.

References

1. Centers for Disease Control and Prevention: Provisional Drug Overdose Death Counts. https://www.cdc.gov/nchs/nvss/vsrr/drug-overdose-data.htm. Accessed 3 Feb 2019
2. Stevens, J.: Internet Stats & Facts for 2019. https://hostingfacts.com/internet-facts-stats/. Accessed 17 Dec 2018
3. Wu, H.C., Luk, R.W.P., Wong, K.F., Kwok, K.L.: Interpreting TF-IDF term weights as making relevance decisions. ACM Trans. Inf. Syst. (TOIS) 26(3), 13 (2008)
4. Suykens, J.A., Vandewalle, J.: Least squares support vector machine classifiers. Neural Process. Lett. 9(3), 293–300 (1999)
5. Kim, Y.: Convolutional neural networks for sentence classification. arXiv preprint arXiv:1408.5882 (2014)
6. Mackey, T.K., Liang, B.A.: Global reach of direct-to-consumer advertising using social media for illicit online drug sales. J. Med. Internet Res. 15(5), e105 (2013)
7. Stroppa, A., di Stefano, D., Parrella, B.: Social media and luxury goods counterfeit: a growing concern for government, industry and consumers worldwide. The Washington Post (2016)
8. Hu, X., Tang, J., Zhang, Y., Liu, H.: Social spammer detection in microblogging. In: Twenty-Third International Joint Conference on Artificial Intelligence (2013)
9. Zheng, X., Zeng, Z., Chen, Z., Yu, Y., Rong, C.: Detecting spammers on social networks. Neurocomputing 159, 27–34 (2015)
10. Agrawal, M., Velusamy, R.L.: R-SALSA: a spam filtering technique for social networking sites. In: 2016 IEEE Students' Conference on Electrical, Electronics and Computer Science (SCEECS), pp. 1–7. IEEE (2016)
11. Jain, G., Sharma, M., Agarwal, B.: Spam detection in social media using convolutional and long short term memory neural network. Ann. Math. Artif. Intell. 85(1), 21–44 (2019)
12. Hull, D.A.: Stemming algorithms: a case study for detailed evaluation. J. Am. Soc. Inf. Sci. 47(1), 70–84 (1996)
13. Salton, G., McGill, M.J.: Introduction to modern information retrieval (1986)
14. Mikolov, T., Chen, K., Corrado, G., Dean, J.: Efficient estimation of word representations in vector space. arXiv preprint arXiv:1301.3781 (2013)
15. Mikolov, T., Sutskever, I., Chen, K., Corrado, G.S., Dean, J.: Distributed representations of words and phrases and their compositionality. In: Advances in Neural Information Processing Systems, pp. 3111–3119 (2013)

16. Zhang, Y., Wallace, B.: A sensitivity analysis of (and practitioners' guide to) convolutional neural networks for sentence classification. arXiv preprint arXiv:1510.03820 (2015)
17. Wu, Y., Skums, P., Zelikovsky, A., Rendon, D.C., Liao, X.: Predicting opioid epidemic by using Twitter data. In: Zhang, F., Cai, Z., Skums, P., Zhang, S. (eds.) ISBRA 2018. LNCS, vol. 10847, pp. 314–318. Springer, Cham (2018). https://doi.org/10.1007/978-3-319-94968-0_30
18. Powers, D.M.: Evaluation: from precision, recall and F-measure to ROC, informedness, markedness and correlation (2011)

Data Analysis and Methodology

A Robustness Analysis of Dynamic Boolean Models of Cellular Circuits

Ariel Bruner and Roded Sharan[✉]

Blavatnik School of Computer Science, Tel Aviv University,
69978 Tel Aviv, Israel
roded@tau.ac.il

Abstract. With ever growing amounts of omics data, the next challenge in biological research is the interpretation of these data to gain mechanistic insights about cellular function. Dynamic models of cellular circuits that capture the activity levels of proteins and other molecules over time offer great expressive power by allowing the simulation of the effects of specific internal or external perturbations on the workings of the cell. However, the study of such models is at its infancy and no large scale analysis of the robustness of real models to changing conditions has been conducted to date. Here we provide a computational framework to study the robustness of such models using a combination of stochastic simulations and integer linear programming techniques. We apply our framework to a large collection of cellular circuits and benchmark the results against randomized models. We find that the steady states of real circuits tend to be more robust in multiple aspects compared to their randomized counterparts.

1 Introduction

Protein-protein interaction (PPI) networks have been mapped for over a decade using techniques such as yeast two hybrid and co-immunoprecipitation. Comprehensive maps are available for multiple species, including yeast and human, with hundreds of thousands of interactions in each (see, e.g., the BioGrid database [1]). As we and others have shown, substantial portions (over 40%) of those maps are spanned by signaling interactions [2]. Thus, protein networks have been extensively used for interpreting genome-wide screens in the context of a phenotype of interest. Previous work in this regard can be broadly categorized into topology-based and logic-based. The first category includes the vast majority of current studies. It aims to characterize a given phenotype by: (i) the location of its affected proteins in a network, identifying network regions, or modules, that are associated with the phenotype [3–10]; and (ii) the network-based attributes of the proteins, such as degree of connectivity and more [11,12].

Electronic supplementary material The online version of this chapter (https://doi.org/10.1007/978-3-030-20242-2_16) contains supplementary material, which is available to authorized users.

Z. Cai et al. (Eds.): ISBRA 2019, LNBI 11490, pp. 183–195, 2019.
https://doi.org/10.1007/978-3-030-20242-2_16

The second, logic-based category aims to provide mechanistic models for the phenotype of interest [13]. While a variety of models can be employed, most prior works focus on logical (Boolean) models or variations thereof [14], thanks to the rich history of these models in the biological domain [15]. These works include Boolean modeling of fundamental systems such as EGFR signaling [16, 17] and MAPK signalling [18]; algorithms to optimize Boolean models against experimental data [19–21]; algorithms to traverse the state space of a model [22–24]; and generalizations of these models to multi-valued [25,26], constraint-based [27,28] and probabilistic [29] ones.

The advantage of the latter logic-based models is that they allow simulating a process of interest under different genetic and environmental perturbations. In recent years, a concentrated community effort has led to the curation of dozens of logic models for a diverse array of cellular processes [3], providing for the first time the opportunity to study the properties of these circuits at large scale.

Since biological processes need to operate consistently in a noisy environment, they must be robust to perturbations, such as stochastic changes in molecular concentrations and protein activity. Extensive research has been done on empirical stability of biological processes [30,31], as well as stability of Boolean models [32–35] and alternative modelling frameworks [36,37].

However, there are few studies that compare the robustness of real networks to those of random networks with similar structural properties. Indeed, random networks can be artificially selected for robustness, affecting their structural properties [33,34]. We are aware of only a single work that directly compares real networks to topologically-similar random ones. Specifically, Daniels et al. [38] compared 67 biological models with topologically- and functionally-similar random models, identifying a node-based sensitivity measure that differs between the two types of models.

Here, for the first time, we compare real and similar random models in terms of their dynamical properties. Since the space of model states is exponentially large, mapping its attractors and basins of attraction is very costly. To tackle this computational problem we design novel integer linear programming (ILP) formulations for model learning and attractor computations. We apply our methodology to a set of 30 curated models covering fundamental biological processes in multiple species with up to 62 nodes. We analyze their steady state (attractor) properties and their robustness to small perturbations. As a benchmark we use randomized models that maintain the topology and logical complexity of the real models. We find that real models tend to be more robust than random models in several aspects that concern the steady state changes in response to intrinsic noise in the system or due to small perturbations to its logic.

2   Preliminaries

For a cell circuit of interest, a Boolean model is a pair (G, F), where $G = (V, E)$ is a directed graph representing the circuit topology with $n = |V|$ nodes and F is a collection of Boolean functions over V. In order to define F let us denote, for each $v_i \in V$, the ordered set $p(v_i) = (v_j : (v_j, v_i) \in E)$, with an arbitrary order over V. Let $f_i : \{0,1\}^{|p(v_i)|} \rightarrow \{0,1\}$ be a Boolean function representing the activity of v_i in terms of its predecessors $p(v_i)$.

An assignment of binary activity levels to each node defines a *state* of the model. We assume a synchronous model, where for each discrete time point t and vertex v_i, $v_{i,t} \in \{0,1\}$ defines the vertex activity at time t, for an arbitrary starting value $v_{i,0}$. Denote $p_t(v_i) = (v_{j,t} : v_j \in p(v_i))$, then the model is updated as follows (for all $v_i \in V$ and $t \geq 0$):

$$v_{i,(t+1)} = f(v_{i,t}) := \begin{cases} v_{i,t} & |p(v_i)| = 0 \\ f_i(p_t(v_i)) & otherwise \end{cases} \tag{1}$$

An *attractor* for G is a simple cycle of network states. Formally, a state path A of length T is a tuple $(V_t : 0 \leq t \leq T)$ where each $V_t = (v_{i,t} : v_i \in V)$ represents a state and (1) holds for each $0 \leq t < T$. We say that A is an attractor if $V_T = V_0$ and for all other t, $V_t \neq V_0$. By definition, every state in a synchronous model is followed by exactly one state, hence every state leads to a single attractor. For a given attractor, we call the set of states that lead to this attractor its *basin of attraction*, and use the notation $s \rightarrow A$ to mean that s is in the basin of attraction of A.

3   Methods

In order to test the robustness of a given model we examine the impact of small perturbations on its dynamics. Specifically, we consider two types of perturbations: (i) changing k bits in the truth tables of one or more of its functions, where k is a small constant; (ii) switching from any state to another state whose Hamming distance to the original is at most k. We study the impact of such changes on the set of model attractors by quantifying for each change the percent of model states that lead to a different attractor in the perturbed model/state.

Since the space of model states is exponentially large, mapping its attractors and basins of attraction is very costly. To tackle this computational problem we combine novel integer linear programming (ILP) formulations for model learning and attractor computations with stochastic simulations for estimating the size of each basin of attraction. To describe our formulation for attractor representation, we first describe a set of sub-programs that form its building blocks.

3.1   State Comparisons

The representation of attractor states requires efficient means to compare them to one another to make sure that they represent a cyclic path of distinct states and that no two attractors share the same state. One way to tackle this problem as well as remove degrees of freedom from solutions is to assign a unique numeric key to each possible network state. The resulting order can then be used to implement equality and inequality constraints.

A natural key for a state is the binary number it represents, but the representation size may exceed the number of bits in a computer word. Instead, we implement this representation by assuming an arbitrary upper bound of M bits per word and splitting the number into $\ell+1$ pieces, where $\ell := \left\lceil \frac{|V|}{M} \right\rceil - 1$. For any two states $a, b \in \{0,1\}^{|V|}$ represented by $x_0, \ldots x_\ell$ and $y_0, \ldots, y_\ell$, respectively, to implement an indicator for an inequality such as $a > b$, we use auxiliary binary variables $(z_0, \ldots, z_\ell)$ and, for convenience, denote $z_{-1} = 0$. We introduce the following constraints for all $0 \leq j \leq \ell$:

$$(2^M + 1)z_j \geq x_j - y_j + z_{j-1}$$
$$(2^M + 1)z_j \leq x_j - y_j + z_{j-1} + 2^M. \tag{2}$$

It holds that $z_0 = 1$ iff $x_0 > y_0$ and for $j \geq 1$:

$$z_j = 1 \iff x_j > y_j \lor (x_j = y_j \land z_{j-1} = 1)$$

This implies that $z_\ell = 1$ iff $a > b$, and we can denote $I_{a>b} = z_\ell$. Using $z_{-1} = 1$ captures a weak ordering, and allows us to implement the indicators $I_{a=b}$ and $I_{a \neq b}$ as well.

We can use these indicators to implement a set membership indicator $I_{a \in A}$ representing whether a state a belongs to a set of states A, with the formulation:

$$\forall a' \in A : I_{a \in A} \geq I_{a = a'}$$
$$I_{a \in A} \leq \sum_{a' \in A} I_{a = a'}$$

3.2   Path Constraints

Let $a_0, \ldots, a_T$ be ordered sets of binary vector variables, representing graph states, such that $\forall t, a_t = (v_{1,t}, \ldots, v_{n,t})$. Let $F(a_t) = (f(v_{1,t}), \ldots, f(v_{n,t}))$, then we can enforce the ordered set of states to represent a path by requiring $a_{t+1} = F(a_t)$ for all $t < T$. We follow the technique used in [39] as detailed below.

For an input node v_i such that $|p(v_i)| = 0$, we only require $v_{i,t+1} = v_{i,t}$. Otherwise, denote by i_j the index of the j'th node in $p(v_i)$. For each binary vector $r \in \{0,1\}^{|p(v_i)|}$, representing a possible input, we create an auxiliary non-negative variable s_r whose value is 0 iff r is the input vector to v_i at time $t+1$, i.e., $r = p_t(v_i)$:

$$s_r = |p(v_i)| - \sum_j \left(1 - r_j + v_{i_j,t}(2r_j - 1)\right)$$
$$v_{i,t+1} \geq f_i(r) - s_r \tag{3}$$
$$v_{i,t+1} \leq f_i(r) + s_r$$

We call this sub-program $PATH(a_0, \ldots, a_T)$. For the case of a length-1 path, we use the simpler expression $a_1 = F(a_0)$ to refer to this sub-program. We can also derive a formulation for requiring the states not to form a path, which we denote by $\overline{PATH}(a_0, \ldots, a_T)$.

3.3   Attractor Learning

We are now ready to present our formulations for attractor learning under various constraints. Using fixed values for the outputs of F, we can use such formulations to compute the attractors of a given model. By substituting the outputs of F with binary variables, we get a formulation whose solution represents a possible model on G and its attractors. For brevity, in the following formulations whenever we write an indicator as a constraint, we require its value to be 1. All formulation depend on specifying in advance upper bounds on the number and length of the model's attractors. For clarity, we assume at first that there are exactly K attractors of length T each. We apply the conditional constraint mechanism below to generalize the formulations to the case that K and T are upper bounds. Let $\{a_{k,t}\}_{1 \leq k \leq K, 0 \leq t \leq T}$ denote ordered sets of binary vector variables representing the K attractors. That is, each $a_{k,t}$ is a vector of n binary variables denoting the state of attractor k at time t. These variables can be populated using the following formulation (with empty objective):

$$\text{Max } \emptyset \text{ s.t.}$$

$$\forall k \leq K : \qquad\qquad PATH(a_{k,0}, \ldots, a_{k,T})$$
$$\forall k \leq K : \qquad\qquad\qquad I_{a_{k,0}=a_{k,T}}$$
$$\forall k \leq K, t > 0 : \qquad\qquad\qquad I_{a_{k,T}>a_{k,t}}$$
$$\forall k < K : \qquad\qquad\qquad I_{a_{k+1,T}>a_{k,T}} \qquad (4)$$

where degrees of freedom in attractor representation are eliminated by forcing the final state of each attractor to be the largest in terms of its associated numeric value, and the final states of different attractors to be ordered as well.

To generalize the program from T, K being exact numbers to upper bounds, we introduce activity variables $w_{k,t}$ for each state $a_{k,t}$, with $w_{k,-1} = 0$ for all k. These activity variables affect which constraints are in effect and what variables are zeroized. Their application to the program requires a conditioning construct which enforces a constraint conditioned on the value of the activity variables. Such conditioning can be constructed by inequality constraints that take effect iff the condition is met. For example, we can formulate the constraint $a \rightarrow b \geq c$, which requires $b \geq c$ iff $a = 1$, by the modified constraint $b \geq c - (1 - a)$. The generalized formulation is as follows:

$$\text{Max } \emptyset \text{ s.t.}$$

$$\forall k, t < T : \qquad (w_{k,t} \wedge w_{k,t+1}) \rightarrow a_{k,t+1} = F(a_{k,t})$$
$$\forall k, t < T : \qquad (w_{k,t} \wedge \neg w_{k,t-1}) \rightarrow I_{a_{k,0}=a_{k,T}}$$
$$\forall k, t > 0 : \qquad\qquad w_{k,t} \rightarrow I_{a_{k,T}>a_{k,t}}$$
$$\forall k < K : \qquad (w_{k,T} \wedge w_{k+1,T}) \rightarrow I_{a_{k+1,T}>a_{k,T}}$$
$$\forall k, t < T - 1 : \qquad\qquad w_{k,t} \leq w_{k,t+1}$$
$$\forall k < K : \qquad\qquad w_{k,T} \leq w_{k+1,T}$$
$$\forall k : \qquad\qquad w_{k,T-1} = w_{k,T}$$
$$\forall k, t : \qquad\qquad \neg w_{k,t} \rightarrow I_{a_{k,t}=0^n} \qquad (5)$$

where 0^n denotes a vector of n zeros and the last constraint results in zeroizing variables that are not in use.

3.4   Robustness to State Perturbations

To quantify robustness, we first consider the sensitivity of a model to small perturbations to its current state, similar to [33, 34, 40]. As attractors represent the steady states or behaviors of the system under study, it is natural to consider the ability of a perturbation to alter the system's behavior. Such perturbations could occur due to the inherent noise in the biological system.

Let (G, F) be a given model with a set ATT of attractors. Given a state a of an attractor A, we consider a perturbations of k of its bits corresponding to a set S of nodes. We denote the perturbed state by $XOR(a, S)$. We say that a is *sensitive* to S if $XOR(a, S) \not\to A$. We say that A is sensitive to S if there exists such $a \in A$.

We evaluate the robustness of a model by a weighted average of its sensitive attractors, where attractors are weighted by their basin size estimate and the result is averaged over multiple choices of S, constrained to have cardinality k. The procedure is as follows:

Algorithm 1. Stochastic state sensitivity score

---

1. Calculate model attractors.
2. Sample a state s and a node set S uniformly at random.
3. Simulate the model from s to derive its attractor A.
4. Sample a state $a \in A$ uniformly at random.
5. Simulate the model from $XOR(a, S)$ to determine A's sensitivity to S.
6. Repeat steps (1-5) 200 times and average the sensitivity results.

As an alternative sensitivity measure, we use an ILP formulation to compute an upper bound rather than an average on sensitivity. That is, we find the maximal weighted average of attractors that are sensitive to a set S of a given cardinality. Denote the relative basin size of attractor $A \in ATT$ by $w_A := |\{s \in \{0,1\}^n : s \to A\}|/2^n$. We can use the stochastic procedure above to estimate w_A for each $a \in ATT$. For the formulation, we further assume that any state reaches its attractor after at most P steps. We denote the set of attractor states by R. We represent the perturbed set S by binary variables $s_1, \ldots, s_n$, and formulate the following program:

$$\text{Max} \sum_{A \in ATT} w_A \cdot I_{b_P^A \in R \setminus A} \text{ s.t.}$$

$$\sum_i s_i \leq k$$

$$\forall A \in ATT : I_{a^A \in A}$$

$$\forall A \in ATT : I_{b_0^A = XOR(a^A, S)}$$

$$\forall A \in ATT : PATH(b_0^A, \ldots, b_P^A)$$

3.5 Robustness to Logic Perturbations

An orthogonal way to measure the robustness of a model is by the effects of perturbing its logical functions on the model's attractors [31,34]. As before, we assume that up to k bits are changed in the truth tables of the Boolean functions.

Let (G, F) be a source model with a set ATT of attractors and let (G, F') be the perturbed model with a set ATT' of attractors. We consider the percent of model states whose associated attractors were eliminated by the change: $\sum_{A \in ATT \setminus ATT'} w_A$. For brevity, we omit the constraint regarding the number of bit changes.

$$\text{Max} \sum_{A \in ATT} w_A \cdot I_A \text{ s.t.}$$

$$\forall A \in ATT : \qquad I_A = 1 \rightarrow \overline{PATH(A)}$$
$$\forall A \in ATT : \qquad I_A = 0 \rightarrow PATH(A)$$

To estimate the average rather than maximum sensitivity to changes, we use a stochastic procedure similar to the above.

4 Results

4.1 Data Retrieval and Implementation Details

We downloaded 76 Boolean models from the CellCollective repository [41], using the truth tables export features. We excluded 3 models with missing information or could not be parsed and 3 acyclic models (hence, no dynamic behavior).

For each model we applied four scoring variants quantifying the sensitivity of the model to state changes (stochastic and ILP-based variants) and attractor elimination due to function changes (stochastic and ILP-based variants). We used 1000 iterations to estimate each model's attractors and their basin sizes. For stochastic measures we used 200 iterations per model. For optimization variants we set the bound on the path length to an attractor to be $P = 1$ (the results were qualitatively similar for $P = 10$). We explored different values of the number k of bit changes. The results followed similar trends when k ranged from 1 to 4, we focus the description on $k = 1$.

We benchmarked the sensitivity of the different models against randomized models in which both topology and logic functions are randomized. Specifically, the topology was randomized while preserving in- and out-degrees using the switching method with $10n$ switch attempts; the Boolean functions were randomized while preserving the number of 1's in their truth tables by permuting table rows. We generated 40 random models for each real model.

We split the processing jobs among five servers. Two with 72 CPUs (2300 MHz) and 258 GB RAM, and the remaining three with 40 CPUs (2800 MHz) and 258 GB RAM. When running the stochastic algorithms we ran iterations in parallel, and let the ILP solver (Gurobi) utilize all cores. Table 1 provides running time statistics. Running times varied considerably between models, thus we set a timeout of 90 min (in addition to ILP problem setup time) for running the different scoring methods on each model and discarded the results for models on which the computation was not completed within the given time bound. Overall, we collected results for 30 of the 70 models.

Table 1. Running times in seconds of the different scoring variants.

Variant	Median	Mean	Max
Model change, ILP	0	1	11
Model change, stochastic	27	83	1917
State change, ILP	14	322	6728
State change, stochastic	6	38	1371

4.2 Real Models are More Robust than Their Randomized Counterparts

We applied our robustness computation pipeline to the 30 models described above[1] and compared the results to those obtained on randomized models. When applying the ILP-based sensitivity analysis of model changes we observed that in almost all cases (for both real and random models) a single bit change was enough to eliminate the entire attractor space of a model. Thus, we computed an alternative score where the change is constrained to a specific node at a time, producing a sensitivity score for each of the model's nodes.

The distributions of the five score variants (3 model-based and 2 node-based) on real and random models are given in Fig. 1. Evidently, real models have lower sensitivity scores than random models across all score variants. To quantify the deviation of the real models from the corresponding distributions of random scores, we used a Wilcoxon signed-rank test, where the score of each real model was paired with the mean score of its random counterparts. For the node-based model perturbation scores, we paired the mean score of vertices in a real model with the mean scores of nodes in its random counterparts.

The comparison results are summarized in Table 2, revealing that in all cases the real models are more robust to changes than their randomized instances. These differences were significant for all score types ($p < 0.05$), and especially for the state changes ($p < 0.005$), as also evident from the box plots in Fig. 2.

---

[1] The source code used for the analysis can be found at github.com/arielbro/attractor_learning, commit hash 83474950c9fc3aa61277d5535a142aad90ff7eed.

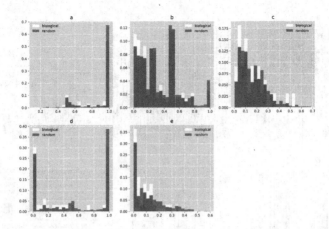

Fig. 1. Distribution of model scores for each score variant, comparing real and random models. (a) ILP per-node scores for model perturbation. (b) Stochastic per-node scores for model perturbation. (c) Stochastic scores for model perturbation. (d) ILP scores for state perturbation. (e) Stochastic scores for state perturbation.

4.3   Stochastic vs. Optimization-Based Sensitivity Analysis

It is interesting to compare the stochastic scores that measure the mean sensitivity to a change vs. the ILP-based optimization scores that measure the extremes. As an example, nearly all models had an optimization score of 1 for attractor elimination, while not for the stochastic scores, suggesting that there is a marked difference between a random perturbation to the logic of a model and a targeted one. For individual nodes, there was a correlation of 0.20 between the stochastic and ILP-based variants. For state perturbations, there was a strong correlation (0.74) between the stochastic and ILP-based variants. These relationships are visualized in Fig. 3. The ILP-based scores are not a perfect upper bound for the stochastic ones since they assume a restricted path length to a modified attractor, but are almost always higher for the same model.

Table 2. Comparison of mean sensitivity scores for real and random models.

Variant	Real	Random	p-value
Model change, ILP (nodes)	0.837	0.863	0.049
Model change, stochastic (nodes)	0.311	0.339	0.017
Model change, stochastic	0.154	0.168	0.013
State change, ILP	0.226	0.416	3.89e−03
State change, stochastic	0.073	0.126	8.35e−04

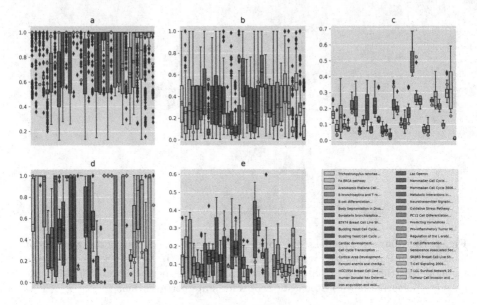

Fig. 2. Box plots showing the distribution of sensitivity scores for each base network. Whiskers denote furthest datapoint within 1.5 IQR from median. Red dots denote the real models, and for per-node scores the mean of real model node scores. (a) ILP per-node scores for model perturbation. (b) Stochastic per-node scores for model perturbation. (c) Stochastic scores for model perturbation. (d) ILP scores for state perturbation. (e) Stochastic scores for state perturbation. (Color figure online)

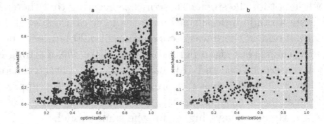

Fig. 3. Scatter plots comparing ILP-based and stochastic-based scores across all models. Real models appear in red and randomized models in blue. (a) Per-node model perturbation scores. (b) State perturbation scores. (Color figure online)

5   Conclusions

We have presented a first analysis of the attractor landscape of Boolean models of cellular circuits and its sensitivity to state and logic changes. Our analysis combined novel ILP formulations with stochastic simulations to overcome the inherent complexity of the problem. We found that real models are more robust than their randomized counterparts with respect to both state and logic changes.

While the analysis could be successfully applied to dozens of circuits with up to 62 nodes each, the optimization problem remains challenging for larger circuits, especially when allowing longer paths from a state to its associated attractor. Another interesting problem the generalization of the methods presented to asynchronous update schemes.

Acknowledgements. RS was supported by a research grant from the Israel Science Foundation (no. 715/18).

References

1. Chatr-Aryamontri, A., et al.: The BioGRID interaction database: 2015 update. Nucleic Acids Res. **43**(Database issue), D470–D478 (2015)
2. Silberberg, Y., Kupiec, M., Sharan, R.: A method for predicting protein-protein interaction types. PLoS ONE **9**, e90904 (2014)
3. Ideker, T., et al.: Discovering regulatory and signalling circuits in molecular interaction networks. Bioinformatics **18**(1), s233–s240 (2002)
4. Vandin, F., Upfal, E., Raphael, B.: Algorithms for detecting significantly mutated pathways in cancer. J. Comput. Biol. **18**(3), 507–522 (2011)
5. Cowen, L., et al.: Network propagation: a universal amplifier of genetic associations. Nat. Rev. Genet. **18**(9), 551–562 (2017)
6. Shachar, R., et al.: A systems-level approach to mapping the telomere length maintenance gene circuitry. Mol. Syst. Biol. **4**, 172 (2008)
7. Yeger-Lotem, E., et al.: Bridging high-throughput genetic and transcriptional data reveals cellular responses to alpha-synuclein toxicity. Nat. Genet. **41**(3), 316–323 (2009)
8. Huang, S., Fraenkel, E.: Integrating proteomic and transcriptional and and interactome data reveals hidden components of signaling and regulatory networks. Sci. Sig. **2**(81), ra40 (2009)
9. Yosef, N., et al.: Toward accurate reconstruction of functional protein networks. Mol. Syst. Biol. **5**, 248 (2009)
10. Dittrich, M., et al.: Identifying functional modules in protein-protein interaction networks: an integrated exact approach. Bioinformatics **24**(13), i223–i231 (2008)
11. Said, M., et al.: Global network analysis of phenotypic effects: protein networks and toxicity modulation in Saccharomyces cerevisiae. Proc. Natl. Acad. Sci. USA **101**(52), 18006–18011 (2004)
12. Jonsson, P., Bates, P.: Global topological features of cancer proteins in the human interactome. Bioinformatics **22**(18), 2291–2297 (2006)
13. Novere, N.L.: Quantitative and logic modelling of molecular and gene networks. Nat. Rev. Genet. **16**(3), 146–158 (2015)
14. Morris, M., et al.: Logic-based models for the analysis of cell signaling networks. Biochemistry **49**(15), 3216–3224 (2010)
15. Kauffman, S.: Metabolic stability and epigenesis in randomly constructed genetic nets. J. Theor. Biol. **22**(3), 437–467 (1969)
16. Samaga, R., et al.: The logic of EGFR/ErbB signaling: theoretical properties and analysis of high-throughput data. PLoS Comput. Biol. **5**(8), e1000438 (2009)

17. Oda, K., et al.: A comprehensive pathway map of epidermal growth factor receptor signaling. Mol. Syst. Biol. **1**, 2005.0010 (2005)
18. Grieco, L., et al.: Integrative modelling of the influence of MAPK network on cancer cell fate decision. PLoS Comput. Biol. **9**(10), e1003286 (2013)
19. Saez-Rodriguez, J., et al.: Discrete logic modelling as a means to link protein signalling networks with functional analysis of mammalian signal transduction. Mol. Syst. Biol. **5**, 331 (2009)
20. Mitsos, A., et al.: Identifying drug effects via pathway alterations using an integer linear programming optimization formulation on phosphoproteomic data. PLoS Comput. Biol. **5**(12), e1000591 (2009)
21. Moignard, V., et al.: Decoding the regulatory network of early blood development from single-cell gene expression measurements. Nat. Biotechnol. **33**(3), 269–276 (2015)
22. Chaouiya, C., Naldi, A., Thieffry, D.: Logical modelling of gene regulatory networks with GINsim. Methods Mol. Biol. **804**, 463–479 (2012)
23. Qiu, Y., et al.: On control of singleton attractors in multiple Boolean networks: integer programming-based method. BMC Syst. Biol. **8**(Suppl. 1), S7 (2014)
24. Dubrova, E., Teslenko, M.: A SAT-based algorithm for finding attractors in synchronous Boolean networks. IEEE/ACM Trans. Comput. Biol. Bioinform. **8**(5), 1393–1399 (2011)
25. Morris, M., et al.: Training signaling pathway maps to biochemical data with constrained fuzzy logic: quantitative analysis of liver cell responses to inflammatory stimuli. PLoS Comput. Biol. **7**(3), e1001099 (2011)
26. Huard, J., et al.: An integrative model links multiple inputs and signaling pathways to the onset of DNA synthesis in hepatocytes. FEBS J. **279**(18), 3290–3313 (2012)
27. Dasika, M., Burgard, A., Maranas, C.: A computational framework for the topological analysis and targeted disruption of signal transduction networks. Biophys J. **91**(1), 382–398 (2006)
28. Vardi, L., Ruppin, E., Sharan, R.: A linearized constraint-based approach for modeling signaling networks. J. Comput. Biol. **19**(2), 232–240 (2012)
29. Gat-Viks, I., et al.: A probabilistic methodology for integrating knowledge and experiments on biological networks. J. Comput. Biol. **13**(2), 165–181 (2006)
30. Alon, U., et al.: Robustness in bacterial chemotaxis. Nature **387**, 913–917 (1997)
31. Li, F., et al.: The yeast cellcycle network is robustly designed. Proc. Natl. Acad. Sci. USA **101**, 4781–4786 (2004)
32. Aldana, M., Cluzel, P.: A natural class of robust networks. Proc. Natl. Acad. Sci. USA **100**, 8710–8714 (2003)
33. Fretter, C., Szejka, A., Drossel, B.: Perturbation propagation in random and evolved Boolean networks. New J. Phys. **11**, 033005:1–033005:13 (2009)
34. Sevim, V., Rikvold, P.: Chaotic gene regulatory networks can be robust against mutations and noise. J. Theor. Biol. **253**, 323–332 (2008)
35. Peixoto, T.: Redundancy and error resilience in Boolean networks. Phys. Rev. Lett. **104**, 048701 (2010)
36. Klemm, K., Bornholdt, S.: Topology of biological networks and reliability of information processing. Proc. Natl. Acad. Sci. USA **102**, 18414–18419 (2005)
37. Lyapunov, A.M.: The general problem of the stability of motion. Int. J. Control **55**, 531–534 (2007)
38. Daniels, B., et al.: Criticality distinguishes the ensemble of biological regulatory networks. Phys. Rev. Lett. **121**, 138102 (2018)

39. Sharan, R., Karp, R.: Reconstructing Boolean models of signaling. J. Comput. Biol. **13**(2), 165–181 (2006)
40. Ghanbarnejad, F., Klemm, K.: Impact of individual nodes in Boolean network dynamics. EPL **99**, 58006 (2012)
41. Helikar, T., et al.: The cell collective: toward an open and collaborative approach to systems biology. BMC Syst. Biol. **6**, 96 (2012)

Graph Transformations, Semigroups, and Isotopic Labeling

Jakob L. Andersen[1], Daniel Merkle[1,2(✉)], and Peter S. Rasmussen[1]

[1] Department of Mathematics and Computer Science,
University of Southern Denmark, 5230 Odense, Denmark
`daniel@imada.sdu.dk`
[2] Department of Systems Biology, Harvard Medical School, Boston, MA 02115, USA

Abstract. The Double Pushout (DPO) approach for graph transformation naturally allows an abstraction level of biochemical systems in which individual atoms of molecules can be traced automatically within chemical reaction networks. Aiming at a mathematical rigorous approach for isotopic labeling design we convert chemical reaction networks (represented as directed hypergraphs) into transformation semigroups. Symmetries within chemical compounds correspond to permutations whereas (not necessarily invertible) chemical reactions define the transformations of the semigroup. An approach for the automatic inference of informative labeling of atoms is presented, which allows to distinguish the activity of different pathway alternatives within reaction networks. To illustrate our approaches, we apply them to the reaction network of glycolysis, which is an important and well understood process that allows for different alternatives to convert glucose into pyruvate.

Keywords: Glycolysis · Isotopic labeling · Hypergraphs ·
Double pushout

1 Introduction

The study of semigroups has provided essential insight into the analysis of decomposability of systems. One of the most important contributions is the Krohn-Rhodes theory (or algebraic automata theory) [17] which allows to decompose any semigroup into simpler components (similar to the Jordan-Holder decomposition for groups). While conceptually very powerful, its application in life sciences is quite limited. A notable exception is a series of scientific contributions to analyse biological systems, more specifically metabolic and gene regulatory networks [10]. Based on semigroup theoretical approaches, subsystems and their hierarchical relations are identified within several biochemical systems, including the lac operon, the Krebs cycle, and the p53-mdm2 genetic regulatory pathway.

Supported in part by the Independent Research Fund Denmark, Natural Sciences, grant DFF-7014-00041.

An essential tool for these studies is the analysis of an associated algebraic structure based on the Krohn-Rhodes decomposition. The hierarchical decomposition is also used as a way of assessing structural complexity changes of biological systems. However, for the Krohn-Rhodes complexity [17] (i.e., the least number of groups in a wreath product) it is still unknown if its computation is even decidable. Furthermore, it is also practically a very difficult problem to solve [9]. Therefore an alternative decomposition approach, the holonomy decomposition, [14] has been promoted. Holonomy decomposition allows to analyse large(r) systems. The decomposition approaches as well as most other methods that aim to analyse metabolic networks (e.g., Flux Balance Analysis (FBA) or Elementary Flux Modes (EFM) [16,19]) are based on a fixed set of abstracted molecular structures and rules (named reactions) that transform molecules into each other. In the decomposition approaches a biological system is represented as a state automata, where subsets of substrates correspond to states in the state set A and a transition function $\delta \colon A \times X \to A$ describes for which "input" a transition is performed. An input can, e.g., encode the presence of a specific enzyme. In FBA, EFM, and similar approaches the central object for the analysis is the stoichiometric matrix of the system, where rows encode compounds and rows encode reactions. The null spaces of the stoichiometric matrix are used in order to characterise the system under consideration.

We follow a more chemically motivated (and direct) modelling approach with several important differences. (i) Being the most natural approach, we model chemical reaction networks as directed multi-hypergraphs [20], where the directed hyper-edges correspond to chemical reactions. (ii) Molecules are not modelled as abstract entities, but as undirected graphs, with vertices representing atoms, and edges representing the chemical bonds between them. (iii) Hypergraphs are generated by methods based on graph transformation [3]. This level of abstraction allows to trace each individual atom through a single or a series of reactions. (iv) We employ a mathematical framework in which pathways are rigorously defined as integer-hyperflows on directed multi-hypergraphs [4].

In the following we introduce a semigroup approach in order to assist in the design of (stable) isotopic labeling experiments. A typical labeling experiment uses a isotope-labelled educt molecule, for instance glucose with a single or several labelled atoms (e.g. carbon-13, ^{13}C, in place of a ^{12}C) at a specific position of the molecule. These compounds are then used for some experiment, for instance the glucose is ingested or by other means transformed into the product molecules. By extracting the product molecules and analysing at which position the labeled atoms ended up (e.g. by using mass spectrometry), the original labeling might have been informative or not, i.e., it might or might not be possible to learn something about the chemical process involved, or to allow for quantification of different pathway alternatives. We present a framework to automatically infer if a specific labeling experiment is informative or not. Besides introducing a theoretical framework, we also aim to support the cross-fertilisation of different scientific communities, including biology, chemistry and theoretical computer science. In order to make such a bridge viable we borrow many techniques from algorithmic engineering [18] as all the methods presented are efficiently implemented and easily accesible to scientists with a limited background in computer science. E.g., as an efficient inference of the automorphism group molecular graphs is essential for our approach, we employ our state-of-the-art algorithmic engineering approach for inference of the automorphism group of a graph [5].

2   Basic Definitions

We will keep our mathematical notation brief. For details wrt. transformation semigroups, we refer the reader, e.g., to [13].

2.1   Semigroup Theory

Definitions. (Group, Semigroup, Monoid, Transformation Semigroup, Orbit). A *group* is a set G together with a operation $\bullet$. The operation takes two elements of G and returns an element of G. A group must satisfy the four group axioms; closure, associativity, identity, and invertibility. A *semigroup* must satisfy only two requirements: (i) (closure) if $g, h \in G$, then $g \bullet h$ must be in G, and (ii) (associativity) for all $g, h, k \in G$, it must hold that $(g \bullet h) \bullet k = g \bullet (h \bullet k)$. If a semigroup includes an identity element it is called a *monoid*. A *transformation* t is a function that maps a set Ω to itself, i.e., $t \colon \Omega \to \Omega$. A *transformation semigroup* on a finite set Ω is a set of transformations which is closed under composition. If every transformation is bijective, then the semigroup is also a *permutation group*. It is well know that any semigroup can be realised as a transformation semigroup of some set. The *orbit* of a point $\omega \in \Omega$ is all the points that can be reached via (semi)group actions. We write it as $\mathrm{Orb}_G(\omega) = \{g(\omega) \mid g \in G\} \subseteq \Omega$. The orbit can also be defined for tuples: let $\overline{\omega} = (\omega_1, \ldots, \omega_k)$ be a group element from Ω^k, then $\mathrm{Orb}_G(\overline{\omega}) = \{(g(\omega_1), \ldots, g(\omega_k)) \mid g \in G\}$. A *generating set* (or generating system) S is a subset of G such that every element of G can be expressed as a finite product of elements of S. We write $G = \langle S \rangle$. If $S = \{s_1, s_2, \ldots, s_k\}$, it is also common to write $G = \langle s_1, s_2, \ldots, s_k \rangle$.

Many problems involving semigroups of a finite set of generators are hard. It can be shown that determining any "sensible" property of such semigroups is even undecidable and, e.g., that membership testing in commutative transformation semigroups is NP-complete [8] (while being polynomial in permutation groups). Despite these results, the computational runtime needed for the results of this contribution will not be dominated by semigroup computations.

2.2   Graph Transformations

Definitions (Molecule, Reaction, Span, Double Pushout Approach, Atom Maps, Derivation, Derivation Graph). A *molecule* is a simple connected graph $G = (V, E)$. Vertices of V are annotated with an attribute describing the atom type and its charge. The edges E are annotated with the bond type, e.g., single bond $(-)$, double bond $(=)$, or aromatic bond. With molecules represented as graphs a chemical reaction from a set of educt graphs to a set of product graphs can naturally be defined as graph transformations. To this end, we employ the *double pushout (DPO) formalism* [7]. In DPO, the rewriting of the educt pattern L into the product pattern R is specified as a *span* $L \leftarrow K \rightarrow R$, where K denotes the subgraph of L that remains unchanged during the rewriting operation (hence it is also a subgraph of R). The rule $L \leftarrow K \rightarrow R$ can be applied to a graph G if and only if (i) the pattern (precondition) L can be embedded in G (by means of a suitable graph morphism m) and (ii) there are objects D and H with so-called pushouts such that the diagram

$$L \xleftarrow{\;\;l\;\;} K \xrightarrow{\;\;r\;\;} R$$
$$\left\downarrow m \right. \qquad \left\downarrow \right. \qquad \left\downarrow \right.$$
$$G \longleftarrow D \longrightarrow H$$

commutes [11]. The objects D and H are guaranteed to be unique if they exist. The graph H is the product obtained by rewriting the educt G with respect to the rule $L \leftarrow K \rightarrow R$ and the matching morphism m. For a graph rewriting rule to be chemical we require in addition that (i) all graph morphisms are injective (i.e., they describe subgraph relations), (ii) the restriction of l and r to the vertices is bijective (ensuring the atoms are preserved), and (iii) that changes in edges (chemical bonds) and charges conserve the total number of electrons. The first two conditions ensure that all reactions are logically reversible [1]. Furthermore, they ensure that given $m \colon L \rightarrow G$, the *atom map* of every reaction is well-defined, i.e., given $m \colon L \rightarrow G$, there is a bijection $\varphi \colon V(G) \rightarrow V(H)$ between the vertices (atoms) of the educt graph G and the product graph H. The third condition captures much of the semantics of chemistry, which views chemical reactions as rearrangements of the electron pairs that form the chemical bonds. We call the application of a graph transformation rule a *direct derivation* from G to H via rule p and morphism m, written $G \xRightarrow{p,m} H$ or $G \xRightarrow{p} H$ if the morphism is unimportant. An example DPO diagram for a chemical reaction called inverse aldol addition is depicted in Fig. 1. Note that for two different morphisms m and m' it might hold $G \xRightarrow{p,m} H$ and $G \xRightarrow{p,m'} H$, i.e., the same rule might transform a set of molecules G into a set of molecules H using different atom maps. A *derivation graph* (DG) is a directed hypergraph $H = (V, E)$ consisting of vertices V and hyperedges E. Each vertex in the DG contains a molecule-graph. A hyperedge $e = (e^+, e^-)$ is a multiset of tail-vertices e^+ to a multiset of head-vertices e^-. In the case of the DG, hyperedges correspond to DPO rule applications which in turn correspond to chemical reactions. The DG might be given or can be computed by starting with an initial set of molecules and then applying rules following some strategy, for instance apply all rules repeatedly, until no new molecules are produced. In [4] a pathway corresponds to an integer hyperflow within a DG. Here, as we are interested if a specific reaction is either active or not within a pathway, we conveniently define a *pathway* to be a (sub)set of hyperedges of a given DG (which could easily be inferred by all edges of an integer hyperflow according to [4], for which there is a non-zero flow).

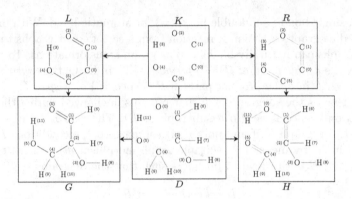

Fig. 1. A chemical example of a DPO diagram with the application of a rule (inverse aldol addition). The graph transformation corresponds to an edge in the derivation graph.

3   The Hypergraph Semigroup Approach

The overarching goal of an isotopic labeling experiment is to be able to distinguish different pathways (or quantify their activity level). In the first step of our approach we construct a hypergraph-semigroup based on a DG that covers different pathway alternatives. This semigroup will be used to compute orbits of atoms. Based on the orbits, we will define a *pathway table* where rows are potentially active pathways to be distinguished, and columns are the informative candidates for atom labeling and the entries are results of orbit calculations.

3.1   Forward Approach

Definitions (Atomic Linearisation, Hypergraph Semigroup). Let n be the total number of all atoms of all molecular graphs in a DG, and let $\Omega = \{1, 2, \ldots, n\}$. An *atomic linearisation* of a DG is an ordered list $[1, 2, \ldots, n]$ of ids corresponding to all the atoms of all molecules[1]. The *hypergraph semigroup* of a derivation graph $H = (V, E)$ is a semigroup $G = \langle S_\Omega \rangle$ acting on Ω, where the generators S_Ω are defined by the union of (partial) transformations and automorphisms, i.e., $S_\Omega = T_\Omega \cup A_\Omega$. More precisely, $T_\Omega = \bigcup_{e \in E} T(e)$ is the union of all atom maps (resulting from a chemical reaction, i.e., from a direct derivation using graph transformation) of all hyperedges in the DG. These atom maps correspond to partial transformations of the semigroup acting on Ω. The generators based on the automorphisms (i.e, the symmetries) of the molecular graphs are defined as $A_\Omega = \bigcup_{v \in V} a_v$, where a_v is a set of generators of the automorphism group of the molecular graph v, i.e., A_Ω describes all symmetries of all molecular graphs in the DG. Note, that $\langle a_v \rangle$ corresponds to a permutation group.

---

[1] Note: The linearisation ids are 1-indexed since they will be used in a semigroup where the tradition is to use the range 1 to n.

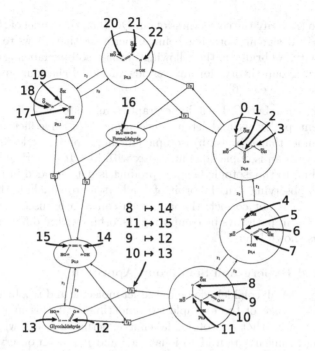

Fig. 2. A DG for the Formose reaction. To reduce clutter we restrict Ω to the 22 carbon atoms in the DG. Integers 1 to 22 correspond to the ids of an atomic linearisation. For the derivation $\{p_{0,2}\} \xrightarrow{p,m} \{p_{0,0}, \text{glycolaldehyde}\}$ (based on an inverse aldol addition) the corresponding partial transformation is $\{8 \mapsto 14, 11 \mapsto 15, 9 \mapsto 12, 10 \mapsto 13\}$. The only symmetry within a molecular graph is found in $p_{0,0}$, i.e., $A_\Omega = a_{p_{0,0}} = \{(14\ 15)\}$.

An example based on a DG for the so-called Formose reaction is depicted in Fig. 2. For modelling details and the 4 graph transformation rules (an aldol-addition reaction, a keto-enol tautomerisation reaction, and their inverses) used to compute the DG see [1]. We restrict Ω to carbon atoms to reduce clutter of the illustration. While every hyperedge in the DG defines a transformation of the hypergraph semigroup, (14 15) is the only automorphism found (in cycle notation) in all molecular graphs (an inclusion of hydrogens would significantly increase the number of automorphisms). We depict the partial transformation $t := \{8 \mapsto 14, 11 \mapsto 15, 9 \mapsto 12, 10 \mapsto 13\}$ and note, that this is also a partial permutation as an inverse is always well defined. This is true for any partial transformation resulting from an edge in the DG, and simply reflects the fact, that any chemical reaction can theoretically be inverted. Additional constraints, e.g., from thermodynamics, might make this inversion very unlikely or impossible in reality, and therefore the corresponding hyperedge might not exist in the DG. In the depicted DG all chemical reactions are indeed reversible (in mathematical terms: all inverses of the partial permutations are also generators of the semigroup), which is usually not the case.

In order to (i) verify the correctness of our DG (i.e., the model of the chemical system) and (ii) design an isotopic labeling experiment that allows to distinguish pathway activity, we bridge in the following wet-lab isotopic labeling experimentation and semigroup theory (for an overview of model checking approaches in Systems Biology, see e.g. [6]).

Forward Approach. Let G be a hypergraph semigroup of a DG covering a set of different pathways and let a single atom k of an educt molecule be labelled. Assume, that a (wet-lab) isotopic labeling experiment leads to a product molecule with an isotopic label at vertex with id i. If $i \notin \mathrm{Orb}_G(k)$, then the pathway leading to the isotopic labelled product is not included in the DG.

While not observing i in the orbit of k allows us to conclude that the DG (i.e., the model) is not correct, this approach can also be used to distinguish activity of different pathways by computing the orbits within different pathways in a DG, as is shown later.

3.2 Inverted Hypergraph Semigroup Approach

Consider Fig. 2, let glycolaldehyde be the educt molecule and $p_{0,6}$ be the product molecule for the sake of the example (we underline that this makes no sense chemically). Assume that an isotopic labeling experiment is run with the atom corresponding to linearisation id 13 is labelled and $p_{0,6}$ is an observed product (with labeled atoms corresponding to linearisation ids 0, 2, and 3). Without even thinking about pathways, it would be desirable to be able to tell if the model is correct – should it even be possible to observe this labeling of $(0, 2, 3)$? Using the inverted hypergraph semigroup, we can answer in this case it is not.

Definition (Inverted Hypergraph Semigroup). For a given DG, let $G = \langle S_\Omega \rangle = \langle T_\Omega \cup A_\Omega \rangle$ be a hypergraph semigroup acting on Ω, where $t \in T_\Omega$ are partial permutations defined by the atom maps of a DG and $a \in A_\Omega$ are permutations describing the automorphisms of the molecular graphs in the DG. The *inverted hypergraph semigroup* G^{-1} of G is a semigroup $G^{-1} = \langle T_\Omega^{-1} \cup A_\Omega \rangle$ where $T_\Omega^{-1} = \{t^{-1} \mid t \in T_\Omega\}$.

Note, that t^{-1} with $t \in T_\Omega$ is not modelling the action of a chemical reaction, but is used in order to infer a precondition that must hold in order to make t applicable. Using the orbit of tuples this leads to the following approach in order to answer the question raised earlier.

Inverted Semigroup Approach. Let G be a hypergraph semigroup of a DG covering a set of different pathways and let a single atom k of an educt molecule be labelled. Assume, that a (wet-lab) isotopic labeling experiment leads to a product molecule with isotopic labels at vertexes $(i_1, \ldots, i_n)$. If $(k, \ldots, k) \notin \mathrm{Orb}_{G^{-1}}((i_1, \ldots, i_n))$ then the pathways leading to the isotopic labeled product is not included in the DG.

The inverted approach underlines an insight that can be used to distinguish pathways: if some atoms have a similar or even identical orbit in two different pathways, using the inverted orbit can potentially disambiguate the two. To this end, we will introduce pathway tables.

Table 1. An example of a pathway table.

Pathway	Atom label		
	0, C	1, C	2, C
1	0, 1, 3, 4	2	1, 3
2	0, 1, 2, 4	3	2, 4
3	1, 2, 3, 4	0	1, 2, 3, 4
4	0, 1, 2, 3, 4		
DG	0, 1, 2, 3, 4	0, 2, 3	0, 1, 2, 3, 4

4   Pathway Tables for Isotopic Labeling Design

Suppose we have a set of potential pathways, P, and we aim to distinguish them. For simplicity, we assume that each pathway has a single educt molecule S and a single product molecule T. Let S_V be the set of atoms in S. Let $A \subseteq S_V$ be the set of atoms that are candidates for isotopic labeling. Assume the DG is also a pathway, corresponding to the union of all pathways under consideration. Consider Table 1 which depicts an automatically inferred $P \times A$ table, where each cell $(p, a) \in P \times A$ contains the subset of $\mathrm{Orb}_{G_p}(a)$ that is contained in the product molecule T (to reduce clutter we omit the subscript, details of the underlying pathways are omitted). The columns are labelled with the local ids (not linearisation ids) of A, referring to atoms of the educt molecule S. The atom type is also shown to ease the reading of the table. The rows are the four potential pathways of P as well as the pathway "DG" which is the union of all pathways. Each table entry is a comma-separated list of local ids (not linearisation ids) that correspond to the orbit of the specific column in the product molecule T, in other words, the part of the orbit that reaches T. The reason that the local (instead of linearisation) ids of molecules are used is convenience for a biologist/chemist when analysing the information.

We will use the following notation to easier relate the ids to where they are coming from: i_S means the carbon with local id i in educt molecule S. For example, 1_S corresponds to the carbon with local id 1 in S, i.e., the column from Table 1 called "1, C". We use a similar notation i_T for ids in product molecule T.

Mass spectrometry (MS) is an analytical standard technique in order to measure masses of chemical compounds. Together with stable isotope labeling experimentation this is an extremely powerful tool for the analysis of (bio)chemical systems. Note, that is much easier and cheaper to detect whether some molecule analysed via MS has zero or one labelled atoms (mass difference) compared to determining where the labelled atoms are located in a molecule (position). In contrast to inferring the mass of a compound (via so-called MS[1] experimentation), the inference of the position of a labelled atom usually requires the fragmentation of the compound (via MS[2] experiments), or the fragmentation of fragments (MS[3]), and so on. The fragmentation process of a compound leads to a series of MS[k] spectra, which can be used as input for inferring the position of a labelled atom. As it is much easier to determine if there are no labels versus if there are some labels, it is preferred to look for mass-distinguishing factors over positions. In the table above, we see such an example. If someone carries out an isotopic labeling experiment where carbon 1_S of S is labelled and not any labels at all are observed in T then we know that pathway 4 cannot have been active. To distinguish pathways 1, 2, and 3 based on a 1_S labeling requires determination of the location of the label in the target.

In the example, consider labeling 0_S (1st column). Each entry has a lot of ids in common with the rest of the entries, but it can still be used. For instance, observing 3_T would exclude pathway 2. The inverted semigroup approach analyses inverted orbits: consider a pair of atoms that pathway 1 and 3 have in common, for instance $(1_T, 3_T)$. Now compute the inverted orbit for this pair in both pathway 1 and 3: $\mathrm{Orb}_{G_{P_1}^{-1}}((1_T, 3_T)) = \{\ldots, (0_S, 2_S), (2_S, 0_S),$ $\mathrm{Orb}_{G_{P_3}^{-1}}((1_T, 3_T)) = \{\ldots, (0_S, 2_S), (2_S, 2_S), (0_S, 0_S), (2_S, 0_S)\}$. The "$\ldots$" represents the part of the orbit not in molecule S, which is not relevant for this part of the analysis. We know that we started with a label only at 0_S, thus both elements of the tuple $(1_T, 3_T)$ must have originated from 0_S. The only of the two pathways where this could have happened is pathway 3 because of the $(0_S, 0_S)$. Therefore, if we ever observe $(1_T, 3_T)$, we can exclude pathway 1. Note, that it is straightforward to remove non-informative columns (where all entries are identical) and to merge identical lines (i.e., non-distinguishable pathways) of a pathway table.

5   A Chemical Example

We have applied our approach to several chemical examples and will present here the analysis of alternative glycolysis pathways. Glycolysis is a process that converts one glucose molecule into two pyruvate molecules. There are two glycolysis types that achieve this: The Entner-Doudoroff (ED) and the Embden-Meyerhoff-Parnass (EMP) pathway [12]. Using the graph transformation rules for glycolysis as presented in [2], it is possible to determine what carbons in glucose are informative in order to distinguish the ED and EMP pathway by labeling. Figure 3 depicts the derivation graph which is an union of the two pathways depicted with green (ED) and blue (EMP) arrows. Figure 4 shows the source molecule (glucose) and the target molecule (pyruvate) and the local atom ids as used in the pathway table.

Table 2 depicts the pathway table of the two pathways as well as their union. The columns correspond to all informative atoms, i.e., all atoms that allow to distinguish different pathways. Ignoring hydrogens, it turns out that there are six atoms in glucose that can be used to distinguish the EMP and ED pathways. For instance, if an isotopic labeling experiment is carried out with a ^{13}C in the place of the normal ^{12}C at position 6 in the glucose molecule, and the mass spectrometry results show only pyruvates with ^{13}C at positions different from 2, then we can conclude that the EMP pathway was not active. The orbit of the oxygens with internal id 0 or 13 (not so 7 or 12) does not include any atoms

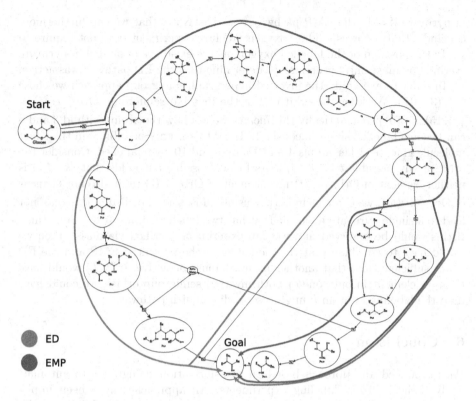

Fig. 3. A combined DG for the ED and EMP pathways. Not all compounds are depicted in order to reduce clutter (e.g., cofactors NADH, ATD, ADP, and water molecules are not depicted).

Fig. 4. Glucose and pyruvate molecules where the local ids can be seen.

Table 2. Pathway table for two pathways (EMP and ED) realising glycolysis.

			Atom label			
Pathway	0, O	6, C	7, O	12, C	13, O	15, O
EMP		2	3	5		0
ED	0	5	0	2	3	4
Both	0	2, 5	0, 3	2, 5	0, 3	0, 4

in pyruvate if only the EMP pathway is active. Note, that we can furthermore conclude, that a corresponding oxygen labeling experiment does not require to verify the position of the labelled oxygen, as the existence of a label in pyruvate (verified by mass only) is enough to determine that the ED pathway was active.

In order to exemplify the inverted hypergraph semigroup approach we chose $t = (3_T, 0_T)$. We infer the orbit of t in the inverted semigroup: $\mathrm{Orb}_{G_{\mathrm{EMP}}^{-1}}(t) = \{(4, 15), (7, 10)\}$. Semantically this means, we ask how the atoms with id 3 and 0 can become labelled simultaneously. In EMP this is only possible if in the educt molecule glucose either atoms 4 and 15, or 7 and 10 were labelled. Consider the orbit of any possible $t = (i_T, j_T)$. As for all possible t it holds that (k_S, k_S) is neither element of $\mathrm{Orb}_{G_{\mathrm{EMP}}^{-1}}(t)$ nor element of $\mathrm{Orb}_{G_{\mathrm{ED}}^{-1}}(t)$ (and also not element of $\mathrm{Orb}_{G_{\mathrm{Both}}^{-1}}(t)$), we can conclude that using an isotopic labelled glucose can not lead to a single pyruvate molecule that has two labelled atoms at the same time. If this would be observed in a wet-lab experiment (verified via mass), then we would have to conclude that the modelling of the presented EMP and the ED is not sufficient, and that another pathway must be active. If (k, k) would have been an element in only one of both inverted semigroup orbits, we could have used the labeling of atom k in glucose to distinguish pathways.

6   Conclusion

We introduced an approach based on transformation semigroups to automatically design isotopic labeling experiments. All approaches have been implemented and will be integrated in an open source framework for chemically inspired graph transformation [3]. A natural next step is to apply the decomposition results known for semigroups. While the modelling approach in [9] is applied on a different level of abstraction (the transformations of a semigroup map chemical compounds onto each other, in contrast to our approach where atoms are mapped), there is an interesting similarity: In [15] local substructures that exhibit symmetry on a network level are found to be permutation groups. Such permutation groups are considered as natural and important subsystems (a "pool of local reversibility"). In our approach local substructures are permutation groups that describe natural symmetries: the automorphisms of any molecular graph will describe a permutation group as a substructure of a semigroup. In our future work we will apply the decomposition approaches to gain a deeper understanding of the atom transition networks by a hierarchical decomposition.

References

1. Andersen, J.L., Flamm, C., Merkle, D., Stadler, P.F.: Inferring chemical reaction patterns using graph grammar rule composition. J. Syst. Chem. 4(1), 4 (2013)
2. Andersen, J.L., Flamm, C., Merkle, D., Stadler, P.F.: 50 shades of rule composition. In: Fages, F., Piazza, C. (eds.) FMMB 2014. LNCS, vol. 8738, pp. 117–135. Springer, Cham (2014). https://doi.org/10.1007/978-3-319-10398-3_9

3. Andersen, J.L., Flamm, C., Merkle, D., Stadler, P.F.: A software package for chemically inspired graph transformation. In: Echahed, R., Minas, M. (eds.) ICGT 2016. LNCS, vol. 9761, pp. 73–88. Springer, Cham (2016). https://doi.org/10.1007/978-3-319-40530-8_5

4. Andersen, J.L., Flamm, C., Merkle, D., Stadler, P.F.: Chemical transformation motifs – modelling pathways as integer hyperflows. IEEE/ACM Trans. Comput. Biol. Bioinform. (2018)

5. Andersen, J.L., Merkle, D.: A generic framework for engineering graph canonization algorithms. In: 2018 Proceedings of the 20th Workshop on Algorithm Engineering and Experiments (ALENEX), pp. 139–153 (2018). https://doi.org/10.1137/1.9781611975055.13

6. Brim, L., Češka, M., Šafránek, D.: Model checking of biological systems. In: Bernardo, M., de Vink, E., Di Pierro, A., Wiklicky, H. (eds.) SFM 2013. LNCS, vol. 7938, pp. 63–112. Springer, Heidelberg (2013). https://doi.org/10.1007/978-3-642-38874-3_3

7. Corradini, A., Montanari, U., Rossi, F., Ehrig, H., Heckel, R., Löwe, M.: Algebraic approaches to graph transformation - part i: basic concepts and double pushout approach. In: Rozenberg, G. (ed.) Handbook of Graph Grammars and Computing by Graph Transformation, chap. 3, pp. 163–245. World Scientific (1997)

8. East, J., Egri-Nagy, A., Mitchell, J.D., Péresse, Y.: Computing finite semigroups. J. Symb. Comput. (2018, in press). Early access online

9. Egri-Nagy, A., Nehaniv, C.L.: Hierarchical coordinate systems for understanding complexity and its evolution, with applications to genetic regulatory networks. Artif. Life **14**(3), 299–312 (2008). https://doi.org/10.1162/artl.2008.14.3.14305

10. Egri-Nagy, A., Nehaniv, C.L., Rhodes, J.L., Schilstra, M.J.: Automatic analysis of computation in biochemical reactions. BioSystems **94**(1–2), 126–134 (2008). https://doi.org/10.1016/j.biosystems.2008.05.018

11. Ehrig, H., Ehrig, K., Prange, U., Taentzer, G.: Fundamentals of Algebraic Graph Transformation. Springer-Verlag, Berlin (2006). https://doi.org/10.1007/3-540-31188-2

12. Flamholz, A., Noor, E., Bar-Even, A., Liebermeister, W., Milo, R.: Glycolytic strategy as a tradeoff between energy yield and protein cost. Proc. Natl. Acad. Sci. **110**(24), 10039–10044 (2013)

13. Ganyushkin, O., Mazorchuk, V.: Classical Finite Transformation Semigroups, vol. 9. Springer, Heidelberg (2009). https://doi.org/10.1007/978-1-84800-281-4

14. Ginzburg, A.: Algebraic Theory of Automata. Academic Press, Cambridge (1968)

15. Nehaniv, C.L., et al.: Symmetry structure in discrete models of biochemical systems: natural subsystems and the weak control hierarchy in a new model of computation driven by interactions. Philos. Trans. R. Soc. A: Math. Phys. Eng. Sci. **373**(2046) (2015). https://doi.org/10.1098/rsta.2014.0223

16. Orth, J.D., Thiele, I., Palsson, B.Ø.: What is flux balance analysis? Nat. Biotech. **28**, 245–248 (2010)

17. Rhodes, J., Nehaniv, C.L., Hirsch, M.W.: Applications of Automata Theory and Algebra. World Scientific, September 2009. https://doi.org/10.1142/7107

18. Sanders, P.: Algorithm engineering – an attempt at a definition. In: Albers, S., Alt, H., Näher, S. (eds.) Efficient Algorithms. LNCS, vol. 5760, pp. 321–340. Springer, Heidelberg (2009). https://doi.org/10.1007/978-3-642-03456-5_22

19. Schuster, S., Hilgetag, C.: On elementary flux modes in biochemical reaction systems at steady state. J. Biol. Syst. **2**(02), 165–182 (1994)

20. Zeigarnik, A.V.: On hypercycles and hypercircuits in hypergraphs. Discrete Math. Chem. **51**, 377–383 (2000)

Iterative Spaced Seed Hashing: Closing the Gap Between Spaced Seed Hashing and k-mer Hashing

Enrico Petrucci[1], Laurent Noé[2], Cinzia Pizzi[1(✉)], and Matteo Comin[1(✉)]

[1] Department of Information Engineering, University of Padova, Padova, Italy
{cinzia.pizzi,comin}@dei.unipd.it
[2] CRIStAL UMR9189, Université de Lille, Lille, France

Abstract. Alignment-free classification of sequences has enabled high-throughput processing of sequencing data in many bioinformatics pipelines. Much work has been done to speed-up the indexing of k-mers through hash-table and other data structures. These efforts have led to very fast indexes, but because they are k-mer based, they often lack sensitivity due to sequencing errors or polymorphisms. Spaced seeds are a special type of pattern that accounts for errors or mutations. They allow to improve the sensitivity and they are now routinely used instead of k-mers in many applications. The major drawback of spaced seeds is that they cannot be efficiently hashed and thus their usage increases substantially the computational time.

In this paper we address the problem of efficient spaced seed hashing. We propose an iterative algorithm that combines multiple spaced seed hashes by exploiting the similarity of adjacent hash values in order to efficiently compute the next hash. We report a series of experiments on HTS reads hashing, with several spaced seeds. Our algorithm can compute the hashing values of spaced seeds with a speedup of 6.2x, outperforming previous methods. Software and Datasets are available at ISSH

Keywords: k-mers · Spaced seeds · Gapped q-gram · Efficient hashing

1 Introduction

In computational biology, sequence classification is a common task with many applications such as phylogeny reconstruction [16], protein classification [20], metagenomic [11,18,21]. Even if sequence classification is addressable via alignment, the scale of modern datasets has stimulated the development of faster alignment-free similarity methods [1,3,4,16,23].

The most common alignment-free indexing methods are k-mer based. Large-scale sequence analysis often relies on cataloguing or counting consecutive k-mers (substring of length k) in DNA sequences for indexing, querying and similarity

© Springer Nature Switzerland AG 2019
Z. Cai et al. (Eds.): ISBRA 2019, LNBI 11490, pp. 208–219, 2019.
https://doi.org/10.1007/978-3-030-20242-2_18

searching. A common step is to break a reference sequence into k-mers and indexing them. An efficient way of implementing this operation is through the use of hash based data structures, e.g. hash tables. Then, to classify sequences are also broken into k-mers and queried against the hash table to check for shared k-mers.

In [17] it has been shown that requiring the matches to be non-consecutive increases the chance of finding similarities and they introduced spaced seeds. They are a modification to the standard k-mer where some positions on the k-mer are set to be "don't care" or wildcard to catch the spaced matches between sequences. In spaced seeds, the matches are distributed so as to maximize the sensitivity, that is the probability to find a local similarity.

Spaced seeds are widely used for approximate sequence matching in bioinformatics and they have been increasingly applied to improve the sensitivity and specificity of homology search algorithms [15, 19]. Spaced seeds are now routinely used, instead of k-mers, in many problems involving sequence comparison like: multiple sequence alignment [5], protein classification [20], read mapping [22], phylogeny reconstruction [16], metagenome reads clustering and classification [2, 8, 21].

In all these applications, the use of spaced seeds, as opposed to k-mers, has been reported to improve the performance in terms of sensitivity and specificity. However, the major drawback is that the computational cost increases. For example, when k-mers are replaced by spaced seeds, the metagenomic classification of reads of Clark-S [21] increases the quality of classification, but it also produces a slowdown of 17x with respect to the non-seed version. A similar reduction in time performance when using spaced seeds is reported also in other applications [2, 20, 22].

The main reason is that k-mers can be efficiently hashed. In fact, the hashing of a k-mer can be easily computed from the hashing of its predecessor, since they share $k - 1$ symbols. For this reason, indexing all consecutive k-mers in a string can be a very efficient process. However, when using spaced seeds these observations do not longer hold. Therefore, improving the performance of spaced seed hashing algorithms would have a great impact on a wide range of bioinformatics applications. The first attempt to address this question was in the Thesis of R. Harris [13], but hard coding was used to speed-up a non linear packing. Recently, we develop an algorithm based on the indexing of small blocks of runs of matching positions that can be combined to obtain the hashing of spaced-seeds [9]. In [6, 10] we proposed a more promising direction, based on spaced seed self-correlation, in order to reuse part of the hashes already computed. We showed how the hash at position i can be computed based on one best previous hash. Despite the improvement in terms of speedup, the number of symbols that need to be encoded in order to complete the hash could still be high. In this paper we solved this problem through: (1) a better way to use previous hashes, maximizing re-use; (2) an iterative algorithm that combines multiple previous hashes. In fact, our algorithm arranges multiple previous hashes in order to recover all $k - 1$ symbols of a spaced seed, so that we only need to encode the new symbol, like with k-mer hashing.

2   Methods: Iterative Spaced Seed Hashing

2.1   Spaced Seed Hashing: Background

A *spaced-seed* Q (or just a seed) is a string over the alphabet $\{1,0\}$ where the 1s correspond to matching positions and 0 to non-matching positions or wildcards, e.g. 1011001. A spaced seed Q can be represented as a set of non negative integers corresponding to the matching positions (1s) in the seed, e.g. $Q = \{0,2,3,6\}$, a notation introduced in [14]. The *weight* of a seed, denoted as $|Q|$, corresponds to the number of 1s, while the *length*, or span $s(Q)$, is equal to $\max(Q)+1$.

Given a string x, the positioned spaced seed $x[i + Q]$ identifies a string of length $|Q|$, where $0 \leq i \leq n-s(Q)$. The positioned spaced seed $x[i + Q]$, also called Q-gram, is defined as the string $x[i + Q] = \{x_{i+k}, k \in Q\}$.

Example 1. Given the seed 1011001, defined as $Q = \{0,2,3,6\}$, with weight $|Q| = 4$ and span $s(Q) = 7$. Let us consider the string $x = AATCACTTG$.

$$x \quad A\ A\ T\ C\ A\ C\ T\ T\ G$$
$$Q \quad 1\ 0\ 1\ 1\ 0\ 0\ 1$$
$$x[0 + Q]\ A\quad T\ C\qquad T$$

The Q-gram at position 0 of x is defined as $x[0 + Q] = ATCT$. Similarly the other Q-grams are $x[1 + Q] = ACAT$, and $x[2 + Q] = TACG$.

In this paper, for ease of discussion, we will consider as hashing function the simple encoding of a string, that is a special case of the Rabin-Karp rolling hash. Later, we will shown how more advanced hashing function can be implemented at no extra cost. Let's consider a coding function from the DNA alphabet $\mathcal{A} = \{A, C, G, T\}$ to a binary codeword, $encode : \mathcal{A} \rightarrow \{0,1\}^{log_2|\mathcal{A}|}$, where $encode(A) = 00, encode(C) = 01, encode(G) = 10$, and $encode(T) = 11$. Following the above example, we can compute the encodings of all symbols of the Q-gram $x[0 + Q]$ as follows:

$$x[0 + Q]\quad A\ T\ C\ T$$
$$encodings\ 00\ 11\ 01\ 11$$

Finally, the hashing value of the Q-gram $ATCT$ is 11011100, that is the merge of the encodings of all symbols using little-endian notation. More formally, a standard approach to compute the hashing value of a Q-gram at position i of the string x is the following function $h(x[i + Q])$:

$$h(x[i + Q]) = \bigvee_{k \in Q} (encode(x_{i+k}) \ll m(k) * log_2|\mathcal{A}|) \tag{1}$$

Where $m(k)$ is the number of matching positions that appears to the left of k. The function m is defined as $m(k) = |\{i \in Q, \text{ such that } i < k\}|$. In other words, given a position k in the seed, m stores the number of shifts that we need to apply to the encoding of the k-th symbols in order to place it into the hashing. The vector m is important for the computation of the hashing value of a Q-gram.

Example 2. In this example, we report an example of hashing value computation for the Q-gram $x[1 + Q]$.

x	A	A	T	C	A	C	T	T	G
Q		1	0	1	1	0	0	1	
m		0	1	1	2	3	3	3	

shifted encodings   00≪0   01≪2 00≪4   11≪6

           <u>00</u>

               <u>0100</u>

                  <u>000100</u>

hashing value   <u>11000100</u>

The above example shows how the hashing value of $x(1 + Q)$ can be computed through the function $h(x[1+Q]) = h(ACAT) = 11000100$. The hashing value of the other Q-gram can be determined with a similar procedure, i.e. $h(x[2+Q]) = h(TACG) = 10010011$. The hashing function $h(\cdot)$ is a special case of the Rabin-Karp rolling hash. However, more advanced hashing functions can be defined in a similar way. For example, the cyclic polynomial rolling hash can be computed by replacing: shifts with rotations, OR with XOR, and the function encode($\cdot$) with a table, where DNA characters are mapped to random integers.

In this paper we want to address the following problem.

Problem 1. Let us consider a string $x = x_0 x_1 \ldots x_i \ldots x_{n-1}$, of length n, a spaced seed Q and a hash function h that maps strings into a binary codeword. We want to compute all hashing values $\mathcal{H}(x, Q)$ for all the Q-grams of x, starting from the first position 0 of x to the last $n - s(Q)$.

$$\mathcal{H}(x, Q) = \langle h(x[0 + Q]), h(x[1 + Q]), \ldots h(x[n - s(Q)]) \rangle$$

To compute the hash of a contiguous k-mer it is possible to use the hash of its predecessor. In fact, given the hashing value at position i, the hashing for position $i + 1$ can be obtained with two operations, a shift and the insertion of the encoding of the new symbol, since the two hashes share $k - 1$ symbols. However, if we consider the case of a spaced seed Q, we can clearly see that this observation does not hold. In fact, in the above example, two consecutive Q-grams, like $x[0 + Q] = ATCT$ and $x[1 + Q] = ACAT$, do not necessarily have much in common. Since the hashing values are computed in order, the idea is to speed up the computation of the hash at a position i by reusing part of the hashes already computed at previous positions. In this paper we present a solution for Problem 1 that maximizes the re-use of previous hashes so that only one symbol needs to be encoded in the new hash, as with k-mers hashing.

2.2 Iterative Spaced Seed Hashing

In the case of spaced seeds, one can reuse part of previous hashes to compute the next one, however we need to explore not only the hash at the previous position, as with k-mers, but the $s(Q) - 1$ previous hashes. A first attempt to solve this problem was recently proposed in [10], where the hash at position i is computed based on one best previous hash. Despite the improvement in terms of speedup with respect to the standard hashing method, the number of symbols that need to be read in order to complete the hash could still be high. In this paper we reduced this value to just one symbol by working in two directions: (1) we devise a better way to use a previous hash, maximizing re-use (2) we propose an iterative algorithm that combines multiple previous hashes.

Let us assume that we want to compute the hashing value at position i and that we already know the hashing value at position $i - j$, with $j < s(Q)$. We can introduce the following definition of $C_{g,j} = \{k \in Q : k + j \in Q \land m(k) = m(k + j) - m(j) + m(g)\}$ as the positions in Q that after j shifts are still in Q with the propriety that k and $k + j$ positions are both in Q and they are separated by $j - g - 1$ (not necessarily consecutive) ones. In other words if we are processing the position i of x and we want to reuse the hashing value already computed at position $i - j$, $C_{g,j}$ represents the symbols, starting at position g of $h(x[i - j + Q])$, that we can keep while computing $h(x[i + Q])$.

Example 3. Let's consider $Q = \{0, 1, 2, 4, 6, 8, 10\}$. If we know the first hashing value $h(x[0 + Q])$ and we want to compute the second hash $h(x[1 + Q])$, the following example show how to construct $C_{0,1}$.

			0	1	2	3	4	5	6	7	8	9	10
k			0	1	2	3	4	5	6	7	8	9	10
Q			1	1	1	0	1	0	1	0	1	0	1
$Q \ll 1$		1	1	1	0	1	0	1	0	1	0	1	
$m(k)$			0	1	2	3	3	4	4	5	5	6	6
$m(k+1)-m(1)+m(0)$		−1	0	1	2	2	3	3	4	4	5	5	
$C_{0,1}$			0	1									

The symbols at positions $C_{0,1} = \{0, 1\}$ of the hash $h(x[1 + Q])$ have already been encoded in the hash $h(x[0 + Q])$ and we can keep them. In order to complete $h(x[1 + Q])$, the number of remaining symbols are $|Q| - |C_{0,1}| = 5$.

In the paper [10] we use only the symbols in $C_{0,j}$, that is g was always 0. As we will see in the next examples, if we are allowed to remove the first g symbols from the hash of $h(x[i - j + Q])$, we can recover more symbols in order to compute $h(x[i + Q])$.

Example 4. Let us consider the hash at position 2 $h(x[2 + Q])$, and the hash at position 0 $h(x[0 + Q])$. In this case we are interested in $C_{0,2}$.

			0	1	2	3	4	5	6	7	8	9	10
k			0	1	2	3	4	5	6	7	8	9	10
Q			1	1	1	0	1	0	1	0	1	0	1
$Q \ll 2$	1	1	1	0	1	0	1	0	1	0	1		
$m(k)$			0	1	2	3	3	4	4	5	5	6	6
$m(k+2)-m(2)+m(0)$	−2	−1	0	1	1	2	2	3	3	4	4		
$C_{0,2}$			0										

Thus, the only position that we can recover is $C_{0,2} = \{0\}$. On the other hand, if we are allowed to skip the first position of the hash $h(x[0 + Q])$ and consider $C_{1,2}$, instead of $C_{0,2}$, we have:

k				0	1	2	3	4	5	6	7	8	9	10
Q				1	1	1	0	1	0	1	0	1	0	1
$Q \ll 2$	1	1	1	0	1	0	1	0	1	0	1		
m(k)				0	1	2	3	3	4	4	5	5	6	6
m(k+2)−m(2)+m(1)	−1	0	1	2	2	3	3	4	4	5	5		
$C_{1,2}$				2		4		6		8			

Where, we can re-use the symbols $C_{1,2} = \{2, 4, 6, 8\}$ of $h(x[0 + Q])$ in order to compute $h(x[2 + Q])$. This example shows how the original definition of C_j in [10], that in this work corresponds to $C_{0,2} = \{0\}$, was not optimal and more symbols could be recovered from the same hash with $C_{1,2} = \{2, 4, 6, 8\}$.

In [10], the hash value at a given position was reconstructed starting from the best previous hash. However, the number of symbols to be inserted to complete the hash could still be high. In this paper we propose a new method that not only consider the best previous hash, but all previous hashes at once. For a given hash to be computed h_i, we devised an iterative algorithm that is able to find a combination of the previous hashes that covers all symbols of h_i, apart from the last one. That is, we can combine multiple hashes in order to recover $|Q| - 1$ symbols of h_i, so that we only need to read the new symbol, like with k-mer hashing.

Let's assume that we have already computed a portion of the hash h_i, and that the remaining symbols are $Q' \subset Q$. We can search the best previous hash that covers the largest number of positions of Q'. To this end, we define the function $BestPrev(s, Q')$ that searches for this best previous hash:

$$BestPrev(s, Q') = argmax_{z \in [0, s-1], k \in [1, s]} |C_{z,k} \cap Q'|$$

This function will return a pair (g, j) that identifies the best previous hash at position h_{i-j} from which, after removing the first g symbols, we can recover $|C_{g,j} \cap Q'|$ symbols. In order to extract these symbols from h_{i-j} we define a mask, $Mask_{g,j}$, that filters these positions. The algorithm iteratively searches the best previous hashes, until all $|Q| - 1$ symbols have been recovered. An overview of the method is shown below:

Our iterative algorithm scans the input string x and computes all hashing values according to the spaced seed Q. In order to better understand the amount of savings we evaluate the algorithm by counting the number of symbols that are read and encoded. First, we can consider the input string to be long enough so that we can discard the transient of the first $s(Q) - 1$ hashes. Let us continue to analyze the spaced seed 11101010101, that corresponds to $Q = \{0, 1, 2, 4, 6, 8, 10\}$. If we use the standard function $h(x[i + Q])$ to compute all hashes, each symbol of x is read $|Q| = 7$ times.

Algorithm 1. Iterative Spaced Seed Hashing

1: Compute $C_{g,k}$ and $Mask(g,k)$ $\forall g, k$;
2: $h_0 :=$ compute $h(x[0 + Q])$;
3: **for** $i := 1$ to $s(Q) - 1$ **do**
4:     $Q' = Q$;
5:     **while** $|Q'| \neq 1$ **do**
6:         $(g, k) = BestPrev(i, Q')$;
7:         **if** $(Q' \cap C_{g,k}) == \emptyset$ **then**
8:             Exit while;
9:         **else**
10:             $h_i := h_i$ OR $((h_{i-k}$ AND $Mask(g,k)) >> k * log_2|\mathcal{A}|)$;
11:             $Q' = Q' - C_{g,k}$;
12:         **end if**
13:     **end while**
14:     **for all** $k \in Q'$ **do**
15:         insert $encode(x_{i+k})$ at position $m(k) * log_2|\mathcal{A}|$ of h_i;
16:     **end for**
17: **end for**
18: **for** $i := s(Q)$ to $|x| - s(Q)$ **do**
19:     $Q' = Q$;
20:     **while** $|Q'| \neq 1$ **do**
21:         $(g, k) = BestPrev(s(Q) - 1, Q')$;
22:         $h_i := h_i$ OR $((h_{i-k}$ AND $Mask(g,k)) >> k * log_2|\mathcal{A}|)$;
23:         $Q' = Q' - C_{g,k}$;
24:     **end while**
25:     insert $encode(x_{i+s(Q)-1})$ at last position of h_i ;
26: **end for**

In the first iteration of our algorithm (lines $= 19$–25) $Q' = Q$ and the best previous hash $BestPrev(s(Q) - 1, Q') = (1, 2)$ is $C_{1,2} = \{2, 4, 6, 8\}$. Thus, while computing h_i we can recover these 4 symbols from h_{i-2}. At the end of the first iteration Q' is updated to $\{0, 1, 10\}$. During the second iteration the best previous hash $BestPrev(s(Q) - 1, Q') = (0, 1)$ is $C_{0,1} = \{0, 1\}$. As above, we can append these two symbols from h_{i-1} to the hash h_i. Now, we have that $Q' = \{10\}$, that is only one symbol is left. The last symbol is read and encoded into h_i, and the hash is complete. In summary, after two iterations all $|Q| - 1$ symbols of h_i have been encoded into the hash, and we only need to read one new symbol from the sequence. Moreover, if one needs to scan a string with a spaced seed and to compute all hashing values, the above algorithm guarantees to minimize the number of symbols to read. In fact, with our algorithm, we can compute all hashing values while reading each symbol of the input string only once, as with k-mers.

3   Results and Discussion

In this section we will present the results of some experiments in which ISSH is compared against two other approaches available in literature: FISH [9] (block-based) and FSH [10] (overlap-based).

3.1   Experimental Settings

We use the same settings as in previous studies [9,10]. The spaced seeds belong to three different types of spaced seeds, according to the objective function used to generate them: maximizing the hit probability [21]; minimizing the overlap complexity [12]; and maximizing the sensitivity [12]. We tested three spaced seeds for each type, all with weight $W = 22$ and length $L = 31$ (see Appendix of [9]). Furthermore, we used other sets of spaced seeds, built with *rashbari* [12], which have weights from 11 to 32 and the same length. The complete list of the spaced seeds used is reported in the Appendix of [9]. The datasets of metagenomic reads to be hashed were taken from previous papers on binning and classification [7,11,23]. All the experiments have been performed on a laptop equipped with an Intel i7-3537U CPU at 2 GHz and 8 GB of RAM.

3.2   Analysis of the Time Performances

The first comparison we present is between the performances of ISSH, FISH and FSH in terms of speedup with respect to the standard hash computation (i.e. applying Eq. 1 to each position). Figure 1 shows the average speedup among all datasets, for each of the spaced seeds $Q1$–$Q9$, obtained by the three different methods.

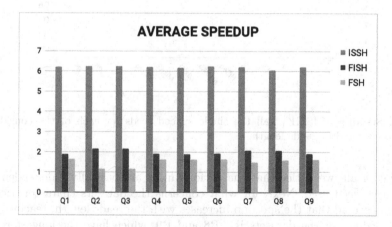

Fig. 1. The average speedup obtaind by ISSH, FISH and FSH with respect to the standard computation.

It can be seen that ISSH is much faster than both FISH and FSH for all the spaced seeds. In terms of actual running time, the standard approach (Eq. 1) requires about 14 min to compute the hashes for a single spaced seed on all datasets. ISSH takes just over 2 min with an average speedup of 6.2. As for the other two approaches, FISH and FSH, they compute the hashes in 6 and 9 min respectively, with an average speedup of 2 (FISH) and 1.5 (FSH).

We also notice that the variation among the speedups, relative to different spaced seeds using the same method, are lower for ISSH, for which the speedups are in the range [6.05–6.25] while for FISH and FSH the range is [1.89–2.16] and [1.18–1.58], respectively. For all the tested methods there is a correlation between the spaced seed structure and the time needed for the computation. FISH depends on the number of blocks of 1s, while both ISSH and FSH depend on the spaced seed self-correlation. ISSH performances are also sensitive to the number of iterations. However, the experiments show that, even if FSH performs a single iteration, the time required to naively compute the hash for all the non-overlapping positions is more than the time required by ISSH to perform more iterations. Moreover, for all the tested spaced seeds the number of iterations needed by ISSH was on average 4.

Figure 2 gives an insight on the performance of ISSH with respect to each spaced seed and each datasets considered.

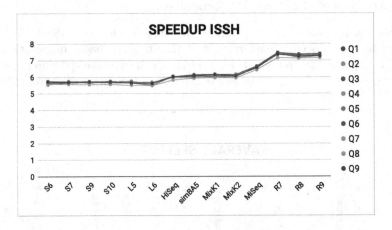

Fig. 2. Speedup of ISSH of all the single spaced seeds for each of the considered datasets, ordered by reads length.

First of all, we notice that the performances are basically independent on the spaced seed used. Next, for what concerns the datasets characteristics, it can be observed that the speedup increases with the reads length, reaching the highest values for the datasets R7, R8 and R9, which have the longest reads. This behavior is expected: when considering longer reads the slowdown caused by the initial transient – in which more than one symbol has to be encoded – is less relevant with respect to the total running time.

In Fig. 3 we report the speedups on each datasets obtained by $Q7$, a typical spaced seed (the other spaced seeds performances are similar) using ISSH, FISH and FSH.

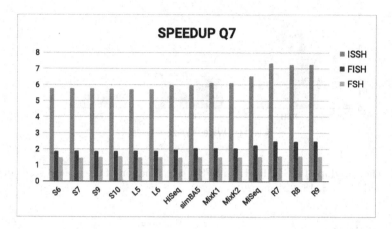

Fig. 3. Details of the speedup on the spaced seed $Q7$ on each datasets, ordered by reads length, using ISSH, FISH and FSH.

All the results are compatible with the above observations: ISSH, if compared to FISH and FSH, allows to compute the hashing values faster for all the datasets. Futhermore, by using ISSH, the improvement on long reads datasets is larger than the improvement obtained with FISH or FSH.

3.3   Effect of Spaced Seeds Weight on Time Performances

The experiments presented here point out the connection between the density of a spaced seed and the speedup. We considered four sets of nine spaced seeds, generated with *rasbhari* [12], with weights 14, 18, 22 and 26 and a fixed length of 31.

In Fig. 4 we compare the average speedup of ISSH, FISH and FSH for these sets of spaced seeds as a function of the weight W. We notice that the speedup grows as the weight increases. This phenomenon is consistent among all the methods we analyzed. It is reasonable to think that such difference is due to how the hashes are computed with the standard method using Eq. 1 (against which all methods are compared), because denser spaced seeds imply hashes with a larger number of symbols that need to be encoded and joined together. Moreover, for ISSH we have that denser spaced seeds have more chances of needing fewer previously calculated hashes to compute each of the $|Q| - 1$ symbols, thus saving further iterations.

Both these effects are emphasized when looking at the actual running times needed by the least dense group ($W = 14$) and by the most dense group ($W = 26$) of spaced seeds. The standard method requires 9.73 and 15.11 min, respectively, while ISSH spends only 2.75 and 2.16 min to perform the same task.

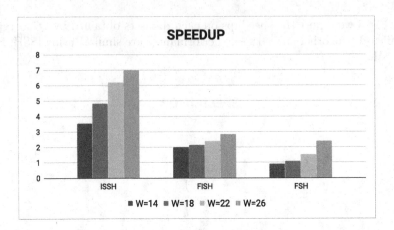

Fig. 4. The speedup of ISSH, FISH and FSH as a function of the spaced seeds density (L = 31 and W = 14, 18, 22, and 26).

4   Conclusions

In this paper we present ISSH (Iterative Spaced Seed Hashing), an iterative algorithm that combines multiple previous hashes in order to maximize the re-use of already computed hash values. The average speedup of ISSH with respect to the standard computation of hash values is in range of [3.5x–7x], depending on spaced seed density and reads length. In all experiments ISSH outperforms previously proposed algorithms. Possible directions of research are the combination of multiple spaced seeds and the investigation of global optimization schemes.

References

1. Apostolico, A., Guerra, C., Landau, G.M., Pizzi, C.: Sequence similarity measures based on bounded hamming distance. Theor. Comput. Sci. **638**, 76–90 (2016)
2. Břinda, K., Sykulski, M., Kucherov, G.: Spaced seeds improve k-mer-based metagenomic classification. Bioinformatics **31**(22), 3584 (2015)
3. Comin, M., Verzotto, D.: Beyond fixed-resolution alignment-free measures for mammalian enhancers sequence comparison. IEEE/ACM Trans. Comput. Biol. Bioinform. **11**(4), 628–637 (2014)
4. Comin, M., Leoni, A., Schimd, M.: Clustering of reads with alignment-free measures and quality values. Algorithms Mol. Biol. **10**(1), 4 (2015)
5. Darling, A.E., Treangen, T.J., Zhang, L., Kuiken, C., Messeguer, X., Perna, N.T.: Procrastination leads to efficient filtration for local multiple alignment. In: Bücher, P., Moret, B.M.E. (eds.) WABI 2006. LNCS, vol. 4175, pp. 126–137. Springer, Heidelberg (2006). https://doi.org/10.1007/11851561_12
6. Girotto, S., Comin, M., Pizzi, C.: Fast spaced seed hashing. In: Proceedings of the 17th Workshop on Algorithms in Bioinformatics (WABI). Leibniz International Proceedings in Informatics, vol. 88, pp. 7:1–7:14 (2017)
7. Girotto, S., Comin, M., Pizzi, C.: Higher recall in metagenomic sequence classification exploiting overlapping reads. BMC Genomics **18**(10), 917 (2017)

8. Girotto, S., Comin, M., Pizzi, C.: Metagenomic reads binning with spaced seeds. Theor. Comput. Sci. **698**, 88–99 (2017)
9. Girotto, S., Comin, M., Pizzi, C.: Efficient computation of spaced seed hashing with block indexing. BMC Bioinform. **19**(15), 441 (2018)
10. Girotto, S., Comin, M., Pizzi, C.: FSH: fast spaced seed hashing exploiting adjacent hashes. Algorithms Mol. Biol. **13**(1), 8 (2018)
11. Girotto, S., Pizzi, C., Comin, M.: MetaProb: accurate metagenomic reads binning based on probabilistic sequence signatures. Bioinformatics **32**(17), i567–i575 (2016)
12. Hahn, L., Leimeister, C.A., Ounit, R., Lonardi, S., Morgenstern, B.: rasbhari: Optimizing spaced seeds for database searching, read mapping and alignment-free sequence comparison. PLoS Comput. Biol. **12**(10), 1–18 (2016)
13. Harris, R.S.: improved pairwise alignment of genomic DNA. Ph.D. thesis, University Park, PA, USA (2007)
14. Keich, U., Li, M., Ma, B., Tromp, J.: On spaced seeds for similarity search. Discret. Appl. Math. **138**(3), 253–263 (2004)
15. Kucherov, G., Noé, L., Roytberg, M.A.: A unifying framework for seed sensitivity and its application to subset seeds. J. Bioinform. Comput. Biol. **4**(2), 553–569 (2006)
16. Leimeister, C.A., Boden, M., Horwege, S., Lindner, S., Morgenstern, B.: Fast alignment-free sequence comparison using spaced-word frequencies. Bioinformatics **30**(14), 1991 (2014)
17. Ma, B., Tromp, J., Li, M.: PatternHunter: faster and more sensitive homology search. Bioinformatics **18**(3), 440 (2002)
18. Marchiori, D., Comin, M.: SKraken: fast and sensitive classification of short metagenomic reads based on filtering uninformative k-mers. In: Proceedings of the 10th International Joint Conference on Biomedical Engineering Systems and Technologies - Volume 3: BIOINFORMATICS, (BIOSTEC 2017), pp. 59–67. INSTICC, SciTePress (2017)
19. Noé, L., Martin, D.E.K.: A coverage criterion for spaced seeds and its applications to support vector machine string kernels and k-mer distances. J. Comput. Biol. **21**(12), 947–963 (2014)
20. Onodera, T., Shibuya, T.: The gapped spectrum kernel for support vector machines. In: Perner, P. (ed.) MLDM 2013. LNCS (LNAI), vol. 7988, pp. 1–15. Springer, Heidelberg (2013). https://doi.org/10.1007/978-3-642-39712-7_1
21. Ounit, R., Lonardi, S.: Higher classification sensitivity of short metagenomic reads with CLARK-S. Bioinformatics **32**(24), 3823 (2016)
22. Rumble, S.M., Lacroute, P., Dalca, A.V., Fiume, M., Sidow, A., Brudno, M.: SHRiMP: accurate mapping of short color-space reads. PLOS Comput. Biol. **5**(5), 1–11 (2009)
23. Wood, D.E., Salzberg, S.L.: Kraken: ultrafast metagenomic sequence classification using exact alignments. Genome Biol. **15**, R46 (2014)

DM-SIRT: A Distributed Method for Multi-tilt Reconstruction in Electron Tomography

Zihao Wang[1,2], Jingrong Zhang[1,2], Xintong Liu[1,3], Zhiyong Liu[1(✉)], Xiaohua Wan[1(✉)], and Fa Zhang[1(✉)]

[1] High Performance Computer Research Center, Institute of Computing Technology, Chinese Academy of Sciences, Beijing 100190, China
{wangzihao,zyliu,wanxiaohua,zhangfa}@ict.ac.cn
[2] University of Chinese Academy of Sciences, Beijing, China
[3] School of Automation, Beijing Institute of Technology, Beijing, China

Abstract. The 'missing wedge' of single tilt in electron tomography introduces severely artifacts into the reconstructed results. To reduce the 'missing wedge' effect, a widely used method is 'multi-tilt reconstruction', which collects projections using multiple different axes. However, as the number of tilt series increases, its computing and memory costs also rises. While the demand to speed up its reconstruction procedure grows, the huge memory requirement from the 3D structure and strong data dependencies from projections heavily limit its parallelization. In our work, we present a new fully distributed multi-tilt reconstruction framework named DM-SIRT. To improve the parallelism of the reconstruction process and reduce the memory requirements of each process, we formulate the multi-tilt reconstruction as a consensus optimization problem and design a distributed multi-tilt SIRT algorithm. To improve the reconstruction resolution, we applied a multi-agent consensus equilibrium (MACE) with a new data division strategy. Experiments show that along with the visually and quantitatively improvement in resolution, DM-SIRT can acquire a 5.4x speedup ratio compared to the raw multi-tilt reconstruction version. It also has 87% decrease of memory overhead and 8 times more scalable than the raw reconstruction version.

Keywords: Cryo-electron Tomography · Parallel computing · Consensus optimization · Multi-tilt reconstruction · TxBR

1 Introduction

Recently, Cryo-electron Tomography (Cryo-ET) is gaining popularity among structural biologists. One great superiority of Cryo-ET is its ability to reveal the three-dimensional (3D) structure of cellular or macromolecular assemblies in near-native state at nanometer scale [1,2]. Electron tomography tilts the samples to generate a series of two-dimensional (2D) projection images (also called tilt

© Springer Nature Switzerland AG 2019
Z. Cai et al. (Eds.): ISBRA 2019, LNBI 11490, pp. 220–231, 2019.
https://doi.org/10.1007/978-3-030-20242-2_19

series) and reconstructs 3D structure from these tilt series. Due to physical limitations of microscopes, the sample is tilted around a single fixed axis with a range from $-70°$ to $+70°$. The absence of projection from orientations from $70°$ to $90°$ and $-70°$ to $-90°$ leads to a 'missing wedge' in Fourier space, which will lead to severe ray artifacts for the reconstructed tomograms.

There are currently two mainstream methods to solve the problem of 'missing wedge'. The first one is based on the single tilt series. These algorithms apply prior constraints or compressed sensing framework to the reconstructed tomogram to compensate the missing wedge such as Discrete Algebraic Reconstruction Technique (DART) [3], Iterative Compressed-sensing Optimized NUFFT Reconstruction (ICON) [4] and Model-Based Iterative Reconstruction (MBIR) [5]. Another kind of method is using multiple tilt series taken by rotating the sample in a plane, which is called multiple axes acquisition strategies. In the popular double axes acquisition, the sample is rotated $90°$ to obtain two tilt series [6]. What's more, the strategy can be extended to 8-tilt series or 16-tilt series by rotating the sample in a series of uniformly sampled orientations It can effectively reduce the 'missing wedge' effect to a 'missing pyramid' of data, or even to a 'missing cone' [7].

IMOD [8], TxBR [9] and AuTom-dualx [10] all provide the complete processing procedure for double axes data. However, as the data size increases, the process of multiple axes data encountered several obstacles. The first one is that the degree of parallelism is limited by the existing parallel strategies. For single tilt data series, one common parallel strategy is to split the reconstructed volume along the tilting axis such as Y-axis. But, as the geometry in TxBR is non-linear and the Y-axis varies while X-axis rotating, splitting reconstructed volume along Y-axis to perform the parallelization is not applicable. For multi-axis data collection, they usually split the reconstructed volume along Z-axis [11]. Due to the physical limitations of electron microscope, the thickness of the sample is quite limited, which can severely restrict the parallel degree. The second obstacle is the huge memory requirement from reconstruction results. Comparing with the single tilt series, the 16-tilt data collection owns almost 2000 projection views. For the iterative method, the updating requires assessing the whole projection series in each iteration, which means each thread needs to access the whole projection series. Taking the 4096*4096 projection series as an example, each thread needs to access 242 GB memory in each iteration. Its memory requirement cannot be fulfilled for most multi-core computers.

To address these challenges of multiple tilts reconstruction in cryo-ET, we present a new fully Distributed Multi-tilt Simultaneous Iterative Reconstruction Technique named DM-SIRT. In the framework, we formulate the multi-tilt reconstruction as a consensus optimization problem. We divide the multi-tilt data into multiple subsets, which will be processed separately. Then we apply a multi-agent consensus equilibrium (MACE) approach [12] to optimize the results of each subset through updating the global result iteratively. This approach has been proved to be converged to the global optimal result. To our knowledge, this is the first multi-tilt reconstruction method based on consensus optimization.

What's more, we use two hierarchies parallel method to reduce the communication overhead and use new data division strategy to prevent the overfitting during reconstruction.

Our proposed method owns some advantages in computing and memory costs. First of all, the distributed data can improve the parallelism of the reconstruction process because we adopted a new data partitioning strategy. Secondly, the strategy reduces the number of processed projection data so it can reduce the memory requirements and adjust the number of projections processed according to the memory of the real environment. Finally, we use two hierarchies parallel method to reduce the communication overhead and improve the scalability of DM-SIRT. Benefiting from these strategies, multi-tilt reconstruction can be done in high efficiency.

The rest of the paper is organized as follows. Section 2 shows backgrounds about the multi-tilt reconstruction and multi-agent consensus equilibrium. Section 3 presents the process and implementation of our new distributed framework DM-SIRT. In Sects. 4 and 5, we test the resolution, time and scalability performance of DM-SIRT by comparing it with widely used methods. Finally, Sect. 6 presents our conclusion.

2   Related Work

2.1   Multi-tilt Reconstruction

The multi-tilt data acquisition is similar to the single axis tomography. In double-tilt also known as dual-axis tomography, the sample will be rotated 90° in the XOY plane and obtain a combination of two tilt series, showed in Fig. 1(a). The multi-tilt series is a direct extension of the single tilt series. According to the number of rotation angles, multi-tilt can be named as double-tilt, 4-tilt, 8-tilt or 16-tilt data showed in Fig. 1(b). Along with the increase of the number of tilt, the reconstruction artifacts caused by the 'missing wedge' will be gradually weakened.

Once the tilt series has been acquired, the first step is the alignment of images within data set. This step can adjust the data to a single global coordinate system to ensure the accuracy of reconstruction [13]. There are some methods of multi-tilt data alignment including IMOD, TxBR and AuTom-dualx. After the process of alignment, the data is ready to be reconstructed. As direct back projection method can not make full use of the relation of multi-tilt data, most current reconstruction methods for multi-axis data are based on the iterative method SIRT [14] such as ADA-SIRT [13], combined dual-axis SIRT [15], and W-SIRT [7].

2.2   Multi-agent Consensus Equilibrium

Consensus Equilibrium. The simplest form of the statistical reconstruction method is:

$$x^* = \underset{x}{argmin} \left\{ f(x) + h(x) \right\} \tag{1}$$

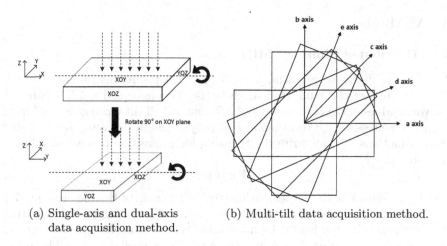

(a) Single-axis and dual-axis data acquisition method.

(b) Multi-tilt data acquisition method.

Fig. 1. Data acquisition of cryo-electron tomography

where f is the data fidelity function (minimizing the difference between reconstruction and real data), h is the regularizing function (suppression of noise data) and x is our reconstruction result. In more general settings, if the data collected from multi-modal data collection or using different fidelity functions, the cost function can be written as

$$minimize \ f(x) = \sum_{i=1}^{N} f_i(x) \tag{2}$$

where variable $x \in \mathbb{R}^n$ and $f_i : \mathbb{R}^n \rightarrow \mathbb{R} \cup \{+\infty\}$. In the consensus optimization, the minimization of the original cost function with the constraint that the separate variables must share a common value.

$$minimize \ f(x) = \sum_{i=1}^{N} f_i(x) \ subject \ to \ x_i = x, \ i = 1, ..., N \tag{3}$$

Buzzard et al. in [12] propose a general framework named Multi-Agent Consensus Equilibrium (MACE) to solve the consensus optimization problem such as Eq. (3). The framework can handle multiple heterogeneous models come from physical sensors or learned from data. Like the Eq. (4) proposed by ADMM [16], it maps Eq. (3) to a auxiliary function to solving the consensus equilibrium. After mapping, it reformulates the optimization function as a fixed point problem and uses iteration framework to achieve convergence. A more detailed description for the framework and convergence proof can be found in [12].

$$F_i(z_i) = \underset{v_i}{argmin} \left\{ f_i(v_i) + \frac{\|v_i - z_i\|^2}{2\sigma^2} \right\} \tag{4}$$

3   Methods

3.1   Distributed Multi-tilt SIRT

To solve the discussed problem, we first analyzed the reconstruction method. Let n denote the total number of voxels in the 3D volume and $\mathbf{x} \in \mathbb{R}^n$ represent the voxel values of the 3D volume. $\mathbf{p} \in \mathbb{R}^m$ contains all tilt series generated from $\mathbf{x}$ and m represents the total number of projections in different angles. Using these definitions, we can write the tomographic acquisition process as a linear equation:

$$W\mathbf{x} = \mathbf{p} \tag{5}$$

The W is defined as the projection matrix. In the matrix W, the element W_{ij} specifies the contribution of voxel $\mathbf{X}_j$ to projection $\mathbf{P}_i$. The algebraic reconstruction methods solve the Eq. (5) by minimizing the norm of the residuals vectors in Eq. (6). Through this, it can find the model $\mathbf{x}$ as similar as possible to the experimentally projections.

$$\mathbf{x} = \underset{x}{argmin} \, \|W\mathbf{x} - \mathbf{p}\|^2 \tag{6}$$

In multi-tilt reconstruction, we often use the family of iterative algebraic reconstruction algorithms like ADA-SIRT, W-SIRT mentioned before. Within one iteration, their updating strategies are similar. For the kth iteration, the concrete updating strategy is:

$$\mathbf{x}^{k+1} = \mathbf{x}^k + \alpha W^T \left(\mathbf{p} - W\mathbf{x}^k \right) \tag{7}$$

W is the projection matrix from all the orientations and $\mathbf{p}$ is projections from all angles. It uses all projection data in one iteration which severely limits degree of parallelism and costs huge memory. Because each tilt is highly independent and has the same optimization target, we can rewrite the Eq. (6) as minimizing the sum norm of each tilt i and add the consensus constraint:

$$f_i(\mathbf{x}_i) = \underset{\mathbf{x}_i}{argmin} \, \|W_i\mathbf{x}_i - \mathbf{p}_i\|^2 \tag{8}$$

$$minimize \; f(\mathbf{x}) = \sum_{i=1}^{N} f_i(\mathbf{x}_i) \; subject \; to \; \mathbf{x}_i = \mathbf{x}, \; i = 1, ..., N \tag{9}$$

Some papers have proved that the SIRT update scheme is guaranteed to converge to a weighted least squares solution [14,17,18]. So we rewrite the Eq. (7) as the optimizer for each tilt proximal map F:

$$\mathbf{x}_i^{k+1} = \mathbf{x}_i^k + \alpha_i W_i^T \left(\mathbf{p}_i - W_i\mathbf{x}_i^k \right) \tag{10}$$

We apply the consensus equilibrium framework for multi-tilt electron tomography reconstruction and parallelize the SIRT to improve efficiency and reduce memory consumption. It is named as DM-SIRT and the detail of the algorithm is shown in Algorithm 1.

Algorithm 1. DM-SIRT: Distributed Multi-tilt Simultaneous Iterative Reconstruction Technique

---

Input: N: the number of different subsets, k: the number of iteration number, **p**: multiple tilt series projection data, W_i: the different tilt angle projection geometry matrix from TxBR, x_0: init value of x^*, $\bar{w}$: the average result of all subsets, v_i^k: the kth iteration result of subset x_i, z_i^k: the kth update result after merged in each iteration.

Output: x^*: The reconstruction volume

1: $k \leftarrow 0$, $\bar{w}^k \leftarrow x_0$, $w_i^k \leftarrow x_0$
2: **while** *not converged* **do**
3:    **for** $i \in 1$ to N **do**
4:       $z_i^k \leftarrow 2\bar{w}^k - w_i^k$
5:       $v_i^k \leftarrow z_i^k + \alpha_i W_i^T \left(\mathbf{p}_i - W_i z_i^k\right)$ {SIRT for subset(inner loop)}
6:       $w_i^{k+1} \leftarrow \rho(2v_i^k - z_i^k) + (1-\rho)w_i^k$ {Mann Iteration in MACE}
7:       $\bar{w}^{k+1} \leftarrow (w_1^k + w_i^k + ... + w_N^k)/N$
8:    **end for**
9:    $k \leftarrow k + 1$
10: **end while**
11: Solution: $x^* \leftarrow \bar{w}^k$ {Consensus solution}

3.2 Two Hierarchies Parallel Strategy

From the above analysis, we parallel DM-SIRT on two hierarchies. As for the first level, projection images are divided according to the multi-tilt projection angle (usually N can be 4, 8 or 16...). We divide the whole angles into N subsets to parallel. In the second level, the reconstructed volume is divided along Z-axis. Because of the geometry in TxBR is non-linear, we must divide the whole volume along Z-axis and use as many processes as possible to calculate the Eq. (7). In general, we adopted the master-slave process hierarchy. Different from the traditional master-slave architecture, to make the best of computing resources, all the processes including master process participate in computing.

Based on the two parallel hierarchies, we also mark processes as two types to reduce the communication overhead between processes and show the relationship in Fig. 2. The first type of process is responsible for the update of variables in the outer loop of DM-SIRT such as z_i, w_i and $\bar{w}$ in Algorithm 1. Each subset only has one process with this type so it can reduce the communication processes while updating these variable. The second type of process is responsible for the update of v^k in the inner loop of DM-SIRT. The communication between different subsets is independent. To reduce memory consumption, we designed a customized data structure, which only sustains the needed by each process. The first type of process using Reduce and Bcast to synchronize with global data. The second type of process using Scatter and Gather to communicate the need data. These strategies can not only improve communication efficiency but also reduce the memory occupancy of each process.

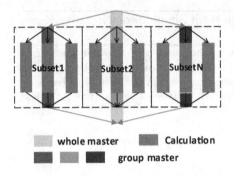

whole master   Calculation

group master

Fig. 2. The two hierarchies parallel strategy for DM-SIRT

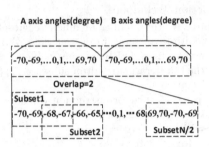

Fig. 3. The data division strategy for DM-SIRT

3.3 Overlapped Data Division Strategy

In the consensus equilibrium method, they often use a round-robin method to divide data. However, it is not suitable for multi-tilt reconstruction. Although we have adjusted multi-tilt data to the same geometry, there is still existing difference between different axis projection. We divided the data in the same tilt into the same subset which can contribute to updating faster in the inner loop of DM-SIRT. Within the scope of the same axis, there are existing more commonness in the adjacent angle. In addition, each subset need contain some overlapped angles to avoid the overfitting in the inner loop of DM-SIRT.

Based on these observations, we group the adjacent angles in the same tilt together. Figure 3 shows our new data division strategy in DM-SIRT and the overlap angle number is set to 2. When dividing a subset, it is prior to ensure that the data of the same tilt is divided together so we first separate the projection angle from A, B axis. After that, dividing the angle of the same tilt, we group the adjacent angles together like subset1 including $-70°$, $-69°$ and subset2 including $-68°$, $-67°$. To avoid overfitting, each subset will get some overlap angles from the next subset like subset1 has $-68°$ and $-67°$ in subset2.

4   Experimental Setup

Datasets. We use the cryo-ET data named EEL-Crosscut were taken using an FEI Titan operated at 300kV from National Center for Microscopy and Imaging Research (NCMIR) and the micrography was produced using a 4096*4096 CCD camera. The tilt series includes two axes a, b and the acquisition method is shown in Fig. 1(b). The tilt angles of the projection images in each axis range from $-60°$ to $60°$ at 1-degree interval. There are 121 images in the tilt series of each axis. The size of each projection image is 4096*4096 with a pixel size of 1.36 nm. In this paper, to ensure all methods can work, we binned the tilt series with factors four to generate a dataset of size is 1024*1024. The tilt series are aligned using TxBR and the size of reconstruction result is 1024*1024*66.

Computing Platform. Experiments were performed on a cluster with 16 working nodes in total. Each node equipped with two 16-core 2.3 GHz AMD Opteron (TM) 6376 processors and 64 GB of RAM. All programs in this paper were implemented and compiled with Intel MPI 5.0.

Experimental Setup. We use four methods as the comparison to analyze the performance of our method. First one is the conventional method named RAW-SIRT which uses full angles to reconstruct the volume. The second is the simple reconstruction method named Filtered-back Projection (FBP) [19]. The third method is our proposed named DM-SIRT. The last method is dividing the projection angle, reconstruct independently without using consensus framework, combining the result directly named DirectCombine-SIRT. All iterative methods use the same relax factor 0.3 [20], and the whole iteration time is set to 100. The inner iteration time of DM-SIRT is set to 10, and outer iteration time is also set to 10 which makes 100 iterations in total equal to the iteration number of RAW-SIRT so that we can ensure the fairness of comparison. We use eight subsets for DM-SIRT and the overlap number is set to 2.

5   Results

In this section, we describe our results of experiments. First of all, we compare the reconstruction result of RAW-SIRT, FBP, DM-SIRT, and DirectCombine-SIRT. Next, we compare the timing and memory performance of RAW-SIRT and DM-SIRT. Finally, we analyze the scalability of DM-SIRT.

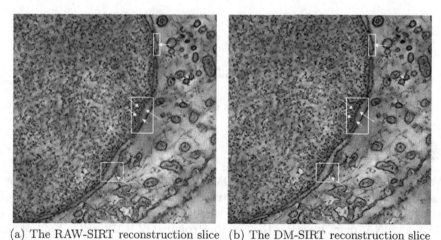

(a) The RAW-SIRT reconstruction slice   (b) The DM-SIRT reconstruction slice

Fig. 4. The reconstruction results using multi-tilt data

5.1   Reconstruction Precision

Figure 4 shows the center slice of the reconstructed volume of the EEL-Crosscut dataset. In the red box, we can find DM-SIRT generate much clear edge information than RAW-SIRT based on the visual assessment. However, more area looks similar so we use the normalized correlation coefficient (NCC) showed in Eq. (11) between the reprojections of different reconstruction methods with the original tilt series (axis A in this paper) to analyze the quantized accuracy.

$$NCC(\mathbf{x_1}, \mathbf{x_2}) = \frac{\sum(\mathbf{x_1} - \mu_{\mathbf{x_1}})(\mathbf{x_2} - \mu_{\mathbf{x_2}})}{\sqrt{\sum(\mathbf{x_1} - \mu_{\mathbf{x_1}})^2}\sqrt{\sum(\mathbf{x_2} - \mu_{\mathbf{x_2}})^2}} \qquad (11)$$

From Fig. 5, we can find that the DM-SIRT performance is the best on the NCC result overall tilt angle. RAW-SIRT as the standard method not performed well in the high angle comparing with DM-SIRT, but in the low angle slightly better than DM-SIRT. Based on our experiments, the accuracy of FBP and DirectCombine-SIRT is worse compared to DM-SIRT and RAW-SIRT which justified the previous discussion that these methods are not usually used due to their bad performance.

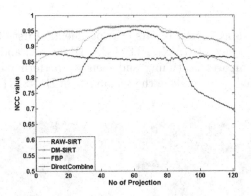

Fig. 5. The NCC comparing with different reconstruction methods

5.2   Performance Results

We test the time of RAW-SIRT and DM-SIRT to quantify the overall performance on a cluster. Table 1 lists reconstruction time at the different number of nodes which each node has 32 cores. In these tests result, we can find the 'Out-Mem' and 'NA' mark indicates the two shortcomings of RAW-SIRT. First of all, RAW-SIRT only can divide the data along Z-axis so the degree of parallelism is limited by Z-axis thickness. The result 'NA' in the last three columns represents that the RAW-SIRT can not scale up to 128 cores because of the Z-axis thickness is 66. The other shortcoming is that each process needs to handle the whole projection data leading to high memory occupancy. However, reducing memory occupancy by increasing the number of process on one node is still not applicable, it will cause memory overflow as shown in the first column in Table 1.

Our method DM-SIRT improves the parallelism by dividing the projection angle so it can scale to the entire cluster with 512 cores. In addition, due to the data division strategy, the memory consumption of each process will also be reduced and the detail data showed in Sect. 5.3. From the Table 1, DM-SIRT achieves 74 min reconstruction time at 512 nodes. It is 5.4 times faster than the RAW-SIRT performance on 64 nodes and is 8 times more scalable than RAW-SIRT.

Table 1. The speedups of DM-SIRT compared to RAW-SIRT

Nodes	2 Nodes	4 Nodes	4 Nodes	8 Nodes	16 Nodes
Cores	64	64	128	256	512
RAW-SIRT (min)	OutMem	403	NA	NA	NA
DM-SIRT (min)	411	411	236	135	74

5.3 Memory Overhead

We analyze the memory overhead in each process. The main memory consumption includes the reconstructed result and the projection data needed by each process. As RAW-SIRT only divides the reconstruct volume, each process needs to handle the whole projection series. We use subset 0 to represent RAW-SIRT method in Fig. 6. The number of the subset is two, four and eight represents the different number of subset adopted by DM-SIRT. Based on the results shown in Fig. 6, with the increase of the subset number, the memory consumption of each process decrease accordingly. When DM-SIRT adopts 8 subsets, it has 87% decrease of memory overhead as comparing with RAW-SIRT.

5.4 Scalability Performance

In the scalability test, we fixed the total number of tasks in all nodes. We tested the scalability with the mentioned reconstruction data size 1024*1024*66. We only increased the number of the process from 64 to 512. From Fig. 7, we can observe that the parallel efficiency decreased to 87% when using 128 processes and decreased further to 70% when using 512 processes. The observed degradation of scalability efficiency is acceptable.

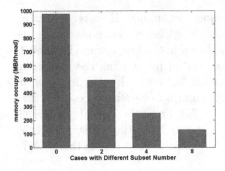
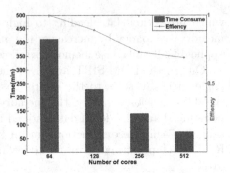

Fig. 6. The memory overhead in different subset

Fig. 7. The scalability performance

6   Conclusion

In this work, we present a new fully distributed multi-tilt reconstruction framework named DM-SIRT. We are the first to formulate the reconstruction as a consensus optimization problem in Cryo-ET. With the help of multi-agent consensus equilibrium approach, we improve the parallelism of the reconstruction process and reduce the memory requirements through reducing the number of projection data which each process needed. We also proposed the two hierarchies parallel method to improve the scalability of DM-SIRT and use overlapped data division strategy to prevent the overfitting during reconstruction. Benefiting from these strategies, multi-tilt reconstruction can be done with the visually and quantitatively improvement in resolution. Experiments also show that our proposed method has a high degree of parallelism, low memory consumption, and high scalability.

Acknowledgments. We acknowledge Albert Lawrence and Sebastien Phan at UCSD for providing the experimental dataset. This research is supported by the Strategic Priority Research Program of the Chinese Academy of Sciences Grant (No. XDA19020400), the National Key Research and Development Program of China (No. 2017YFE0103900 and 2017YFA0504702), Beijing Municipal Natural Science Foundation Grant (No. L182053), the NSFC projects Grant (No. U1611263, U1611261 and 61672493) and Special Program for Applied Research on Super Computation of the NSFC-Guangdong Joint Fund (the second phase).

References

1. Lučić, V., Rigort, A., Baumeister, W.: Cryo-electron tomography: the challenge of doing structural biology in situ. J. Cell Biol. **202**(3), 407–419 (2013)
2. Grotjahn, D.A., Chowdhury, S., Xu, Y., McKenney, R.J., Schroer, T.A., Lander, G.C.: Cryo-electron tomography reveals that dynactin recruits a team of dyneins for processive motility. Nat. Struct. Mol. Biol. **25**(3), 203 (2018)
3. Batenburg, K.J., et al.: 3D imaging of nanomaterials by discrete tomography. Ultramicroscopy **109**(6), 730–740 (2009)

4. Deng, Y., Chen, Y., Zhang, Y., Wang, S., Zhang, F., Sun, F.: ICON: 3D reconstruction with "missing-information" restoration in biological electron tomography. J. Struct. Biol. **195**(1), 100–112 (2016)
5. Yan, R., Venkatakrishnan, S.V., Liu, J., Bouman, C.A., Jiang, W.: MBIR: a cryo-electron tomography 3D reconstruction method that effectively minimizes missing wedge artifacts and restores missing information. bioRxiv (2018). 355529
6. Penczek, P., Marko, M., Buttle, K., Frank, J.: Double-tilt electron tomography. Ultramicroscopy **60**(3), 393–410 (1995)
7. Phan, S., et al.: 3D reconstruction of biological structures: automated procedures for alignment and reconstruction of multiple tilt series in electron tomography. Adv. Struct. Chem. Imaging **2**(1), 8 (2017)
8. Kremer, J.R., Mastronarde, D.N., McIntosh, J.R.: Computer visualization of three-dimensional image data using IMOD. J. Struct. Biol. **116**(1), 71–76 (1996)
9. Lawrence, A., Bouwer, J.C., Perkins, G., Ellisman, M.H.: Transform-based back-projection for volume reconstruction of large format electron microscope tilt series. J. Struct. Biol. **154**(2), 144–167 (2006)
10. Han, R., et al.: Autom-dualx: a toolkit for fully automatic fiducial marker-based alignment of dual-axis tilt series with simultaneous reconstruction. Bioinformatics **35**(2), 319–328 (2018)
11. Zhang, J., Wan, X., Zhang, F., Ren, F., Wang, X., Liu, Z.: A parallel scheme for three-dimensional reconstruction in large-field electron tomography. In: Basu, M., Pan, Y., Wang, J. (eds.) ISBRA 2014. LNCS, vol. 8492, pp. 102–113. Springer, Cham (2014). https://doi.org/10.1007/978-3-319-08171-7_10
12. Buzzard, G.T., Chan, S.H., Sreehari, S., Bouman, C.A.: Plug-and-Play unplugged: optimization-free reconstruction using consensus equilibrium. SIAM J. Imaging Sci. **11**(3), 2001–2020 (2018)
13. Arslan, I., Tong, J.R., Midgley, P.A.: Reducing the missing wedge: high-resolution dual axis tomography of inorganic materials. Ultramicroscopy **106**(11–12), 994–1000 (2006)
14. Gilbert, P.: Iterative methods for the three-dimensional reconstruction of an object from projections. J. Theor. Biol. **36**(1), 105–117 (1972)
15. Haberfehlner, G., Serra, R., Cooper, D., Barraud, S., Bleuet, P.: 3D spatial resolution improvement by dual-axis electron tomography: application to tri-gate transistors. Ultramicroscopy **136**, 144–153 (2014)
16. Boyd, S., Parikh, N., Chu, E., Peleato, B., Eckstein, J., et al.: Distributed optimization and statistical learning via the alternating direction method of multipliers. Found. Trends® Mach. Learn. **3**(1), 1–122 (2011)
17. Gregor, J., Benson, T.: Computational analysis and improvement of SIRT. IEEE Trans. Med. Imaging **27**(7), 918–924 (2008)
18. Sorzano, C., et al.: A survey of the use of iterative reconstruction algorithms in electron microscopy. BioMed Research International (2017)
19. Herman, G.T., Frank, J.: Computational Methods for Three-dimensional Microscopy Reconstruction. Springer, New York (2014). https://doi.org/10.1007/978-1-4614-9521-5
20. Van der Sluis, A., van der Vorst, H.A.: SIRT-and CG-type methods for the iterative solution of sparse linear least-squares problems. Linear Algebr. Appl. **130**, 257–303 (1990)

PeakPass: Automating ChIP-Seq Blacklist Creation

Charles E. Wimberley[✉] and Steffen Heber[✉]

NC State University, Raleigh, NC 27606, USA
{cewimber,sheber}@ncsu.edu

Abstract. ChIP-Seq blacklists contain genomic regions that frequently produce artifacts and noise in ChIP-Seq experiments. To improve signal-to-noise ratio, ChIP-Seq pipelines often remove data points that map to blacklist regions. Existing blacklists have been compiled in a manual or semi-automated way. In this paper we describe PeakPass, an efficient method to generate blacklists, and present evidence that blacklists can increase ChIP-Seq data quality. PeakPass leverages machine learning and attempts to automate blacklist generation. PeakPass uses a random forest classifier in combination with genomic features such as sequence, annotated repeats, complexity, assembly gaps, and the ratio of multi-mapping to uniquely mapping reads to identify artifact regions. We have validated PeakPass on a large dataset and tested it for the purpose of upgrading a blacklist to a new reference genome version. We trained PeakPass on the ENCODE blacklist for the hg19 human reference genome, and created an updated blacklist for hg38. To assess the performance of this blacklist we tested 42 ChIP-Seq replicates from 24 experiments using the Relative Strand Correlation (RSC) metric as a quality measure. Using the blacklist generated by PeakPass resulted in a statistically significant increase in RSC over the existing ENCODE blacklist for hg38 – average RSC was increased by 50% over the ENCODE blacklist, while only filtering an average of 0.1% of called peaks.

Keywords: ChIP-seq · Classification · Quality control · Blacklist

1 Introduction

ChIP-Seq is an experimental technique to determine protein-DNA interaction sites within the genome. During a ChIP-Seq experiment, fragments of DNA crosslinked to proteins are enriched by antibodies that bind to a protein of interest, and then isolated and sequenced. The resulting sequence reads are mapped to a reference genome, and clusters of mapped reads are used to infer potential protein binding sites. Typically, ChIP-Seq is used to identify epigenetic alterations such as histone modifications, and the binding sites of transcription factors and other DNA-binding proteins.

ChIP-Seq blacklists contain genomic regions that frequently produce artifacts and noise in ChIP-Seq experiments. To improve data quality, ChIP-Seq pipelines

© Springer Nature Switzerland AG 2019
Z. Cai et al. (Eds.): ISBRA 2019, LNBI 11490, pp. 232–243, 2019.
https://doi.org/10.1007/978-3-030-20242-2_20

often remove data points that map to blacklist regions. Artifact regions are frequently associated with errors in the underlying genome assembly, sequence repeats, or increased genomic variability, but often the exact cause of the observed artifacts remains elusive [1]. Genome-wide blacklists are a key part of ChIP-Seq pipelines, and they are usually created using a time-consuming, semi-automated process [2]. A recently published regression model based on mappability features has the potential to simplify this procedure, but so far, a validation using ChIP-Seq experiments is still missing [3,4]. Our PeakPass algorithm uses a supervised learning approach: starting from artifact regions derived from the ENCODE blacklist, and a large set of genomic features we apply feature selection in combination with classification to produce a new genome-wide blacklist. Subsequently, we evaluate the quality of our blacklist using a large number of ChIP-Seq experiments.

In our validation experiments we apply different blacklist and control treatments to various ChIP-Seq replicates and measure their effects on quality. One quality metric of particular interest is the RSC signal-to-noise estimate, which is related to unmappable positions within the genome [5]. RSC shows up in stranded cross-correlation plots as the ratio of correlation at the fragment length (signal) to correlation at read length (noise). A correlation spike at the read length is known as the "phantom peak", and represents suboptimal RSC. Ramachandran and colleagues have shown that controlling for alignability during cross-correlation analysis reduces the phantom peak [6]. Carroll and colleagues found that plotting cross-correlation of only duplicated reads results in both read length and fragment length correlation peaks. However, cross-correlation of reads in blacklist regions results in only a read length peak [5]. Therefore, the reduction of the phantom peak height after blacklist treatment is one metric to assess the effectiveness of a blacklist.

In the remaining parts of this paper, we describe our PeakPass approach and illustrate how PeakPass can be used to update a previous blacklist (ENCODE hg19 blacklist) to a new genome version (hg38). Finally, we compare the newly generated blacklist with the corresponding ENCODE hg38 blacklist by computing RSC on multiple ChIP-Seq datasets downloaded from the ENCODE database [7]. Average RSC increased by 50% over the ENCODE hg38 blacklist when applying the PeakPass blacklist. We also compare the repeat composition of the blacklists, and measure how many peaks from the ENCODE ChIP-Seq sets intersect with the new blacklist regions.

2   Methods

Starting from a small training set of artifact regions, PeakPass uses supervised learning (classification) to generate a blacklist that covers the entire genome. To identify a suitable classifier, we compared the performance of 6 algorithms: random forest [8], K-Nearest Neighbor (KNN) [9], Support Vector Machines (SVM) [10] with radial and linear kernels, Neural Networks [11], and Naive Bayes [12] using a blacklist from an earlier human reference genome version (hg19).

Each algorithm was tested on its ability to predict hg19 blacklist regions set aside from the training data; the random forest classifier outperformed all competitors (see the Results section) and was adopted as the PeakPass classification engine. Subsequently, we used PeakPass to predict a blacklist for the most recent genome version hg38. Many assembly errors and gaps in hg19 have been fixed in hg38. The new blacklist for hg38 was tested for its impact on ChIP-Seq quality metrics and called peaks.

We start by dividing the hg19 reference genome into windows of size 1kbp. Subsequently, features were computed for each window as shown in Fig. 1. If 70% or more of the window overlaps with an ENCODE blacklist region the window is labeled as a blacklist region in the learning dataset, otherwise the window is treated as a "normal" region. The resulting data set consists of 8,517 blacklist regions and 2,566,611 normal regions.

Blacklisted Region

Chromosome						
Window #	Window 1	Window 2	Window 3	Window 4	Window 5	Window 6
Feature 1	12	13	2	3	20	15
Feature 2	A	B	A	A	B	B
...	...	...	...	...	...	...
Class	Normal	Normal	Blacklist	Blacklist	Normal	Normal

Fig. 1. Windowed regions of the genome are used to compute machine learning features and labels for supervised learning. Windows that significantly overlap with a blacklist region are labeled as "blacklist" class windows.

PeakPass uses a variety of genomic features to predict artifact regions. We use two alignability based features (similar to DangerTrack [3]): alignability averaged across each window and the number of positions in the window with alignability below 0.1. Alignability measures how often a certain k-mer sequence aligns to the genome. For a specific genomic interval of length k the alignability is defined as the inverse of the number of different matches of the interval in the genome, see Derrien et al. for a more detailed description [13]. Similar to the Kundaje's blacklist candidate generator, PeakPass also uses ratio of multi-mapping loci to uniquely-mapping loci. Repeat-based features include the number of Repeat-Masker [14] repeats that intersect the window and the number of softmasked base pairs. In addition, several complexity-based features (the number of unique 4-mers, number of 2-mer tandem repeats, and A, T, C, and G content), distance to the nearest gene, and distance to the nearest assembly gap are used.

Several features such as unique 3-mers, number of loci with alignability over 0.9, and monomer repeats were removed from the feature set because they were highly correlated with other features, and model performance did not decrease after their removal. In addition, we decided to not use some features because they were not available for certain genome assemblies. These include nearest assembly gap type and gap size. The annotated repeat-types also differed widely between organisms, so this feature was excluded as well. The final list of features we used is shown in Table 1.

Table 1. PeakPass features and feature importance using random forest's out-of-bag feature importance estimate. The importance score of a feature is the mean decrease in accuracy when scrambling this feature in out-of-bag data.

Feature	Importance score
Distance to nearest assembly gap	89.0
Frequency of unique 4-mers in window	44.2
Frequency of softmasked base pairs in window	38.0
Distance to nearest gene	35.5
Frequency of loci with alignability <0.1	19.8
G content of window	17.9
C content of window	17.5
Average alignability on window	17.2
Frequency of 2-mer tandem repeats in window	17.2
T content of window	15.4
A content of window	14.6
Frequency of repeats measured by RepeatMasker	14.2
Ratio of multi-mapping to uniquely-mapping reads in window	13.3

We split the labeled data derived from the ENCODE hg19 blacklist into independent training and test data sets. The training data set was further subdivided into independent parts for hyperparameter tuning (via 10-fold cross-validation, parameters were selected to maximize AUCROC, tuning was performed with the caret library for R [15]), model training, and model comparison (via multiple iterations of sub-sampling). Training sets were undersampled to 2000 observations of each class label. Using the test data set we evaluated the performance of the best classifier. All data sets were disjoint on both window location and blacklist item, this means that windows from the same blacklist item or genomic location cannot appear in both a training and corresponding validation or test set. Finally, we used the fully trained algorithm to predict a new blacklist for hg38.

Because Kundaje's blacklist indicates that some regions are blacklisted due to the presence of certain repeats types, the repeat composition of the PeakPass blacklist was measured [2]. Each blacklist was intersected with the set of Repeat-Masker repeats, and the frequency of each repeat type was computed. In order to normalize repeat frequencies between blacklists of different sizes, we report repeat frequency per kilobase. The results are shown in Table 6.

Next, the blacklists were compared to determine their effect on Relative Strand Correlation (RSC) and Normalized Strand Coefficient (NSC). We selected 42 replicates from 24 transcription factor ChIP-Seq experiments from ENCODE that exhibited a phantom peak and evaluated them under different blacklist treatments. We used the ENCODE cross-correlation implementation to measure RSC. Additional details about the composition of the ChIP-Seq data are shown in Table 2.

Table 2. A variety of different transcription factor ChIP-Seq experiments were used to validate the PeakPass blacklist. Blacklists are not biased toward a specific protein of interest or cell line.

Property	Number of distinct observations
Replicates	42
Experiments	24
Protein target	22
Cell line	12
Laboratory	9

Replicates from the following ENCODE experiment accession IDs were used:

- ENCSR920BLG
- ENCSR000BRU
- ENCSR000BNO
- ENCSR286PCG
- ENCSR000FBC
- ENCSR000FAZ
- ENCSR000EZW
- ENCSR945NSF

- ENCSR519QAA
- ENCSR051DXE
- ENCSR000DON
- ENCSR000DNQ
- ENCSR000DNN
- ENCSR892DRK
- ENCSR000AOA
- ENCSR549NPZ

- ENCSR384LYW
- ENCSR000DSZ
- ENCSR000EXZ
- ENCSR000EUZ
- ENCSR000EVX
- ENCSR000EWI
- ENCSR000DNI
- ENCSR000DKR

Table 3 shows library complexity and PCR bottle-necking quality metrics for the ChIP-Seq datasets we used. We selected these metrics due to their association with ChIP-Seq artifacts. Quality thresholds and rating terminology used in the table below are taken from ENCODE [16].

Finally, the blacklists are intersected with the peaks called by the ENCODE pipeline to measure number of the peaks that are filtered by the blacklists. IDR thresholded peak sets were downloaded from the ENCODE database when available. We report the percentage of called peaks that intersect with the PeakPass blacklist, as these peaks would be filtered by the blacklist.

PeakPass consists of a set of R, Python, and Bash scripts that can be executed in a Unix environment. The PeakPass algorithm, instructions, and the PeakPass hg38 blacklist can be downloaded from our GitHub repository: https://github.com/ewimberley/peakPass.

Table 3. The quality of a replicate may influence the effectiveness of blacklist treatment on that replicate. Therefore, we have listed the quality of the replicates used. The NRF (Non Redundant Fraction) is the number of reads after duplicate removal divided by the number of reads before duplicate removal. PBC1 (PCR Bottlenecking Coefficient 1) is the number of mapped loci with 1 uniquely mapped read divided by the number of loci with one or more uniquely mapped reads.

Metric	Rating	Number of observations
NRF	Poor (0–0.5)	4 (9%)
	Moderate (0.5–0.8)	15 (36%)
	High (<0.8)	23 (55%)
PBC1	Severe (0–0.5)	4 (9%)
	Moderate (0.5–0.8)	9 (22%)
	Mild (0.8–0.9)	11 (26%)
	None (>0.9)	18 (43%)

3   Results

Table 4 compares model performance for different learning algorithms using 32 randomly sampled training and validation sets. Sensitivity and specificity are measured at a threshold of 0.5. We used AUC precision/recall as our metric for model selection because of its good performance on imbalanced data.

Table 4. Performance of the different classification algorithms. Classifier performance values are computed on our validation datasets, whereas performance of the final model is based on the test dataset. The standard deviation of each performance metric is shown in parenthesis next to the mean of the measurement.

Algorithm	Sensitivity	Specificity	AUC P/R	AUCROC
Random forest	0.920 (0.038)	0.966 (0.007)	0.627 (0.061)	0.988 (0.003)
KNN	0.895 (0.035)	0.947 (0.011)	0.531 (0.041)	0.970 (0.011)
SVM (Radial)	0.842 (0.047)	0.963 (0.009)	0.445 (0.085)	0.977 (0.005)
ANN	0.901 (0.056)	0.931 (0.029)	0.299 (0.152)	0.951 (0.021)
SVM (Linear)	0.913 (0.036)	0.948 (0.011)	0.201 (0.060)	0.976 (0.005)
Naïve Bayes	0.906 (0.036)	0.926 (0.018)	0.095 (0.030)	0.960 (0.011)
Final model	0.978	0.963	0.858	0.995

Random forest was selected as the best performing model, and was bench-marked on the test set (286,446 windows of 1kbp each). Our random forest has 2000 trees with a maximum of 10 nodes each. The final model performance on the testing dataset is shown in the last row of the table above. The rest of the results are based on a blacklist created with this model.

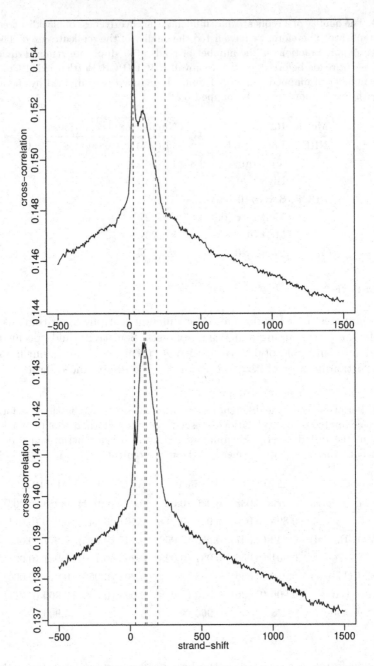

Fig. 2. Strand Cross-Correlation plots for 2 different treatments of the ENCFF900RPG replicate from the ENCODE database. Top: No treatment with RSC 0.708. Maximum correlation is at the phantom peak. Bottom: The PeakPass blacklist with RSC 1.405. Maximum correlation is at the fragment length. Blue vertical lines indicate phantom peak estimates, and red vertical lines indicate fragment length estimates. (Color figure online)

Our blacklist is considerably larger than the ENCODE blacklist: it contains 5,078 regions, with an average length of 14,500bp. In comparison, the ENCODE blacklist for hg38 contains 38 regions with an average length of 450bp. Blacklist regions cluster around assembly gaps, GRC incidents, anomalous mappability regions, centromeres, and telomeres.

After investigating cross-correlation plots of various data sets, we observed that PeakPass greatly improves RSC, but has little impact on NSC. The Peak-Pass treatment decreases both minimum and maximum cross-correlation. The cross-correlation curve itself appears smoother after PeakPass treatment. Example strand cross-correlation plots with different treatments are shown in Fig. 2.

RSC measurements after treatment with different blacklists are shown in Fig. 3. The ENCODE blacklist slightly increases the average RSC. By comparison, the PeakPass blacklist increases the average RSC by 50%.

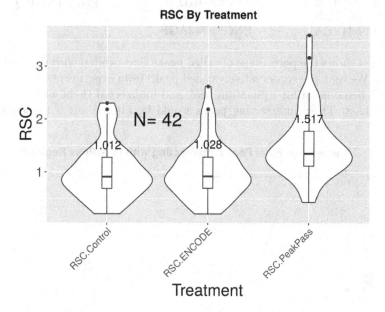

Fig. 3. This violin plot shows the distribution of RSC scores for 42 hg38 ChIP-Seq replicates by treatment. The hg38 ENCODE blacklist has little impact on RSC compared to the PeakPass blacklist. Average RSC for each treatment is shown above its box.

To determine the statistical significance of this effect, we performed a paired Holm's post-hoc test [17]. Our hypotheses are that replicates treated with the ENCODE blacklist have a higher RSC value than without, and that PeakPass treatment results in a higher RSC value than either ENCODE treatment or no treatment. Table 5 contains the p-values for each hypothesis. PeakPass showed a significant improvement over the ENCODE blacklist and the control.

For the replicates listed below, treatment with PeakPass pushes the replicate over the ENCODE recommended RSC threshold of 0.8 [18]. After PeakPass treatment, these replicates could now be used in subsequent analysis steps. Publications using these replicates may want to attempt to reproduce their results

Table 5. P-values from Holm's pairwise hypothesis test. The treatment in each row is hypothesized to have a greater RSC value than the treatments listed in the columns.

	No treatment	ENCODE blacklist treatment
After ENCODE treatment	0.074	N/A
After PeakPass treatment	6e-11	6e-11

using the PeakPass blacklist, and investigate differentially called peaks in order to reduce false positive peak calls.

- ENCFF000XOH
- ENCFF038NOW
- ENCFF061RFH
- ENCFF369TPH
- ENCFF534VEA
- ENCFF629GVZ
- ENCFF693TLQ
- ENCFF701IFH
- ENCFF706HJC
- ENCFF743AGP
- ENCFF877JAH
- ENCFF900RPG
- ENCFF919IIQ

Figure 4 shows the percentage of called peaks that overlap with the PeakPass blacklist. We took the conservatively called peaks from experiments that contain replicates demonstrating a phantom peak and intersected them with the Peak-Pass blacklist. These intersecting peaks would be filtered out if the PeakPass

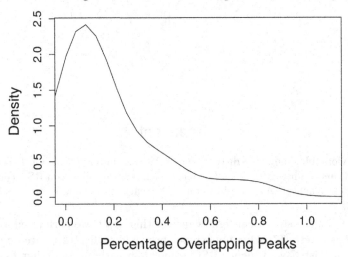

Fig. 4. Percentage of final ChIP-Seq peaks intersecting with PeakPass blacklist regions for 25 conservatively filtered peak sets. Note that one outlier with 10% overlap (the peak set with ENCODE id ENCFF452WYE) does not appear on this graph due to the zoom level. This experiment appears to be an outlier in terms of low quality. The RSC for the replicates from this experiment were less than 0.25, and less than 0.5 after blacklist treatment.

blacklist was used. For most replicates, approximately 0.1% of peaks would have been filtered by PeakPass.

Blacklist regions are associated with repeat elements [2]. The number of repeats per kilobase was measured for both the ENCODE blacklist and the PeakPass predicted blacklist as shown in Table 6. Four of the top five repeats from the hg39 ENCODE blacklist are centromeric repeats, some of which are also enriched in the hg19 ENCODE blacklist [19]. The PeakPass blacklist contains fewer repeats per kilobase, but contains a larger variety of different repeat types (93 different repeat types in the PeakPass blacklist vs 15 different repeat types in the ENCODE blacklist).

Table 6. Relative frequencies (per kbp) of the 5 most frequent repeats in ENCODE and PeakPass blacklists for hg38.

ENCODE blacklist	Frequency/Kbp	PeakPass blacklist	Frequency/Kbp
All	1.4671	All	0.62
(GGAAT)n	0.2934	L1/LINE	0.1490
(TCCAT)n	0.2347	Alu/SINE	0.1208
ALR/Alpha	0.2347	Simple Repeats	0.0937
(GAATG)n	0.1173	MER	0.0411
(GAATGGAATC)n	0.1173	ALR/Alpha	0.0364

4 Conclusion

ChIP-Seq experiments are often plagued by genomic regions that produce artifacts and noise. The underlying cause of this phenomenon is often unclear and therefore, quality control is an important step in every ChIP-seq experiment. Our PeakPass algorithm uses classification to generate genome-scale blacklists that try to protect against such problematic genomic regions. Due to the extreme class imbalance, this task is challenging for most algorithms and we have compared six different classifiers to identify a suitable algorithm. The random forest classifier has outperformed the other classifiers in most of the assessed metrics, and it has shown the best performance for classifying genomic regions as blacklist regions using the features we selected. Hence, we have selected this algorithm as PeakPass' classification engine.

To evaluate the quality of PeakPass predictions we have generated a blacklist for the current version of the human genome (hg38) and compared it to a corresponding ENCODE blacklist. The PeakPass blacklist is considerably larger and increases RSC above the ENCODE blacklist for hg38 by approximately 50%. This increase is statistically significant. However, not all datasets benefit equally from blacklisting. We have observed, not surprisingly, that data sets of lower quality show the highest improvements.

Both the PeakPass blacklist and ENCODE blacklist contain an enrichment of different types of genomic repeats. While the ENCODE contains mostly satellite and centromeric repeats, the PeakPass blacklist contains more retrotransposons such as L1/LINE and Alu/SINE repeats. These types of retrotransposons are frequent in the human genome [20].

There are several avenues of future research. We plan to investigate alternative quality metrics such as the Standardized Standard Deviation (SSD) and percentage of duplicated reads within blacklist regions [5], tune and validate Peak-Pass on histone ChIP-Seq experiments, and apply PeakPass to create blacklists for additional species. Another intriguing research direction focuses on a better understanding of the underlying causes of artifact regions. We hypothesize that alignability problems, as well as highly polymorphic genomic regions are often responsible for the spurious ChIP-Seq peaks.

Acknowledgments. This work was supported in part by the National Science Foundation (grant no. IOS1355019). We thank Robert G. Franks and Miguel Flores-Vergara (both NC State University) for extremely valuable discussions and advice. We are grateful for the ENCODE datasets we used for validating PeakPass. These data were produced by: Michael Snyder, Richard Myers, Sherman Weissman, Xiang-Dong Fu, Kevin Struhl, Bradley Bernstein, John Stamatoyannopoulos, Peggy Farnham, and Vishwanath Iyer.

References

1. Degner, J., et al.: Effect of read-mapping biases on detecting allele-specific expression from RNA-sequencing data. Bioinformatics **25**(24), 3207–3212 (2009)
2. Kundaje, A.: A comprehensive collection of signal artifact blacklist regions in the human genome. http://mitra.stanford.edu/kundaje/akundaje/release/blacklists/hg19-human/hg19-blacklist-README.pdf. Accessed 28 Mar 2019
3. Dolgalev, I., Sedlazeck, F., Busby, B.: DangerTrack: A scoring system to detect difficult-to-assess regions. F1000Research. **6**(443) (2017)
4. Wimberley, C.: PeakPass: a machine learning approach for ChIP-Seq blacklisting. Master's thesis, North Carolina State University (2018)
5. Carroll, T.S., Liang, Z., Salama, R., Stark, R., de Santiago, I.: Impact of artifact removal on ChIP quality metrics in ChIP-seq and ChIP-exo data. Front. Genet. **5**, 75 (2014)
6. Ramachandran, P., Palidwor, G., Porter, C., Perkins, T.: MaSC: mappability-sensitive cross-correlation for estimating mean fragment length of single-end short-read sequencing data. Bioinformatics **29**(4), 444–450 (2013)
7. The ENCODE Project Consortium: An integrated encyclopedia of DNA elements in the human genome. Nature **489**(7414), 57–74 (2012)
8. Ho, T.: Random decision forests. In: Proceedings of the Third International Conference on Document Analysis and Recognition, vol. 1, pp. 278–282 (1995)
9. Fix, E., Hodges, J.: Discriminatory analysis nonparametric discrimination: consistency properties (1951)
10. Cortes, C., Vapnik, V.: Support-vector networks. Mach. Learn. **20**(3), 273–297 (1995)

11. Farley, B., Clark, W.: Simulation of self-organizing systems by digital computer. Trans. IRE Prof. Group Inf. Theory **4**(4), 76–84 (1954)
12. John, G., Langley, P.: Estimating continuous distributions in Bayesian classifiers. In: Proceedings of the Eleventh Conference on Uncertainty in Artificial Intelligence, pp. 338–345 (2013)
13. Derrien, T., et al.: Fast computation and applications of genome mappability. PLoS One **7**(1), e30377 (2012)
14. Smit, A., Hubley, R., Green, P.: RepeatMasker Open-4.0 (2013-2015). http://www.repeatmasker.org
15. Kuhn, M.: Building predictive models in R using the caret package. J. Stat. Softw. **28**(5), 1–26 (2008)
16. The ENCODE Project Consortium: Transcription Factor ChIP-seq Data Standards and Processing Pipeline. https://www.encodeproject.org/chip-seq/transcription_factor/
17. Holm, S.: A simple sequentially rejective multiple test procedure. Scand. J. Stat. **6**(2), 65–70 (1979)
18. Landt, S., et al.: ChIP-seq guidelines and practices of the ENCODE and modEN-CODE consortia. Genome Res. **22**(9), 1813–1831 (2012)
19. Altemose, N., Miga, K.H., Maggioni, M., Willard, H.F.: Genomic characterization of large heterochromatic gaps in the human genome assembly. PLOS Comput. Biol. **10**(5), e1003628 (2014)
20. Kojima, K.: Human transposable elements in Repbase: genomic footprints from fish to humans. Mobile DNA. **9**(2) (2018)

Maximum Stacking Base Pairs: Hardness and Approximation by Nonlinear LP-Rounding

Lixin Liu[1], Haitao Jiang[2,5(✉)], Peiqiang Liu[3,5], Binhai Zhu[4], and Daming Zhu[2]

[1] School of Software, Shandong University, Jinan 250100, Shandong, China
201500301266@mail.sdu.edu.cn
[2] School of Computer Science and Technology, Shandong University,
Qingdao 266237, Shandong, China
{htjiang,dmzhu}@sdu.edu.cn
[3] School of Computer Science and Technology,
Shandong Technology and Business University, Yantai 264006, Shandong, China
liupeiqiang@gmail.com
[4] Gianforte School of Computing, Montana State University,
Bozeman, MT 59717, USA
bhz@montana.edu
[5] Co-innovation Center of Shandong Colleges and Universities:
Future Intelligent Computing, Yantai 264005, Shandong, China

Abstract. Maximum stacking base pairs is a fundamental combinatorial problem from RNA secondary structure prediction under the energy model. The basic maximum stacking base pairs problem can be described as: given an RNA sequence, find a maximum number of base pairs such that each chosen base pair has at least one parallel and adjacent partner (i.e., they form a stacking). This problem is NP-hard, no matter whether the candidate base pairs follow the biology principle or are given explicitly as input. This paper investigates a restricted version of this problem where the base pairs are given as input and each base is associated with at most k (a constant) base pairs. We show that this restricted version is still APX-hard, even if the base pairs are weighted. Moreover, by a nonlinear LP-rounding method, we present an approximation algorithm with a factor $\frac{32(k-1)^3 e^3}{8(k-1)e-1}$. Applying our algorithms on the simulated data, the actual approximation factor is in fact much better than this theoretical bound.

1 Introduction

An Ribonucleic acid (RNA) is single-stranded and can be viewed as a sequence of nucleotides (also known as bases, denoted by A, C, G and U). It plays an important role in regulating genetic and metabolic activities according to the central dogma of biology. To understand the biological functions of RNAs elaborately, we should know their structures at first. The *primary* structure of an RNA strand is formed by the order of the nucleotides. An RNA folds into a three

© Springer Nature Switzerland AG 2019
Z. Cai et al. (Eds.): ISBRA 2019, LNBI 11490, pp. 244–256, 2019.
https://doi.org/10.1007/978-3-030-20242-2_21

dimensional structure by forming hydrogen bonds between nonconsecutive bases that are complementary, such as the Watson-Crick pairs C-G and A-U and the wobble pair G-U. The three-dimensional arrangement of the atoms in the folded RNA molecule is the *tertiary* structure; the collection of base pairs in the tertiary structure is the *secondary* structure. Actually, the secondary structure can tell us where there are additional connections between the bases, and where the RNA molecule could be folded. In the paper [1], the author claimed that "the folding of RNA is hierarchical, since secondary structure is much more stable than tertiary folding", which results in that, the tertiary folding would obey the secondary structure mostly. Since the 3-dimensional structure determines the function of the RNA to some extent, predicting the secondary structure of RNA becomes a key problem to study RNA in a larger and deeper scope.

In 1978, Nussinov et al. [2] began considering the computational study of RNA secondary structure prediction, but this problem is still not well solved yet. The biggest impediment is the existing of pseudoknots, which is composed of two interleaving base pairs provided when we arrange the RNA sequence in a linear order.

Lyngsø and Pedersen [8] have proven that determining the optimal secondary structure possibly with pseudoknots is NP-hard under special energy functions. And Akutsu [9] has shown that it remains NP-hard, even if the secondary structure requires to be planar. There are a lot of positive works where there are no pseudoknot. [2–7] have computed the optimal RNA secondary structure in $O(n^3)$ time and $O(n^2)$ space by the method of dynamic programming. Akutsu in [9], Rivas and Eddy in [10], and Uemura et al. in [11] have presented a polynomial-time algorithm when the types of pseudoknots is limited.

To predict secondary structures with pseudoknots, most research focus on the base pairs individually. The nearest neighbor energy model was studied popularly [8,9,13,14]: the energy of each base pair depends not only on its two bases but also on the other adjacent base pairs. According to the Tinoco model [12]: an RNA structure can recursively be decomposed into loops with independent free energy; the energy of each loop is an affine function in the number of unpaired bases and the number of interior base pairs. The only type of loops without unpaired bases are formed by two adjacent and parallel base pairs, which is called a *stacking*; the negative energy of such stackings stabilizes the RNA structure. In [14], Lyngsø initiated the study for the maximum stacking base pairs problem. He showed this problem to be NP-hard, and devised a polynomial-time exact algorithm over a fixed-size alphabet Σ and with a subset of $\Sigma \times \Sigma$ of legal pair types. Unfortunately, this algorithm has very high complexities of $\Omega(n^{80})$ time and $\Omega(n^{80})$ space even for the canonical alphabet $\{A, C, G, U\}$.

Among all the above results, the base pairs are given implicit, that is, under some fix biology principle, e.g., Watson-Crick base pairs: A-U and C-G, any such two bases can form a base pair. As an alternative, the set of candidate base pairs may be given explicitly as input, because there could be additional conditions from comparative analysis which prevent two bases forming a pair. It would generalize the maximum stacking base pairs problem with explicit base pairs, so the problem remains NP-hard. Jiang [15] improved the approximation factor for the maximum stacking base pairs problem with explicit base pairs to 5/2. Zhou et al. [17] further improved the approximation factor to 7/3 by a *local search* method. Once the candidate base pairs are taken as the input for this problem, naturally, we can put restriction and generalization on it. Similar to the research on graph problems, one basic restriction is to bound the degree of each base, that is, we require each base to associate with at most a constant of k candidate base pairs. This problem is called k-*MSBP*. Also, in light that the optimal secondary structure has the minimum negative energy, we generalize this problem to the weighted version, where we give each base pair a weight (representing energy) and the problem becomes computing a maximum weight stacking base pairs, this problem is called k-*MWSBP*. So far as we know, there is no results on k-*MWSBP*.

The main contributions of this paper are: (1) We show that k-*MSBP* is APX-hard for $k \geq 2$; (2) We devise an approximation algorithm with a factor of $\frac{32(k-1)^3 e^3}{8(k-1)e-1}$ for k-*MWSBP* (and k-*MSBP*) by the nonlinear LP-rounding method. For k-*MSBP*, although the approximation factor in [15] and [17] are better, the time complexity is as high as $O(n^{14})$, while our algorithm takes linear time besides solving a linear program. Moreover, our simulations show a much better practical performance compared with this theoretical bound.

2   Preliminaries

Let $S = s_1 s_2 \cdots s_n$ be an RNA sequence of n bases. A *base pair* is a pair of two nonconsecutive bases, say s_i, s_j, where $|i - j| > 1$, and is denoted by (s_i, s_j). The degree of a base s_i is the number of base pairs that are associated with s_i. Two base pairs are *compatible* if they do not share a common base. A secondary structure of S is a set of mutually compatible base pairs (s_{i_1}, s_{j_1}), (s_{i_2}, s_{j_2}), ..., (s_{i_r}, s_{j_r}). Two base pairs, such as (s_i, s_j) and (s_{i+1}, s_{j-1}) are mutually *adjacent*. A *stacking* is constituted by two mutually adjacent base pairs.

A feasible secondary structure $FS(S)$ of an RNA sequence S fulfills that if (s_i, s_j) is a base pair in $FS(S)$, then either (s_{i+1}, s_{j-1}) or (s_{i-1}, s_{j+1}) or both are base pairs in $FS(S)$, which implies that a feasible secondary structure is composed of helices.

Now we present the formal definition of the problems studied in this paper.

Definition 1. *Maximum Stacking Base Pairs with Degree Bounded, k-MSBP.*
Input: *An RNA sequence S, and a set of candidate base pairs BP, where the degree of each base is bounded by k.*
Output: *A feasible secondary structure $FS(S)$ such that the number of the base pairs is maximized.*

Definition 2. *Maximum Weighted Stacking Base Pairs with Degree Bounded, k-MWSBP.*
Input: *An RNA sequence S, and a set of candidate base pairs BP, where the degree of each base is bounded by k.*

Output: *A feasible secondary structure FS(S) such that the total weight of the base pairs is maximized.*

k-MSBP is a special case of *k-MWSBP* with all base pairs having a weight of 1. In the next section, we will prove that *k-MSBP* is APX-hard which implies that *k-MWSBP* is also APX-hard. Note that an approximation algorithm for *k-MWSBP* also works on *k-MSBP*.

3   Hardness Results

In this section, we will show that *k-MSBP* is APX-hard by a reduction from the Maximum Independent Set Problem on Cubic Graphs (3-MIS).

Theorem 1. *It is NP-hard to approximate 3-MIS within $\frac{140}{139} - \epsilon$ [18].*

For the sake of simplicity, we just prove that 2-MSBP is NP-hard and then APX-hard, which means *k*-MSBP is also APX-hard for all $k \geq 2$.

Given a cubic graph $G = (V, E)$ as an input for 3-MIS, for each vertex $v \in V$, we construct an RNA subsequence of 32 bases: $A_v^1, A_v^2, \ldots, A_v^{18}, U_v^1, \ldots, U_v^{14}$, and 18 candidate base pairs: $(A_v^1, U_v^5), (A_v^2, U_v^4), (A_v^3, U_v^2), (A_v^4, U_v^1), (A_v^7, U_v^8), (A_v^8, U_v^7), (A_v^{11}, U_v^{14}), (A_v^{12}, U_v^{13}), (A_v^{15}, U_v^{11}), (A_v^{16}, U_v^{10}), (A_v^4, U_v^9), (A_v^5, U_v^8), (A_v^6, U_v^4), (A_v^7, U_v^3), (A_v^8, U_v^{12}), (A_v^9, U_v^{11}), (A_v^{10}, U_v^7), (A_v^{11}, U_v^6)$. See Fig. 1 for an example. There are two feasible secondary structures of this RNA subsequence: the first 10 base pairs (which is maximum, see the solid matching edges in Fig. 1) and the last 8 base pairs (see the dotted matching edges). The RNA sequence RG is the concatenation of all the subsequences which are split by peg bases.

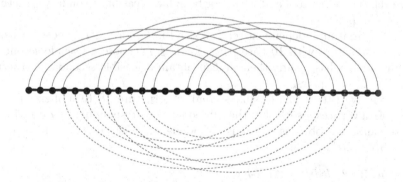

Fig. 1. The RNA subsequence and base pairs corresponding to a vertex.

We can orient the edges of G in such a way that each vertex has at most two incoming edges and at most two outgoing edges. This can be done as follows: iteratively find edge-disjoint cycles in G, and in each cycle, orient the edges to form a directed cycle. The remaining edges form a forest. For each tree in the forest, choose one of its nodes of degree one to be the root, and orient all edges in the tree away from the root. This orientation will clearly satisfy the desired properties.

For each vertex v, we name A_v^1, A_v^3 as two incoming interfaces, A_v^{12}, A_v^{16} as two outgoing interfaces. Initially all interfaces are free. Let (u, v) be an edge from u to v in G.

Construct new base pairs as follows:

1. If A_u^{12} is a free outgoing interface of u and A_v^1 is a free incoming interface of v. Then delete the two bases pairs: (A_v^1, U_v^5), (A_v^2, U_v^4), and make two new base pairs: (A_u^{12}, U_v^5), (A_u^{13}, U_v^4).
2. If A_u^{12} is a free outgoing interface of u and A_v^3 is a free incoming interface of v. Then delete the four bases pairs: (A_v^3, U_v^2), (A_v^4, U_v^1), (A_v^4, U_v^9), (A_v^5, U_v^8), and make four new base pairs: (A_u^{12}, U_v^2), (A_u^{13}, U_v^1), (A_u^{13}, U_v^9), (A_u^{14}, U_v^8).
3. If A_u^{16} is a free outgoing interface of u and A_v^1 is a free incoming interface of v. Then delete the two bases pairs: (A_v^1, U_v^5), (A_v^2, U_v^4), and make two new base pairs: (A_u^{16}, U_v^5), (A_u^{17}, U_v^4).
4. If A_u^{16} is a free outgoing interface of u and A_v^3 is a free incoming interface of v. Then delete the four bases pairs: (A_v^3, U_v^2), (A_v^4, U_v^1), (A_v^4, U_v^9), (A_v^5, U_v^8), and make four new base pairs: (A_u^{16}, U_v^2), (A_u^{17}, U_v^1), (A_u^{17}, U_v^9), (A_u^{18}, U_v^8).

Lemma 1. *Let G be a cubic graph on N vertices. Then, there exists an independent set of size l in G if and only if there exists a feasible secondary structure of size $8N + 2l$.*

Proof. From our construction, each RNA subsequence can have a feasible secondary structure with either 10 base pairs or 8 base pairs. The crucial observation is that, if there is an edge (u, v) between two vertices in G, then the RNA subsequence corresponding to u and v can not both have feasible secondary structures with 10 base pairs.

So, if there is an independent set I of size l in G, for $u \in I$, choose a feasible secondary structure with 10 base pairs, for $u \notin I$, choose a feasible secondary structure with 8 base pairs, then we can obtain a feasible secondary structure with $8N + 2l$ base pairs.

Conversely, if there is a feasible secondary structure FS(RG) of size f, let I consist of all vertices u such that the subsequence corresponding to u contributes 10 base pairs, it is obvious that I is an independent set, and $f = 8(N - |I|) + 10|I| = 8N + 2|I|$.   □

Theorem 2. *2-MSBP is APX-hard.*

Proof. Note that the maximum stacking base pairs instance we construct from an instance of 3-MIS is actually an 2-MSBP. Let I be an instance of 3-MIS and $OPT(I)$ be its optimal solution. Let $f(I)$ be an instance of 2-MSBP constructed from I and $OPT(f(I))$ be its optimal solution. Let y' be a solution of $f(I)$ and $g(y')$ be the corresponding solution of I. The reduction is an L-reduction since it fulfills the following two conditions:

1. $|OPT(f(I))| = 8N + 2OPT(I) \leq 34 \cdot OPT(I)$, since $OPT(I) \geq N/4$.

2. $|OPT(I)| - |g(y')| = (|OPT(f(I))| - 8N)/2 - (|y'| - 8N)/2 = (|OPT(f(I))| - |y'|)/2$.

This completes the proof.   □

4   Approximation Algorithms for k-MWSBP by LP-Rounding

In this section, we will design an approximation algorithms for k-MWSBP by a nonlinear LP-rounding method. Firstly, we formulate k-MWSBP as a 0–1 Integer Linear Program (ILP). Let $S = s_1 s_2 \cdots s_n$ be an RNA sequence of n bases, let BP be the set of candidate base pairs. For each base pair (s_i, s_j), we define a 0–1 variable $x_{i,j}$, if (s_i, s_j) is chosen into the feasible secondary structure, then $x_{i,j} = 1$, otherwise $x_{i,j} = 0$.

ILP-(1):

$$MAX. \quad \sum_{(s_i, s_j) \in BP} \omega_{i,j} x_{i,j}$$

$$S.T. \quad \sum_{j=1}^{n}(x_{i,j} + x_{j,i}) \leq 1, \ for \ i = 1, ..., n \quad (1)$$

$$x_{i,j} - (x_{i-1,j+1} + x_{i+1,j-1}) \leq 0, \ for \ i \leq j \quad (2)$$

$$x_{i,j} \in \{0, 1\}$$

Constraints (1) guarantee that the chosen base pairs are mutually compatible. Constraints (2) require that each chosen base pair must have at least one adjacent partner. Relaxing ILP-(1) to the linear programming formulation.

LP-(2):

$$MAX. \quad \sum_{(s_i, s_j) \in BP} \omega_{i,j} x_{i,j}$$

$$S.T. \quad \sum_{i=1}^{n}(x_{i,j} + x_{j,i}) \leq 1, \ for \ j = 1, ..., n$$

$$x_{i,j} - (x_{i-1,j+1} + x_{i+1,j-1}) \leq 0, \ for \ i \leq j$$

$$0 \leq x_{i,j} \leq 1$$

Algorithm 1. Nonlinear LP-rounding

1: Solving LP-(2) and obtain an optimal solution $x_{i,j} = x_{i,j}^*$.

2: Rounding Strategy: $\Pr(x'_{i,j}=1)=1 - e^{-a\sqrt{x_{i,j}^*}}$.

3: Chose the base pair (s_i, s_j) if and only if $x'_{i,j}=1$.

Theorem 3. *Algorithm-1 is an approximation algorithm for 2-MWSBP with an expected factor of $\frac{32e^3}{8e-1}$.*

Proof. To obtain a feasible secondary structure, every chosen base pair must be compatible with each other, we say that such base pairs are effective. Let $A_{i,j}$ be the event that the base pair (s_i, s_j) is effective. Assume that there are three such base pairs: (s_i, s_j), (s_k, s_i), (s_j, s_l). To make (s_i, s_j) effective, it requires $x'_{i,j} = 1$, $x'_{k,i} = 0$, and $x'_{k,l} = 0$. Thus,

$$Pr(A_{i,j}) = (1 - e^{-a\sqrt{x^*_{i,j}}}) \times e^{-a\sqrt{x^*_{k,i}}} \times e^{-a\sqrt{x^*_{j,l}}}$$

$$\geq (1 - e^{-a\sqrt{x^*_{i,j}}}) \times e^{-2a\sqrt{1-x^*_{i,j}}}$$

$$\geq c\sqrt{x^*_{i,j}}$$

where c is a constant, to be determined later. To make an effective base pairs (s_i, s_j) chosen into the feasible secondary structure, it also requires at least one of (s_{i-1}, s_{j+1}) and (s_{i+1}, s_{j-1}) to be effective. Let $B_{i,j}$ be the event that the base pair (s_i, s_j) take part in constituting stacking and let $z_{i,j}$ be an 0–1 variable, where $z_{i,j} = 1$ if $B_{i,j}$ happens, and $z_{i,j} = 0$ if not.

$$Pr(z_{i,j} = 1) = Pr(A_{i,j})[1 - (1 - Pr(A_{i-1,j+1}))(1 - Pr(A_{i+1,j-1}))]$$

$$\geq c\sqrt{x^*_{i,j}} \cdot [1 - (1 - c\sqrt{x^*_{i-1,j+1}})(1 - c\sqrt{x^*_{i+1,j-1}})]$$

$$\geq c\sqrt{x^*_{i,j}} \left\{ 1 - [\frac{2 - c(\sqrt{x^*_{i-1,j+1}} + \sqrt{x^*_{i+1,j-1}})}{2}]^2 \right\}$$

Since $\sqrt{x^*_{i-1,j+1}} + \sqrt{x^*_{i+1,j-1}} \geq \sqrt{x^*_{i-1,j+1} + x^*_{i+1,j-1}}$, and by constraint (2), we have $x^*_{i-1,j+1} + x^*_{i+1,j-1} \geq x^*_{i,j}$, then, $\sqrt{x^*_{i-1,j+1}} + \sqrt{x^*_{i+1,j-1}} \geq \sqrt{x^*_{i,j}}$.

$$Pr(z_{i,j} = 1) \geq c\sqrt{x^*_{i,j}} \cdot [1 - (\frac{2 - c\sqrt{x^*_{i,j}}}{2})^2]$$

$$\geq c\sqrt{x^*_{i,j}} \cdot (c\sqrt{x^*_{i,j}} - \frac{c^2 x^*_{i,j}}{4})$$

$$\geq x^*_{i,j} \cdot (c^2 - \frac{c^3}{4})$$

Let APP denote the size of the output solution of Algorithm 1, OPT denote the size of the optimal solution, which is also the optimal solution of ILP-(1), and $OPT(LP)$ denote the optimal solution of LP-(2). Obviously, $OPT(LP) \geq OPT$ then we have,

$$E(APP) = E(\sum_{(s_i,s_j)\in BP} \omega_{i,j} \cdot z_{i,j})$$

$$= \sum_{(s_i,s_j)\in BP} \omega_{i,j} Pr(z_{i,j} = 1)$$

$$\geq \sum_{(s_i,s_j)\in BP} \left[\omega_{i,j} \cdot x^*_{i,j} \cdot (c^2 - \frac{c^3}{4})\right]$$

$$= (c^2 - \frac{c^3}{4}) \cdot \sum_{(s_i,s_j)\in BP} \omega_{i,j} x^*_{i,j}$$

$$= (c^2 - \frac{c^3}{4}) \cdot OPT(LP)$$

$$\geq (c^2 - \frac{c^3}{4}) \cdot OPT$$

Let $t = \sqrt{x^*_{i,j}}$, the function $F(t) = \frac{(1-e^{-at})e^{-2a\sqrt{1-t^2}}}{t}, (0 < a \leq 1, 0 \leq t \leq 1)$, reaches its minimum value, when t trends to 0:

$$\lim_{t\to 0} F(t) = \lim_{t\to 0} \frac{(1 - e^{-at})e^{-2a\sqrt{1-t^2}}}{t} = \frac{a}{e^{2a}}.$$

By setting $a = \frac{1}{2}$ and $c = \frac{1}{2e}$, we obtain the best approximation ratio of $\frac{32e^3}{8e-1}$ for 2-MWSBP.

Theorem 4. *Algorithm 1 is an approximation algorithm for k-MWSBP with an expected factor of $\frac{32(k-1)^3 e^3}{8(k-1)e-1}$.*

Proof. The difference between k-MWSBP and 2-MWSBP is the degree of bases. In an k-MWSBP instance, a base pair (s_i, s_j) is not compatible with (s_{t_1}, s_i) , $(s_{t_2}, s_i), \ldots, (s_{t_{k-1}}, s_i)$ and $(s_j, s_{l_1}), (s_j, s_{l_2}), \ldots, (s_j, s_{l_{k-1}})$. Then the probability that (s_i, s_j) is effective is

$$Pr(A_{i,j}) = (1 - e^{-a\sqrt{x^*_{i,j}}}) \cdot e^{-a\sqrt{x^*_{t_1,i}}} \ldots e^{-a\sqrt{x^*_{t_{k-1},i}}} \cdot e^{-a\sqrt{x^*_{j,l_1}}} \ldots e^{-a\sqrt{x^*_{j,l_{k-1}}}}$$

by constraint (1), $x^*_{t_1,i} + \ldots + x^*_{t_{k-1},i} + x^*_{i,j} \leq 1$ and $x^*_{i,j} + x^*_{j,l_1} + \ldots + x^*_{j,l_{k-1}} \leq 1$. Thus,

$$Pr(A_{i,j}) = (1 - e^{-a\sqrt{x^*_{i,j}}}) \cdot e^{-a\sqrt{x^*_{t_1,i}}} \ldots e^{-a\sqrt{x^*_{t_{k-1},i}}} \cdot e^{-a\sqrt{x^*_{j,l_1}}} \ldots e^{-a\sqrt{x^*_{j,l_{k-1}}}}$$

$$\geq (1 - e^{-a\sqrt{x^*_{i,j}}}) \times e^{-(2k-2)a\sqrt{1-x^*_{i,j}}}$$

$$\geq c\sqrt{x^*_{i,j}}$$

The probability that (s_i, s_j) takes part in constituting stacking is

$$Pr(z_{i,j} = 1) = Pr(A_{i,j})[1 - (1 - Pr(A_{i-1,j+1}))(1 - Pr(A_{i+1,j-1}))]$$

$$\geq c\sqrt{x^*_{i,j}} \cdot [1 - (1 - c\sqrt{x^*_{i-1,j+1}})(1 - c\sqrt{x^*_{i+1,j-1}})]$$

$$\geq c\sqrt{x^*_{i,j}} \left\{ 1 - [\frac{2 - c(\sqrt{x^*_{i-1,j+1}} + \sqrt{x^*_{i+1,j-1}})}{2}]^2 \right\}$$

$$\geq c\sqrt{x^*_{i,j}} \cdot [1 - (\frac{2 - c\sqrt{x^*_{i,j}}}{2})^2]$$

$$= c\sqrt{x^*_{i,j}} \cdot (c\sqrt{x^*_{i,j}} - \frac{c^2 x^*_{i,j}}{4})$$

$$\geq x^*_{i,j} \cdot (c^2 - \frac{c^3}{4})$$

Let $t = \sqrt{x^*_{i,j}}$, the function $F(t) = \frac{(1-e^{-at})e^{-(2k-2)a\sqrt{1-t^2}}}{t}, (0 < a \leq 1, 0 \leq t \leq 1)$, reaches its minimum value, again when t trends to 0:

$$\lim_{t\to 0} F(t) = \lim_{t\to 0} \frac{(1 - e^{-at})e^{-(2k-2)a\sqrt{1-t^2}}}{t} = \frac{a}{e^{(2k-2)a}}.$$

By setting $a = \frac{1}{2k-2}$ and $c = \frac{1}{(2k-2)e}$, we obtain the best approximation ratio of $\frac{32(k-1)^3 e^3}{8(k-1)e-1}$ for k-MWSBP. □

5   Simulations

In this section, we show some experiments on randomly generated simulated data. In the simulated data, the length of the RNA sequences ranges from $n = 100$ to $n = 1000$, we choose three values for k: $k = 2$, $k = 3$, $k = 4$. For comparison, besides running the LP-rounding approximation algorithm, we also run the ILP-(1) to obtain the optimal solutions (though when n gets large, the running time gets really high). The performance are summarized as follows.

5.1   Performance Evaluation

For $k = 2$, the experimental results are shown in Table 1. As what is stated in Theorem 2, the approximation factor for 2-MWSBP is about 31. From the experimental results in Table 1, the actual approximation factor is about 4.8, which is much better than the theoretical bound.

For $k = 3$, the experimental results are shown in Table 2. Similarly, the actual approximation factor is about 6.07. Again, the experimental results show much better performance compared with the theoretical results.

For $k = 4$, the experimental results are shown in Table 3. While the practical approximation factor fluctuates more in the case, the average approximation

Table 1. Values of optimal solution (OPT(I)), approximation solution (APP(I)) and the approximation factor, when $k = 2$.

n	OPT(I)	APP(I)	Approximation ratio
100	52	13	4.00
200	113	25	4.52
300	163	31	5.26
400	225	39	5.77
500	313	73	4.29
600	358	69	5.19
700	424	94	4.51
800	455	111	4.10
900	543	130	4.18
1000	632	170	3.72

factor is about 7.44. From our experimental results, we can conclude that the actual performance of our algorithm is much better than the corresponding theoretical bound, the reason is probably due to that the theoretical result is base on the worst-case analysis.

5.2 Runtime Analysis

As discussed before, solving the integer linear program ILP-(1) takes quite some time when n gets large. Hence, we compare the running time of solving the ILP and our LP-rounding approximation algorithm. The results are summarized in Table 4 and Fig. 2.

Table 2. Values of optimal solution (OPT(I)), approximation solution (APP(I)) and the approximation factor, when $k = 3$.

n	OPT(I)	APP(I)	Approximation ratio
100	80	22	3.64
200	170	31	5.48
300	235	32	7.34
400	340	53	6.42
500	475	58	8.19
600	534	81	6.59
700	640	128	5.00
800	708	131	5.40
900	813	123	6.61
1000	955	158	6.04

Table 3. Values of optimal solution (OPT(I)), approximation solution (APP(I)) and the actual approximation factor, when $k = 4$.

n	OPT(I)	APP(I)	Approximation ratio
100	103	22	4.68
200	229	59	3.88
300	332	39	8.51
400	454	37	12.27
500	642	60	10.70
600	707	121	5.84
700	846	128	6.61
800	924	122	7.57
900	1108	166	6.67
1000	1252	163	7.68

Table 4. The running time (seconds) of solving the ILP and the LP-rounding approximation algorithm.

n	$k = 2$		$k = 3$		$k = 4$	
	ILP	LP-rounding	ILP	LP-rounding	ILP	LP-rounding
100	3.11	3.78	3.46	3.90	3.62	3.22
200	11.63	11.91	11.85	11.69	11.69	11.78
300	23.84	25.08	23.90	24.77	24.10	24.91
400	47.26	46.40	46.43	45.88	47.24	46.23
500	147.24	78.16	216.74	78.57	324.22	79.60
600	135.81	117.65	145.16	119.15	354.60	117.45
700	207.78	170.22	209.67	170.00	1403.68	169.49
800	243.78	199.27	216.44	198.71	276.14	199.02
900	258.93	217.81	4096.56	216.48	48420.51	217.05
1000	2642.49	243.57	6683.80	243.70	73952.46	247.81

As shown in Fig. 2, solving the ILP takes much more time as n increases, while the running time of our approximation algorithm is very stable. This is probably due to that the ILP solver takes exponential time, while the approximation algorithm takes polynomial time.

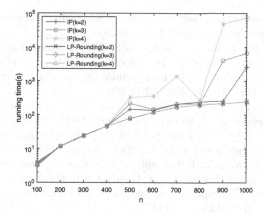

Fig. 2. The plot graph of the running times of the ILP and LP-rounding algorithms.

6   Concluding Remarks

In this paper, we studied a restricted version of the maximum stacking base pairs problem, which originates from RNA secondary structure prediction. Regardless of whether the base pairs are weighted or not, we show that this problem is APX-hard, when the degree of each base is bounded by a constant k. Also, we design the first approximation algorithm for this problem with a factor of $\frac{32(k-1)^3 e^3}{8(k-1)e-1}$ for k-MWSBP by a nonlinear LP-rounding method. Our experimental results indicate a much better performance compared with this theoretical approximation factor and our algorithm is much faster than the exponential time of solving ILP. How to improve the approximation factor for k-MWSBP is an interesting open problem.

Acknowledgments. This research is supported by NSF of China under grant 61872427, 61732009 and 61628207, by NSF of Shandong Provence under grant ZR201702190130. Haitao Jiang is also supported by Young Scholars Program of Shandong University. Peiqiang Liu is also supported by Key Research and Development Program of Yantai City (2017ZH065) and CERNET Innovation Project (No. NGII20161204).

References

1. Tinoco Jr., I., Bustamante, C.: How RNA folds. J. Mol. Biol. **293**, 271–281 (1999)
2. Nussinov, R., Pieczenik, G., Griggs, J.R., Kleitman, D.J.: Algorithms for loop matchings. SIAM J. Appl. Math. **35**(1), 68–82 (1978)
3. Nussinov, R., Jacobson, A.B.: Fast algorithm for predicting the secondary structure of single-stranded RNA. Proc. Natl. Acad. Sci. USA **77**, 6309–6313 (1980)
4. Zuker, M., Stiegler, P.: Optimal computer folding of large RNA sequences using thermodynamics and auxiliary information. Nucleic Acids Res. **9**, 133–148 (1981)
5. Zuker, M., Sankoff, D.: RNA secondary structures and their prediction. Bull. Math. Biol. **46**, 591–621 (1984)

6. Sankoff, D.: Simultaneous solution of the RNA folding, alignment and protosequence problems. SIAM J. Appl. Math. **45**, 810–825 (1985)

7. Lyngsø, R.B., Zuker, M., Pedersen, C.N.S.: Fast evaluation of interval loops in RNA secondary structure prediction. Bioinformatics **15**, 440–445 (1999)

8. Lyngsø, R.B., Pedersen, C.N.S.: RNA pseudoknot prediction in energy based models. J. Comput. Biol. **7**(3/4), 409–427 (2000)

9. Akutsu, T.: Dynamic programming algorithms for RNA secondary structure prediction with pseudoknots. Discrete Appl. Math. **104**(1–3), 45–62 (2000)

10. Rivas, E., Eddy, S.R.: A dynamic programming algorithm for RNA structure prediction including pseudoknots. J. Mol. Biol. **285**(5), 2053–2068 (1999)

11. Uemura, Y., Hasegawa, A., Kobayashi, S., Yokomori, T.: Tree adjoining grammars for RNA structure prediction. Theor. Comput. Sci. **210**(2), 277–303 (1999)

12. Tinoco Jr., I., et al.: Improved estimation of secondary structure in ribonucleic acids. Nat. New Biol. **246**, 40–42 (1973)

13. Ieong, S., Kao, M.-Y., Lam, T.-W., Sung, W.-K., Yiu, S.-M.: Predicting RNA secondary structure with arbitrary pseudoknots by maximizing the number of stacking pairs. J. Comput. Biol. **10**, 981–995 (2003)

14. Lyngsø, R.B.: Complexity of pseudoknot prediction in simple models. In: Díaz, J., Karhumäki, J., Lepistö, A., Sannella, D. (eds.) ICALP 2004. LNCS, vol. 3142, pp. 919–931. Springer, Heidelberg (2004). https://doi.org/10.1007/978-3-540-27836-8_77

15. Jiang, M.: Approximation algorithms for predicting RNA secondary structures with arbitrary pseudoknots. IEEE/ACM Trans. Comput. Biol. Bioinform. **7**(2), 323–332 (2010)

16. Berman, P.: A $d/2$ approximation for maximum weight independent set in d-Claw free graphs. Nordic J. Comput. **7**, 178–184 (2000)

17. Zhou, A., Jiang, H., Guo, J., Zhu, D.: A new approximation algorithm for the maximum stacking base pairs problem from RNA secondary structures prediction. In: Gao, X., Du, H., Han, M. (eds.) COCOA 2017. LNCS, vol. 10627, pp. 85–92. Springer, Cham (2017). https://doi.org/10.1007/978-3-319-71150-8_7

18. Berman, P., Karpinski, M.: On some tighter inapproximability results (extended abstract). In: Wiedermann, J., van Emde Boas, P., Nielsen, M. (eds.) ICALP 1999. LNCS, vol. 1644, pp. 200–209. Springer, Heidelberg (1999). https://doi.org/10.1007/3-540-48523-6_17

Greedy Partition Distance Under Stochastic Models - Analytic Results

Sagi Snir[(✉)]

Department of Evolutionary Biology, University of Haifa, 3498838 Haifa, Israel
ssagi@research.haifa.ac.il

Abstract. Gene partitioning is a very common task in genomics, based on several criteria such as gene function, homology, interactions, and more. Given two such partitions, a metric to compare them is called for. One such metric is based on multi symmetric difference and elements are removed from both partitions until identity is reached. While such a task can be done accurately by a maximum weight bipartite matching, in common settings in comparative genomics, the standard algorithm to solve this problem might become impractical. In previous works we have studied the universal pacemaker (UPM) where genes are clustered according to mutation rate correlation, and suggested a very fast and greedy procedure for comparing partitions. This procedure is guaranteed to provide a poor approximation ratio of 1/2 under arbitrary inputs.

In this work we give a probabilistic analysis of this procedure under a common and natural stochastic environment. We show that under mild size requirements, and a sound model assumption, this procedure returns the correct result with high probability. Furthermore, we show that in the context of the UPM, this natural requirement holds automatically, rendering statistical consistency of this fast greedy procedure. We also discuss the application of this procedure in the comparative genomics rudimentary task of gene orthology where such a solution is imperative.

1 Introduction

Gene partitioning and clustering is ubiquitous in comparative, evolutionary genomics [8,11,22]. Partition is done based on functions, mutation rates, interaction networks, and more. Clustering is commonly performed by statistical, geometric approaches, where *kmeans* is among the most popular [7,12]. Nevertheless, different approaches, different parameters, data quantities, all may lead to differences in partitioning. The latter calls for a metric over the set of partitions. Under such metric, we can define distances and determine whether two partitions are close or far apart and what is the likelihood (p-value) of obtaining such a distance by chance.

One such metric is based on the symmetric difference measure between sets. Elements from the ground set on which the partitions are defined, are

Supported in part by the VolkswagenStiftung grant, project VWZN3157, and the Israel Science Foundation (ISF).

© Springer Nature Switzerland AG 2019
Z. Cai et al. (Eds.): ISBRA 2019, LNBI 11490, pp. 257–269, 2019.
https://doi.org/10.1007/978-3-030-20242-2_22

removed from both partitions, until identity is reached. Such an identity induces a matching (subsets bijection) under which subsets (parts) from both partitions are matched (where, in case of unequal sized partitions, unmatched parts are matched to empty sets), and the symmetric difference between the matched sets is calculated. In [6] Gusfield noted that this task can be casted as an *assignment problem* [3,10] and hence be solved by a maximum flow in a bipartite graph in time $O(mn + n^2 \log n)$ [1]. As the graphs at hand are complete, this turns to be $\theta(n^3)$ and for biological instances of several thousands nodes (e.g. bacterial genomes harbour around 5000 genes) this solution is impractical. A possible remedy to the latter is the use of heuristic approaches that may return sub optimal solutions. These solutions return matchings that imply the removal of excessive number of elements (compared to optimal solutions). Therefore the accuracy of the matching is crucial. In [16] we used a very simple $O(|E| \log |V|)$-time greedy heuristic for the problem, that we denoted *Greedy PartDist*. While the algorithm was shown to provide a poor 1/2-approximation ratio in the general case [15], it exhibited good performance in practice (let alone the asymptotic speedup of almost an order of magnitude).

Motivated by this latter result, in this work we set to explain analytically this improved performance of the Greedy PartDist algorithm. The problem of finding the partition distance is presented as a *convex recoloring* problem [14,14] where element's color indicates original partition membership and we require the target, output, partition to be colored convexly, that is, each part is monochromatic and distinct from all other parts. We define an underlying random setting resembling a wide array of biological settings where errors in clustering follow a random distribution. The analysis is conveyed via the *universal pacemaker* (UPM) realm in which we first tackled this clustering problem [16,17]. While the handling is generic and is not confined to any specific context, the UPM setting confers more intuition to the analysis, and we briefly describe it here. The UPM concept was introduced in the context of molecular evolution to account for variation in gene's mutation rate [18–20] as opposed to the Molecular Clock (MC) [9,23] model, where gene mutation rate is constant. Under the UPM model, genes belonging to the same pacemaker change their mutation rate in correlation. The difference between the two models is depicted in Fig. 2. This gene correlation induces a partition over the gene set where part/subset membership is defined by pacemaker membership which is based either on the level of evolutionary rate or rate correlation [4,5,16].

Our first result is general and shows that when elements have even a slight tendency (i.e. constant, independent of the problem's size), to maintain their original membership (or equivalently, color), it is enough to have even a small (logarithmic in the problem's size) number of elements in each part. We first prove this by conditioning all parts to be of equal magnitude, and subsequently relax this condition with additional, yet constant, increase in the average part size. Then, we return to the UPM setting and show that our mild assumption holds automatically. This in turns furnishes the desirable property of *statistical consistency* of the Greedy PartDist algorithm under the UPM model. Finally, as a further research direction, we describe the application of the algorithm to the fundamental task in comparative genomic of determining gene orthology clustering.

Comment: Due to space considerations, several proofs were omitted and will appear in the journal version.

2 The Evolutionary Model

As the central part of this work is conveyed via an evolutionary setting, we provide here a brief description of this model. An evolutionary tree is a tree $T = (V, E)$ where the set of species are mapped to the leaves of T and the edges represent ancestry relationships. Each edge j is associated with a *time period* $\{t_j\}$ that indicates the time between ancestor to the respected descendant (see Fig. 1(a)). All genes evolve along T by acquiring mutations proportionally to the time along the various edges. As all genes evolve on T in an identical manner, and since we are concerned only in the actual time periods, henceforth we will identify these time periods with tree edges and completely ignore the topological information of T.

Under the molecular clock (MC) each gene g_i tends to evolve at an intrinsic rate r_i that is constant along time but deviates randomly along the time periods. Let $r_{i,j}$ be the *actual* (or *observed*) rate of gene i at period j. Then $r_{i,j} = r_i e^{\alpha_{i,j}}$ where $0 < e^{\alpha_{i,j}}$ is a multiplicative error factor. The number of mutations in gene g_i along period j is hence $\ell_{i,j} = r_{i,j} t_j$, commonly denoted as the *branch length* of gene g_i at period j. Throughout the text, we will reserve the letters i and j (and their derivatives) to index genes and periods respectively (eg. g_i and t_j).

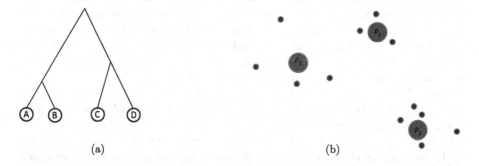

(a)   (b)

Fig. 1. (a) A phylogenetic (evolutionary) tree over the species $\{A, B, C, D\}$. (b) A scheme of a spatial representation of three pacemakers (red big balls) and their associated genes (small blue balls) centered around them (Color figure online)

We now extend this model to the UPM model that include a pacemaker, accelerating or decelerating a gene g_i, relative to its intrinsic rate r_i. Formally, a *pace maker* (or simply PM) P_k is a set of τ *paces* $\beta_{k,j}$, $1 \leq j \leq \tau$ where β_j

is the relative pace of PM during time period j. We reserve k to index PMs. Under the UPM model, a gene g_i that is *associated* with PM P_k has actual rate at period j: $r_{i,j} = r_i e^{\alpha_{i,j}} e^{\beta_{k,j}}$. In Fig. 2 we demonstrate the difference between the two models - the MC (left) and the UPM (right). In both models, the same ratio between the genes' evolutionary rate is maintained at all times, only that under the UPM rates change according to the pacemaker's rate whereas under the MC there is no pacemaker and rates are constant.

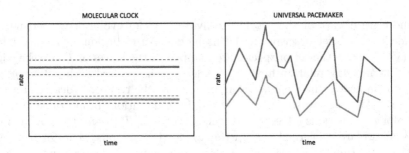

Fig. 2. Left: two genes of different intrinsic rates maintain the same rate along all lineages of a tree. Right: the genes change their intrinsic rate but by the same proportion. In both cases, the same ratio (constant relative rate) is kept at all times.

We assume that every gene is associated with some PM and let $PM(g_i)$ be the PM of gene g_i. Then the latter defines a partition over the set of genes G, where genes g_i and $g_{i'}$ are in the same part if $PM(g_i) = PM(g_{i'})$ (see Fig. 1(b) for illustration). while this is not essential to the current work, It is important to note that gene rates, as well as pace makers paces, are hidden and that we only see for each gene g_i, its set of edge lengths $\ell_{i,j}$. Additionally, the presence of two genes in the same part (PM) does not imply anything about their magnitude of rates, rather on their unison of rate divergence. In [16] we defined and studied *PM Partition identification Problem* that aims at reconstructing the original gene partition based on observed edge lengths solely.

Nevertheless, we can exploit some statistical random structure on the given setting that was observed in nature [21]. This randomness will provide us with signal to distinguish between the objects. By [21] we say that for all genes g_i and periods j, $\alpha_{i,j} \sim N(0, \sigma_G^2)$, and for all PMs P_k and periods j, $\beta_{k,j} \sim N(0, \sigma_P^2)$. In [17] we have provided sufficient conditions to reconstruct the pacemaker partition, in terms of these parameters (the PM and gene variances, and the number of genes/PMs).

3   The Partition Distance Algorithm

While the previous discussion dealt with the partition induced by the PMs and our ability reconstruct it, sometimes we are unable to correctly reconstruct the original partition, whether for reasons of insufficient data, unexpected statistical errors, and more. This calls for the task to assess the quality of reconstruction, or more generally, measure the *partition distance*, defined as the minimal removal of elements from both partitions until they are identical. The task is inherently different than before as we are only concerned with returning a correct answer. One way to look at it is as a recoloring problem [13,14]. The PM identity is a color $c \in C$ so that every gene is colored with the color of its original PM. The *coloring* function C maps a gene g_i to its PM (equivalent to $PM(g_i)$ above) and $C^{-1}(P)$ returns the set of genes of (or associated with) PM P. Given another partition $\mathcal{P}'$ defined over a colored element set, where the coloring is as defined by P, the goal is to recolor the least number of elements in $\mathcal{P}'$ such that every part is monochromatic and the colors of every two parts are distinct. Therefore, henceforth we will use the notions of PM identity and a color interchangeably. In [6] Gusfield noted that the partition distance problem can be casted as an *assignment problem* [3,10] and hence be solved by a maximum flow in a bipartite graph in time $O(mn + n^2 \log n)$ [1]. In the graph $G = (\mathcal{P}, \mathcal{P}', E, w)$, each node $u \in \mathcal{P}$ is a part (PM) in the original partition, and a node $v \in \mathcal{P}'$ is a part in the partition returned by the classifier. For $e = (u, v) \in E$, $w(e)$ is the size of the intersection $C^{-1}(u) \cap C^{-1}(v)$. Note that in $\mathcal{P}$, each part u is monochromatic as it corresponds to an original PM.

In [16] a very simple $O(|E| \log |V|)$ heuristic for the problem is provided. The algorithm, denoted here *Greedy PartDist*, simply sorts the edges in the bipartite graph and then recursively chooses the heaviest edge to the matching, removes all edges adjacent to the chosen edge, and continues in the recursion. This algorithm was shown by Preis to be a 1/2-approximation even for the general case [15] (non bipartite). Here we show that under our probabilistic model this algorithm returns the correct result (as opposed to an approximated one). We first provide the algorithm.

Greedy PartDist $(\mathcal{P}, \mathcal{P}')$:

- Construct the bipartite graph $B(\mathcal{P}, \mathcal{P}', E, w)$ with $w(p, p') = s(p, p')$ for every $p \in \mathcal{P}$ and $p' \in \mathcal{P}'$ s.t. $s(p, p') > 0$
- sort E according to w by descending order and use it as a stack to fetch elements.
- $M_G \leftarrow \emptyset$
- while E is not empty
    - $(p_0, p_0') \leftarrow pop(E)$
    - $M_G \leftarrow M_G \cup (p_0, p_0')$
    - Remove all (p_0, p_j) and (p_i, p_0') items from E
- match arbitrarily zero-degree elements from $\mathcal{P}$ with zero-degree elements from $\mathcal{P}'$ and add to M_G
- Return M_G

3.1   Probabilistic Analysis of Algorithm Greedy PartDist

Before we portray the main result of this section, we explain the setting. We assume an initial assignment of equal number of genes per each PM. This assumption is natural but also necessary as we cannot expect that if some PM has a single gene (or very few genes), that gene will with high probability choose that PM. Next, we assume a constant probability α in which a gene chooses its new part, and this constant is determined by the variances σ_G and σ_P that are also constant parameters in our setting. Therefore, this is the main theorem of this part:

Theorem 1. *Given m PMs and n genes such that $n = \Omega(m \log m)$. Assume all PMs are associated with the same number of genes, and each gene (even slightly) prefers its own PM with probability $0 < \alpha \le 1$, then for m large enough, Algorithm PartDist returns the correct result.*

Proof. The proof follows by first showing under which condition Algorithm PartDist is correct, and then shows this condition holds with high probability. We start with a useful definition.

Definition 1. *We say that a color d is correctly clustered if most of the d-colored genes in $\mathcal{P}$, choose their original color (PM) in $\mathcal{P}'$. We say that a colored partition $\mathcal{P}$ is correctly clustered by another partition $\mathcal{P}'$ if every color in $\mathcal{P}$ is correctly clustered.*

In [16] it was shown that if every color is correctly clustered, then Greedy PartDist returns the correct result.

Lemma 1. *Assume all PMs have the same number of genes - n/m. Assume every gene chooses its own color with probability $0 < \alpha \le 1$ and with probability $1 - \alpha$ chooses uniformly any PM (including its own). Then with probability at least $1 - 2me^{-\frac{\alpha^2}{16}\frac{n}{m}}$ every color is correctly clustered.*

Proof. In our model here every node in $\mathcal{P}$, as well as in $\mathcal{P}'$ corresponds to a single color, and all colors are distinct; that is, there is a bijection $C : V(\mathcal{P}) \to \mathcal{C}$, and the same bijection holds also for $V(\mathcal{P}')$. A gene g_i with some original color (i.e. in $\mathcal{P}$), chooses a new color in $\mathcal{P}'$ and moves to that new node (without changing its original color). Consider a specific color d and let v_d be the node corresponding to the d-color in the partition $\mathcal{P}'$. For a color c and a node v in $\mathcal{P}'$, denote by $C_c^{-1}(v)$ the set of c-genes (i.e. with original color/PM c) that chose the PM represented by v in $\mathcal{P}'$ (or equivalently, the weight of the edge linking between the c-PM in $\mathcal{P}$ to v in $\mathcal{P}'$). Therefore, $C_d^{-1}(v_d)$ is the set of d-colored genes that chose their original color. For a node $v \in \mathcal{P}'$ and a color c, we denote v as the *majority node* for c if $|C_{c(v)}^{-1}| \ge |C_{c(v')}^{-1}|$ for any other $v' \in \mathcal{P}'$. By our assumptions, there are n/m d-colored genes, therefore each retains the d color with probability p where

$$p = \alpha + (1-\alpha)/m, \tag{1}$$

and the expected number of genes retaining color d is,

$$\mathrm{E}\left[C_d^{-1}(v_d)\right] = \frac{n}{m}\left(\alpha + \frac{1-\alpha}{m}\right). \tag{2}$$

We note that we are interested only in the number of genes with original color d who retained d as a color. The final number of genes with color d may be larger for genes with original color other than d who chose d.

We now pursue the following technique to prove a lower bound on the probability for correct clustering. For a given color d (equivalently part or PM), we set a threshold t and require that no color $d' \neq d$ is chosen by at least t d-genes, and that at least t d-genes retain their color. As this implies correct clustering, its probability serves as a lower bound for the latter.

We follow the above idea by bounding from above the probability of the complementary case. The following observation that is given without a proof, formalizes that.

Observation 1. *If there exists $v' \in \mathcal{P}'$ other than v_d, such that $|C_d^{-1}(v')| > |C_d^{-1}(v_d)|$ (or in words, v_d is not the majority for the d-genes), then for any t, either (a) $|C_d^{-1}(v_d)| < t$ or (b) $|C_d^{-1}(v')| > t$.*

In other words, if neither events (a) or (b) occur, then v_d is the majority node for color d. We now show that events (a) and (b) from Observation 1 occur in low probability. For this we need to set a concrete value for t and the value we choose is $\frac{n}{m}\left(\alpha/2 + \frac{1-\alpha}{m}\right)$. Then, events (a) and (b) from Observation 1 become the following condition.

Condition 1. *It is a sufficient condition for correct clustering for a color d, if both sub conditions below do not occur.*

a. $|C_d^{-1}(v_d)| < \frac{n}{m}\left(\alpha/2 + \frac{1-\alpha}{m}\right)$, *or*
b. $|C_d^{-1}(v')| > \frac{n}{m}\left(\alpha/2 + \frac{1-\alpha}{m}\right)$.

The following requirement for $1/m < \alpha$ is sound as for small m's the probability of some $v \neq v_d$ to be the majority node for color d is relatively high and we require α, the color retention preference, to be relatively high.

Claim. Assume $1/m < \alpha$. Then, Conditions (1.a) and (1.b) occur with probability at most $2e^{-\frac{\alpha^2}{16}\frac{n}{m}}$.

The proof of this Claim 3.1 is technical and involved and therefore deferred to the journal version.

As there are m PMs, the probability that events (a) and (b) from Observation 1 occur for some PM is bounded by $2me^{-\frac{\alpha^2}{16}\frac{n}{m}}$. This completes the proof of Lemma 1.

Corollary 1. *If there are m PMs and $n = \Omega(m \log m)$ genes, then with probability $1 - o(1)$ Algorithm Greedy PartDist returns the correct result.*

Proof. Let $n = c_1 m \log m$ where c_1 is a constant. Then by Claim 3.1, a single PM is not correctly clustered with probability at most $2e^{-\frac{\alpha^2}{16}\frac{n}{m}}$. As there are m PMs, by the union bound, the probability that some PM is not correctly clustered is bounded by:

$$2me^{-\frac{\alpha^2}{16}\frac{n}{m}}$$
$$= 2me^{-\frac{\alpha^2}{16}\frac{c_1 m \log m}{m}}$$
$$= 2m^{1-c_1\frac{\alpha^2}{16}}$$
$$= o(1),$$

for $c_1 > \frac{16}{\alpha^2}$.

This completes the proof of Theorem 1.

3.2   Extension to Fully Random Model

We now extend the model above to a fully random model. Recall that in the derivation above (Sect. 3.1), we assumed all PMs possess the same number of genes - n/m. We now extend the model to a fully random model. Under this model, a gene not only chooses its target PM, rather also its source (host) PM and therefore its color. Now, the uniform assumption is not valid anymore and we should account for variable number of genes under any color. To cope with this situation, we give a bound on the number of genes under each PM and incorporate that bound into our analysis. For the sake of simplicity, we assume a uniform distribution over the PMs, that is, a gene chooses each one of the m colors with probability $p = 1/m$, and we note that this can trivially be expanded to any other constant probability set $[p_i]_{i \le m}$ where $\sum_i p_i = 1$. Under this setting, the number of genes under each PM follow a standard multinomial distribution and the expected number of genes under the uniform distribution is $\mathrm{E}\left[|C^{-1}(P)|\right] = n/m$ as our previous assumption (Sect. 3.1). We want now to guarantee that the number of genes under each PM P is at least $(1-\varepsilon)n/m$. while this can be done accurately by the cumbersome inclusion/exclusion principle, we can get a good enough bound using the Chernoff lower bound, Theorem [2, Thm A.1.13] .

$$\Pr\left[|C^{-1}(P)| < (1-\varepsilon)\frac{n}{m}\right]$$
$$= \Pr\left[|C^{-1}(P)| - \mathrm{E}\left[|C^{-1}(P)|\right] < -\varepsilon\frac{n}{m}\right]$$
$$< e^{-\varepsilon^2\left(\frac{n}{m}\right)^2/\left(2\frac{n}{m}\right)}$$
$$= e^{-\frac{\varepsilon^2}{2}\left(\frac{n}{m}\right)}. \tag{3}$$

Again, to account this is correct for all PMs, we use the union bound and obtain

Observation 2. *Under the fully stochastic model, all PMs have at least* $(1-\varepsilon)\frac{n}{m}$ *genes, with probability at least*

$$1 - me^{-\frac{\varepsilon^2}{2}\left(\frac{n}{m}\right)}. \tag{4}$$

It remains now only to incorporate this bound, i.e. the lower bound on the minimum number of genes at each part (PM, color) into our bounds for correct clustering. A close look, with analysis similar to the one before, rather without the assumption of constant size partitions, reveals that we just need to adjust the exponent in the equation by multiplying with $(1 - \varepsilon)$ as n/m is replaced by $(1 - \varepsilon)n/m$. This yields

$$e^{-\frac{(1-\varepsilon)}{2}\frac{n\alpha^2}{m}} \tag{5}$$

and

$$e^{-(1-\varepsilon)\left(\frac{\alpha}{16}\frac{n}{m}\right)}. \tag{6}$$

After combining both (5) and (6) we obtain

$$e^{-\frac{(1-\varepsilon)}{2}\frac{n\alpha^2}{m}} + e^{-(1-\varepsilon)\left(\frac{\alpha}{16}\frac{n}{m}\right)}$$
$$\leq 2e^{-(1-\varepsilon)\frac{\alpha^2}{16}\frac{n}{m}}. \tag{7}$$

Again we multiply by m to account for m PMs, leading to

Observation 3. *Under the fully stochastic model, all PMs are correctly clustered with probability at least*

$$1 - 2me^{-(1-\varepsilon)\frac{\alpha^2}{16}\frac{n}{m}}. \tag{8}$$

Finally, we need to account that both events, that each color (PM) is chosen by at least $1 - \varepsilon$ genes, and that each color is retained, occur with high probability. It is easy to see that this entails multiplying Eqs. (4) and (8) and keeping the product close to one. Hence we obtain:

Corollary 2. *For m number of PMs large enough, and $O(m \log m)$ genes, correct clustering is obtained with high probability under the fully random model.*

The proof is omitted.

4   Specialization to the Universal Pacemaker Setting

The analysis in the previous section, although expressed in the UPM jargon, was general and theoretical and assumed a constant, even small, α for which each gene prefers to retain its original color. However, under our working model of the UPM, we can relax that rigid assumption and instead rely on the model to provide for such a natural preference. Therefore, for the UPM model we can state a significantly stronger assertion. We now show that such an α indeed exists under the UPM model, i.e. a gene indeed prefers to stay closer to its PM, and provide some rough lower bound to it.

Lemma 2. *Under the UPM model, at every time period t_j, there is a constant, i.e. independent of n and m, positive probability for a gene to prefer its own PM.*

Proof. Recall that under this model, a gene sets its rate at some period, around its own PM, i.e. the value (pace) of the PM at that given period. We aim at translating this property into our model. As this part is fairly involved, Fig. 3 illustrates the situation pictorially. Consider two PMs P and P' with paces $\beta_{k,j}$ and $\beta_{k',j}$ respectively, evolving τ periods, and let g_i be a gene associated with PM P - meaning its values (rates, $r_{i,j}$) distribute normally around $\beta_{k,j}$. We concentrate now in the first period. Let d_1 be the average of the PMs values at the first period, and WLOG assume $\beta_{k,1} > \beta_{k',1}$, that is $d_1 = \beta_{k',1} + \frac{1}{2}(\beta_{k,1} - \beta_{k',1})$. Also let d_2 be its antipodal point with respect to $\beta_{k,1}$ - specifically $d_2 = \beta_{k,1} + \frac{1}{2}(\beta_{k,1} - \beta_{k',1})$. Now, it is easy to see that every point p_1 from d_1 toward (and beyond) $\beta_{k',1}$ (i.e. $(-\infty, d_1)$) is closer to $\beta_{k',1}$ than to $\beta_{k,1}$ and if g_i will fall at this interval, i.e. $r_{i,1} \in (-\infty, d_1)$, it will end up closer to P' than to P at this first period. However, for each such point p_1, there is an equiprobable point p_2 on the opposite side of $\beta_{k,1}$ and right to d_2 that is closer to $\beta_{k,1}$ and that cancels its reciprocal point. The opposite however, is not true. There are the points along the interval between d_1 and $\beta_{k,1}$. These are all closer to $\beta_{k,1}$ and have no

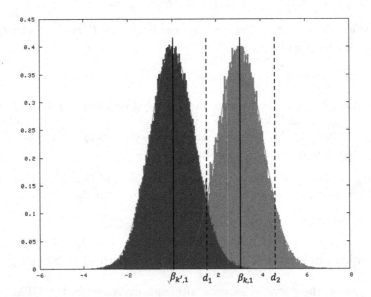

Fig. 3. Consistency of the Greedy PartDist algorithm under the UPM Model: in the figure we see two PMs, red and blue with $\beta_{k,1} = 3$, $\beta_{k',1} = 0$ respectively, and hence d_1 is set in the middle, i.e. $d_1 = 1.5$ and d_2 is symmetric with respect to $\beta_{k,1}$, i.e. $d_2 = 4.5$. Without loos of generality, original genes' color is red and therefore choose coordinates (values) according to the red distribution, i.e. normal $\beta_{k,1} = 3$. All genes with values in the interval (d_1, d_2) will choose the red color (PM), and yet have no equiprobable interval that prefers the blue PM. (Color figure online)

reciprocals that are closer to $\beta_{k',1}$, on the contrary - their antipodal points are in the interval $(\beta_{k,1}, d_2]$ and are also closer to $\beta_{k,1}$. Therefore, the probability of g_i (i.e. $r_{i,1}$) falling in that interval $(d_1, b_{k,1}]$ is

$$\frac{1}{2} - \Phi(d_1 - \beta_{k,1}), \tag{9}$$

where Φ is the normal density function.

Now, the value in Eq. (9) is always non negative and constant as it depends only on the difference $\beta_{k,1} - \beta_{k',1}$, since $d_1 < \beta_{k,1}$, yielding $\Phi(d_1 - \beta_{k,1}) < 1/2$.

The argumentation above is demonstrated in Fig. 3 where we set $\beta_{k',1} = 0$, $\beta_{k,1} = 3$, and hence $d_1 = 1.5$ and $d_2 = 4.5$. Original genes' color is red and therefore choose coordinates (values) according to the red distribution, i.e. normal around $\beta_{k,1} = 3$. All genes with values in the interval (d_1, d_2) will choose the red color (PM), and yet have no equiprobable interval that prefers the blue PM.

Now, Lemma 2 refers to a single time period. However since this holds for every period j (although with different $\beta_{k,1}$ and $\beta_{k',1}$), the probability of falling closer to P increases with every period. Therefore, the probability of g_i falling in the interval $[d_1, \beta_{k,1}]$ serves as a lower bound for α.

We can now state our main claim for this part. An algorithm A is considered *statistically consistent* under a (statistical) model M if the more data A is given (or "sees") from M, the greater the probability to return the correct answer, and as a special case, this probability tends to 1 at the infinity. The above discussion leads to the following conclusion.

Corollary 3. *Algorithm Greedy PartDist is statistically consistent under the UPM stochastic model.*

5   Concluding Remarks and Further Research

In this work we have provided a probabilistic study of the *greedy PartDist* algorithm, that runs a fast, greedy matching on a bipartite graph. We specialized on the universal pacemaker paradigm but the analysis is generic and can apply to a host of probabilistic settings and partitioning. While the algorithm provides a relatively poor guarantee of $1/2$ under a general setting, we here prove that under a relatively sound assumptions, it returns the correct result. Moreover, we show that in a natural stochastic setting, this assumption holds with high probability.

We start by describing the UPM setting where genes are partitioned according to their pacemaker that in turn controls their mutation rate changes. Subsequently we turn to a seemingly unrelated problem of computing distance between partitions and cast it as a *recoloring problem*. We show that under a stochastic setting and a mild natural assumption, the algorithm *greedy PartDist* returns the correct result. This is done in two stages where in the first stage we assume for simplicity constant size parts, and we later relax this assumption to a fully random model. Finally we revert to the UPM setting and show that the natural assumption from the previous section, holds. This in turn yields a strong conclusion of *statistical consistency* regarding the greedy PartDist algorithm under the UPM model.

For further research and a proof for the utility and generality of the result, we would mention that the same need for such an analysis arose in a different setting of *gene orthology detection*, where orthologs between two genomes are sought. Here similarity-based bipartite matching is employed to detect orthologs. For the size of the problem's inputs, the fast greedy PartDist algorithm is a sound compromise. The result presented here has a promising role also in that latter context.

Acknowledgments. We would like to thank Raphy Yuster for very helpful discussion and comments on the probabilistic part of this manuscript. Thanks also to Eugene Koonin and Yuri Wolf for inspiring the question. Part of this work was done while the author was visiting the National Center for Biotechnology Information (NCBI) at the NIH, USA, and the Computational Genomics Summer Institute (CGSI) at UCLA, USA.

References

1. Ahuja, R.K., Magnanti, T.L., Orlin, J.B.: Network Flows. Prentice Hall, Englewood Cliffs (1993)
2. Alon, N., Spencer, J.H.: The Probabilistic Method, 3rd edn. Wiley, New York (2008)
3. Cook, W.J., Cunningham, W.H., Pulleyblank, W.R., Schrijver, A.: Combinatorial Optimization. Wiley, New York (1998)
4. Duchene, S., Ho, S.Y.W.: Mammalian genome evolution is governed by multiple pacemakers. Bioinformatics **31**, 2061–2065 (2015)
5. Duchene, S., Ho, S.Y.W.: Using multiple relaxed-clock models to estimate evolutionary timescales from DNA sequence data. Mol. Phylogenet. Evol. **77**, 65–70 (2014)
6. Gusfield, D.: Partition-distance: a problem and class of perfect graphs arising in clustering. Inf. Process. Lett. **82**(3), 159–164 (2002)
7. Hartigan, J.A., Wong, M.A.: A k-means clustering algorithm. Appl. Stat. **28**, 100–108 (1979)
8. Hurst, L.D., Pál, C., Lercher, M.J.: The evolutionary dynamics of eukaryotic gene order. Nat. Rev. Genet. **5**(4), 299 (2004)
9. Kimura, M.: Molecular evolutionary clock and the neutral theory. J. Mol. Evol. **26**, 24–33 (1987)
10. Lawler, E.L.: Combinatorial Optimization: Networks and Matroids. The University of Michigan, Ann Arbor (1976)
11. Lee, J.M., Sonnhammer, E.L.L.: Genomic gene clustering analysis of pathways in eukaryotes. Genome Res. **13**(5), 875–882 (2003)
12. Lloyd, S.P.: Least squares quantization in PCM. IEEE Trans. Inf. Theory **28**, 129–137 (1982)
13. Moran, S., Snir, S.: Efficient approximation of convex recolorings. J. Comput. Syst. Sci. **73**(7), 1078–1089 (2007)
14. Moran, S., Snir, S.: Convex recolorings of strings and trees: definitions, hardness results and algorithms. J. Comput. Syst. Sci. **74**(5), 850–869 (2008)

15. Preis, R.: Linear time 1/2-approximation algorithm for maximum weighted matching in general graphs. In: Meinel, C., Tison, S. (eds.) STACS 1999. LNCS, vol. 1563, pp. 259–269. Springer, Heidelberg (1999). https://doi.org/10.1007/3-540-49116-3_24

16. Snir, S.: On the number of genomic pacemakers: a geometric approach. Algorithms Mol. Biol. **9**, 26 (2014). Extended abstract appeared in WABI 2014

17. Snir, S.: Bounds on identification of genome evolution pacemakers. J. Comput. Biol. (2019)

18. Snir, S., Wolf, Y., Koonin, E.: Universal pacemaker of genome evolution. PLoS Comput. Biol. **8**, e1002785 (2012)

19. Snir, S., Wolf, Y., Koonin, E.: Universal pacemaker of genome evolution in animals and fungi and variation of evolutionary rates in diverse organisms. Genome Biol. Evol. **6**(6), 1268–1278 (2014)

20. Wolf, Y., Snir, S., Koonin, E.: Stability along with extreme variability in core genome evolution. Genome Biol. Evol. **5**(7), 1393–1402 (2013)

21. Wolf, Y.I., Novichkov, P.S., Karev, G.P., Koonin, E.V., Lipman, D.J.: The universal distribution of evolutionary rates of genes and distinct characteristics of eukaryotic genes of different apparent ages. Proc. Natl. Acad. Sci. **106**(18), 7273–7280 (2009)

22. Yi, G., Thon, M.R., Sze, S.-H.: Identifying clusters of functionally related genes in genomes. Bioinformatics **23**(9), 1053–1060 (2007)

23. Zuckerkandl, E., Pauling, L.: Molecules as documents of evolutionary history. J. Theor. Biol. **8**(2), 357–66 (1965)

Author Index

Printed in the United States
By Bookmasters